The Handbook of Environmental Chemistry

Volume 1 Part C

Edited by O. Hutzinger

The Natural Environment and the Biogeochemical Cycles

With Contributions by
G. G. Choudhry, E. T. Degens, M. Ehrhardt,
R. D. Hauck, S. Kempe, L. W. Lion,
A. Spitzy, P. J. Wangersky

With 55 Figures

Springer-Verlag
Berlin Heidelberg GmbH 1984

Professor Dr. Otto Hutzinger
University of Bayreuth
Chair of Ecological Chemistry and Geochemistry
Postfach 3008, D-8580 Bayreuth
Federal Republic of Germany

Library of Congress Cataloging in Publication Data

Main entry under title: The Natural environment and the biogeochemical cycles. (The Handbook of environmental chemistry; v. 1, pt. A–C)
Includes bibliographical references and index.
1. Biogeochemical cycles. 2. Environmental chemistry. I. Craig, P. (Peter), 1944–. II. Series.
QD31.H335 vol. 1, pt. A–C 574.5′22s [574.5′222] 80-16608 [QH344]

ISBN 978-3-662-15235-5 ISBN 978-3-540-38829-6 (eBook)
DOI 10.1007/978-3-540-38829-6

Originally published by Springer-Verlag Berlin Heidelberg New York in 1984.
Softcover reprint of the hardcover 1st edition 1984

2154/3140-543210

Preface

Environmental Chemistry is a relatively young science. Interest in this subject, however, is growing very rapidly and, although no agreement has been reached as yet about the exact content and limits of this interdisciplinary discipline, there appears to be increasing interest in seeing environmental topics which are based on chemistry embodied in this subject. One of the first objectives of Environmental Chemistry must be the study of the environment and of natural chemical processes which occur in the environment. A major purpose of this series on Environmental Chemistry, therefore, is to present a reasonably uniform view of various aspects of the chemistry of the environment and chemical reactions occurring in the environment.

The industrial activities of man have given a new dimension to Environmental Chemistry. We have now synthesized and described over five million chemical compounds and chemical industry produces about hundred and fifty million tons of synthetic chemicals annually. We ship billions of tons of oil per year and through mining operations and other geophysical modifications, large quantities of inorganic and organic materials are released from their natural deposits. Cities and metropolitan areas of up to 15 million inhabitants produce large quantities of waste in relatively small and confined areas. Much of the chemical products and waste products of modern society are released into the environment either during production, storage, transport, use or ultimate disposal. These released materials participate in natural cycles and reactions and frequently lead to interference and disturbance of natural systems.

Environmental Chemistry is concerned with *reactions in the environment*. It is about distribution and equilibria between environmental compartments. It is about reactions, pathways, thermodynamics and kinetics. An important purpose of this Handbook is to aid understanding of the basic distribution and chemical reaction processes which occur in the environment.

Laws regulating toxic substances in various countries are designed to assess and control risk of chemicals to man and his environment. Science can contribute in two areas to this assessment; firstly in the area of toxicology and secondly in the area of chemical exposure. The available concentration ("environmental exposure concentration") depends on the fate of chemical compounds in the environment and thus their distribution and reaction behaviour in the environment. One very important contribution of Environmental Chemistry to the above mentioned toxic substances laws is to develop laboratory test methods, or mathematical correlations and models that predict the environmental fate of new chemical compounds. The third purpose of this Handbook is to help in the basic understanding and development of such test methods and models.

The last explicit purpose of the Handbook is to present, in concise form, the most important properties relating to environmental chemistry and hazard assessment for the most important series of chemical compounds.

At the moment three volumes of the Handbook are planned. Volume 1 deals with the natural environment and the biogeochemical cycles therein, including some background information such as energetics and ecology. Volume 2 is concerned with reactions and processes in the environment and deals with physical factors such as transport and adsorption, and chemical, photochemical and biochemical reactions in the environment, as well as some aspects of pharmacokinetics and metabolism within organisms. Volume 3 deals with anthropogenic compounds, their chemical backgrounds, production methods and information about their use, their environmental behaviour, analytical methodology and some important aspects of their toxic effects. The material for volume 1, 2 and 3 was each more than could easily be fitted into a single volume, and for this reason, as well as for the purpose of rapid publication of available manuscripts, all three volumes were divided in the parts A and B. Publisher and editor hope to keep materials of the volumes one to three up to date and to extend coverage in the subject areas by publishing further parts in the future. Readers are encouraged to offer suggestions and advice as to future editions of "The Handbook of Environmental Chemistry".

Most chapters in the Handbook are written to a fairly advanced level and should be of interest to the graduate student and practising scientist. I also hope that the subject matter treated will be of interest to people outside chemistry and to scientists in industry as well as government and regulatory bodies. It would be very satisfying for me to see the books used as a basis for developing graduate courses on Environmental Chemistry.

Due to the breadth of the subject matter, it was not easy to edit this Handbook. Specialists had to be found in quite different areas of science who were willing to contribute a chapter within the prescribed schedule. It is with great satisfaction that I thank all 52 authors from 8 countries for their understanding and for devoting their time to this effort. Special thanks are due to Dr. F. Boschke of Springer for his advice and discussions throughout all stages of preparation of the Handbook. Mrs. A. Heinrich of Springer has significantly contributed to the technical development of the book through her conscientious and efficient work. Finally I like to thank my family, students and colleagues for being so patient with me during several critical phases of preparation for the Handbook, and to some colleagues and the secretaries for technical help.

I consider it a privilege to see my chosen subject grow. My interest in Environmental Chemistry dates back to my early college days in Vienna. I received significant impulses during my postdoctoral period at the University of California and my interest slowly developed during my time with the

National Research Council of Canada, before I could devote my full time to Environmental Chemistry, here in Amsterdam. I hope this Handbook may help deepen the interest of other scientists in this subject.

O. Hutzinger

Preface to Parts C of the Handbook

Parts C of the three series
- The Natural Environment and the Biogeochemical Cycles (Vol. 1)
- Reactions and Processes (Vol. 2)
- Anthropogenic Compounds (Vol. 3)

are now either available or in press. During their preparation it became obvious that further parts will have to follow to present the respective subject matters in reasonably complete form.

The publisher and editor have further agreed to expand the Handbook by three new series: Air Pollution, Water Pollution and Environmental Trace Analysis.

Again, I thank all authors as well as collaborators at the Springer Publishing House for their cooperation and help. Thanks are also due to many environmental chemists and reviewers in particular for their critical comments and their positive reception of the Handbook.

Bayreuth, December 1983 Otto Hutzinger

Contents

Marine Gelbstoff

M. Ehrhardt

The Surface of the Ocean

L. W. Lion

Atmospheric Nitrogen. Chemistry, Nitrification, Denitrification, and their Interrelationships

R. D. Hauck

List of Contributors

Dr. Ghulam Ghaus Choudhry
Laboratory of Environmental
and Toxicological Chemistry
University of Amsterdam
Nieuwe Achtergracht 166
NL-1018 WV Amsterdam,
The Netherlands

Prof. Dr. E. T. Degens
Universität Hamburg
SCOPE/UNEP
International Carbon Center
D-2000 Hamburg 13
Federal Republic of Germany

Dr. Manfred Ehrhardt
Institut für Meereskunde
an der Universität Kiel
Düsternbrooker Weg 20
D-2300 Kiel 1
Federal Republic of Germany

Dr. Roland D. Hauck
Division of Agricultural Development
Tennessee Valley Authority
Muscle Shoals, AL 35660, USA

Dr. Stephan Kempe
Universität Hamburg
SCOPE/UNEP
International Carbon Center
D-2000 Hamburg 13
Federal Republic of Germany

Prof. Dr. Leonard W. Lion
Cornell University
Department of
Environmental Engineering
Hollister Hall
Ithaca, NY 14853, USA

Dr. A. Spitzy
Universität Hamburg
SCOPE/UNEP
International Carbon Center
D-2000 Hamburg 13
Federal Republic of Germany

Prof. Dr. Peter J. Wangersky
Department of Oceanography
Dalhousie University
Halifax, Nova Scotia B3H 4J1
Canada

Humic Substances
Structural Aspects, and Photophysical, Photochemical and Free Radical Characteristics

G. G. Choudhry

Laboratory of Environmental and Toxicological Chemistry
University of Amsterdam, 1018 WV Amsterdam, The Netherlands*

Background

Definitions

The organic matter of soils, peats and waters consists of a mixture of plant and animal products in various stages of decomposition together with substances synthesized biologically and/or chemically from the breakdown products as well as microorganisms and small animals [3, 50]. This organic matter is usually divided into two groups:

(i) Nonhumic substances, and
(ii) humic substances.

Nonhumic substances include a large number of relatively simple compounds of known structures and belonging to well-known groups such as carbohydrates, proteins, peptides, amino acids, fats, waxes, resins, pigments, and other low-molecular weight organic substances. In general, these compounds are relatively easily attacked by microorganisms in the soil and thus have a relatively short survival rate. The bulk of the organic matter in most soils and waters consists of humic substances. These are amorphous, brown or black, hydrophilic, acidic, polydisperse substances.

Based on their solubility in alkali and acid, humic substances (HS's) are usually divided into three major fractions:

(i) humic acid (HA), which is soluble in dilute alkaline solution (e.g. 0.5 N NaOH) but is precipitated by acidification (e.g. by addition of 6 M HCl to the alkaline extract, pH 2);
(ii) fulvic acid (FA), which is the humic fraction remaining in the aqueous acidified solution (i.e. it is soluble in both acid and base); and
(iii) humin, which cannot be extracted by dilute base and acid.

* Present address: Pesticide Research Laboratory, Department of Soil Science, University of Manitoba, Winnipeg, Manitoba, Canada R3T 2N2

The insolubility of the third fraction is ascribed to the firmness with which it combines with inorganic soil and water constituents [22, 50]. The HS's containing partially decomposed plant and animal residues are also referred to as humus: the latter can be removed from the HS's with acetyl bromide [3].

Odén [36] has removed the alcohol-soluble portion of the first fraction (HA) as hymatomelanic acid; it is, however, doubtful whether such a sub-division is desirable [3].

Occurrence and Origin

Humic substances occur in agricultural soils, forest soils, mountain soils, peats, coals, oil fields and in volcanic ash. Humus has also been found to exist in the aquatic environments namely in: sea waters, river waters, lake waters, fish ponds, fresh and natural waters, ground waters, waste water sludges and estuaries. These natural macromolecules also occur in: shallow-sea sediments, carbonate and noncarbonate sediments and surface sediments, carbonate and noncarbonate sediments and surface sediments of lakes. Their existence also has been noticed in blood serum, blood proteins and marine algae (references cited in Choudhry [12]).

The synthesis of humus is based on plant residues, the most important compounds in this respect being lignin, carbohydrates and proteins [20]. According to Flaig [19] 50–60% carbohydrates, 1–3% proteins, 10–30% lignins and some phenolic compound participate in the humification processes. Among these groups of organics, lignins are considered to be the most important, because carbohydrates and proteins have a higher rate of decomposition compared to lignin.

Structural Aspects

Data available suggest that, structurally, the three humic fractions are similar to each other, but that they differ in molecular weight, ultimate analysis and functional group content [50]. The chemical structure as well as reactions of humic substances have been the subject of numerous investigations for over 200 years, yet much remains to be learned about these materials. The main task that confronts researchers in this field today is to develop a valid concept of their chemical structures [49].

Molecular Weights

Molecular weights ranging from a few hundred to several millions have been reported for humic substances. Among the methods that have been used are those measuring the number-average, $\bar{M}_n$, (osmotic pressure, cryoscopic, diffusion, isothermal distillation), the weight-average, $\bar{M}_w$, (viscosity, gel filtration) and the z-average, $\bar{M}_z$ (sedimentation) molecular weights. In general, molecular weights of HS's measured by number- and weight-average methods are lower than those determined by z-average method, but there are also wide discrepancies among values within each method. For instance, $\bar{M}_w$'s (weight-average molecular weights) of the order of 300 to >200,000 have been reported for soil HA's [4, 17, 39]. For HA's and FA's extracted from marine sediments $\bar{M}_w$'s were of the range from 700 to 2,000,000 [40, 41] and for HA's extracted from natural waters $\bar{M}_w$'s and $\bar{M}_n$'s ranging from <700 to >26,000 have been reported [20].

Elemental Composition

Humic substances contain carbon, hydrogen, nitrogen, sulfur and oxygen as major elements; C and O being predominant. Schnitzer [48] has reported on the elementary analyses of HA's and FA's both extracted from soils formed under widely differing geographic and pedalogic conditions such as those prevailing in the arctic, the cool temperate, subtropical and tropical climatic zones. The ranges of elemental content for HA's are: C=53.8–58.7%, O=32.8–38.3%, H=3.2–6.2%, N=0.8–4.3% and S=0.1–1.5%; while those for FA's are: C=40.7–50.6%, O=39.7–49.8%, H=3.8–7.0%, N=0.9–3.3% and S=0.1–3.6%. Thus, on the average HA's contain 10% more C but 10% less O than do FA's. According to Khan [25], the elemental analysis of soil humins is of the same order of magnitude as that for HA's. Compared to soil humic compounds, HS's in waters generally contain considerably less C and N (as a mean 43% and 1.1%, respectively in case of water), whereas H is 5.5% [20]. But Stuermer and Payne [61] concluded that sea-water FA's possess relatively lower O content, higher N content and higher H/C ratios as compared to the terrestrial ones.

Functional Groups

The major oxygen-containing functional groups in humic substances are carboxyls, hydroxyls and carbonyls. The means of ranges of the analytical data for these functional groups in HA's and FA's extracted from soils formed under different conditions, which were calculated by Schnitzer [48], are as follows: total acidity=6.7, 10.3 meq./g; COOH=3.6, 8.2 meq./g; phenolic OH=3.9, 3.0 meq/g; alcoholic OH=2.6, 6.1 meq./g; quinonoid and ketonic C=O=2.9, 2.7 meq./g; and OCH_3=0.6, 0.8 meq./g; respectively. The total acidity equals the sum of COOH+ phenolic OH groups. It is obvious that the total acidity, and expecially the COOH content, of "model" FA's are considerably higher than those of "model" HA's and moreover the FA's are richer in alcoholic OH groups. Likewise, Khan [25] has reported the following data for contents (in meq/g) of the major oxygen - containing functional groups in soil humins: total acidity=5.0–5.9, carboxyl=2.6–3.8, phenolic OH=2.1–2.4, carbonyl=4.8–5.7, and methoxyl=0.3–0.4.

In the past, the presence of quinone groups in the structure of natural humic macromolecules and their quantitative determination have been receiving much attention in the literature [12]. For example, several earlier infrared (IR) spectroscopic studies of derivatives of soil FA's and HA's, conducted by various researchers, did not yield evidence of quinoid groups in the form of peaks in the 1660 and 1680 cm^{-1} regions of the spectra. On the other hand, studies with coal and peat humic acids have afforded such IR evidence through spectra of derivatives obtained by specific method (references cited in Choudhry [12]). Mathur [33] reported that the IR spectra of soil methylated FA's and methylated HA's preparations show a distinct peak at 1670 cm^{-1} which is characteristic for the valence frequencies of quinones. Furthermore, it has been reliably shown by Maximov and Glebko [34] that $>C=O$ functional groups, being chiefly of quinonoid origin and readily subject to reversible redox transformations, are contained in the molecules of natural humic acids. The number of quinone groups increases with the transition from soil HA's (e.g. 1.05 m eq/g) through peat HA's (e.g. 1.26 m eq/g) to the HA's of brown coal (e.g. 2.04 m eq/g) and weathered (oxidized) bituminious coal (e.g. 3.27 m eq/g).

NMR Spectrometry

The study of nuclear magnetic resonance (NMR) spectra of humic substances has not received much attention so far. Utilized solvents are $CDCl_3$, CCl_4, deutrated dimethyl sulfoxide (DMSO-d_6), aqueous solutions of HCl (DCl) and NaOH (NaOD) and D_2O; whilst techniques include 60, 100, 270 MHz ^{1}H-NMR and 20–25 MHZ ^{13}C-NMR (Fourier transform) [12]. Recently, Lakatos and Meisel [30] have reported that in the ^{1}H-NMR spectra of FA's in a mixture of $CDCl_3$ and DMSO-d_6, in addition to the signal at chemical shift, $\delta=4.2$ ppm, further lines could be observed in the range of methyl and methylene groups of aliphatic hydrogens at $\delta=0.9$ and 1.4 ppm. The resonance signal at $\delta=4.2$ ppm probably points to some lactone structure, since after alkaline hydrolysis at 105 °C for 7 hours, the signal disappears. Likewise, the presence of aliphatic and aromatic protons and lactone ($\delta=4.2$ ppm) structure in the ^{1}H-NMR spectra of a DMSO-d_6 solution of bogpeat HA's are also known [38].

Stuermer and Payne [61] have investigated seawater- and terrestrial FA's with carbon-13 nuclear magnetic resonance (^{13}C-NMR) and ^{1}H-NMR. Significant differences exist between the ^{13}C-NMR spectra of both type-FA's in 0.01N HCl (48 mg/ml) at 44 °C. The seawater fulvic acid spectrum shows more predominant resonances in 10–60 ppm region, characteristic of aliphatic moieties. Resonances in the 110–160 ppm region of the spectra characteristic of aromatic, heteroaromatic and olefinic constituents are slightly more predominant in the terrestrial FA spectrum. In the ^{1}H-NMR spectrum of Sargasso Sea FA in D_2O (100 mg/ml) at 35 °C, the relative areas of the broad resonances at $\delta=1.0$–1.7 ppm (aliphatic protons), $\delta=1.7$–2.5 ppm (protons on carbons adjacent to functional groups such as carbonyls, aromatic or double bonds) and $\delta=7.3$–7.9 (aromatic protons) are 15: 10:1 [61].

It was concluded by Grant [21] that ^{1}H-NMR absorptions of Podzol humus found in DMSO-d_6 or deutero-pyridine solutions are much more distinct than the spectra (obtained in D_2O) of aqueous extracts and indicate the presence of $CH_3(CH_2)_n$, various environments of $CH_2(CO)$, CH_2-NH-, carbohydrates, H-C-O and usually to a minor extent some aromatic groups. 72% of the H-C groups of the humic fractions occurs in $(CH_2)_n\,CH_3$ structures, with two superimposed groups of peaks, one from (average) $n=7$ at $\tau=8{,}69$ and the other from $n\simeq 20$ at $\tau=8{,}70$ [21]. Wilson and his coworkers' [65] results indicate that the values of parameters $\acute{f}_a$ (the fraction of aromatic carbon plus carboxylic carbon to total carbon) and H^*_{ar} (the fraction of protons which is aromatic) of humic materials extracted from Wakanui silt loam are 0.65 and 0.16, respectively. The combination of a low H^*_{ar} and high $\acute{f}_a$ value recorded for loam extract, shows that solution contains aromatic material which is polyclic or heavily substituted monocyclic.

Quite recently, Ruggiero and his collaborators [44] have carried out ^{1}H-NMR (100 mg/ml in DMSO-d_6) and ^{13}C-NMR (100 mg/ml in 0.1N NaOH) studies on FA's and HA's derived from an Andosol (Mount Vuture, Italy), with a special emphases on the importance of aromatic structures. ^{1}H-NMR spectra of all organic fractions present considerable absorption between 7.4 and 8.8 ppm, due to the presence of aromatic protons. On integerating the ^{1}H-NMR spectra of carefully dried DMSO-d_6 solutions of all samples gave the following quantitative information about the percentage of aromatic protons with respect to the total amount of protons present in the humic substances:FA's 20%, HA's 19%; oxidatively degraded with peracetic acid method at 40 °C FA's $<12\%$ and HA's 14%. The percentages obtained are significant enough to show the importance of aromatic structures in the humic fractions [44].

Products Obtained by Degradation and Physical Separation

For complex materials such as humic and fulvic acids, degradation is often useful in obtaining information on their chemical structures. The degradative methods applied on humic substances are mainly of three types: oxidative, reductive and biological [12]. In general, oxidative degradation has been more successful than reductive degradation since HS's, already contain considerable amounts of oxygen and are difficult to reduce.

More than 150 monomeric products of HS's obtained by degradative and other methods, have been identified. These include three major types of compounds: aliphatics, substituted benzenes and polycyclic aromatic hydrocarbons [12]. The first class consists of *n*-alkane, *n*-alkanoic acids, isoalkanoic acids, fatty acids, alkanedicarboxylic acids like malonic acid, succinic acid etc., 1,2,3-propane tricarbocylic acid, citric acid and 1,2,3,4-butanetetracarboxylic acid. Substituted benzenes include ethyl benzenesulfonate (*1*), ethyl benzylsulfonate (*2*), *p*-toluenesulfonamides (*3,4*), benzene mono- and polycarboxylic acids (*5–16*), polyphenols (*17–20*), polyhydroxytoluenes (*21–23*), hydroxy- (and methoxy-) benzaldehydes (*24–27* etc.), hydroxy- (and methoxy-) benzoic acids (*28–33* etc.), hydroxybenzene polycarboxylic acids (*34–40* etc.), polycarbohydroxy pehnyl acetic acids (*41–45*), tetracarbohydroxybenzyl alcohols (*46–48*), guaiacyl – and syringyl – propionic acids (*49, 50* respectively), dibutyl – and dioctyl phthalates (*51, 52* respectively) and monoalkyl benzenes. The third class of products consists of compounds like naphthalenes (*53–56*), anthracenes (*57–59*), phenanthrenes (*60–62*), 2,3-benzofluorene (*63*) 1,2-benzofluorene (*64*), fluoranthene (*65*), 1,2-benzanthracene (*66*), chrysene (*67*), triphenylene (*68*), pyrene (*69*), methyl pyrenes (*70, 71*), perylene (*72*), 1,2-benzopyrene (*73*), 3,4-benzopyrene (*74*), 1,12-benzoperylene (*75*), coronene (*76*), naphta (2′,3′:1,2) pyrene (*77*) and carbazole (*78*) (references cited in Choudhry [12]). Most of these polyclic aromatic hydrocarbons result from reductive degradation of humic substances. In addition, dehydrodiveratric acid (*79*) also appears as oxidative degradation product of methylated- FA's, -HA's and – humin [26]. In the case

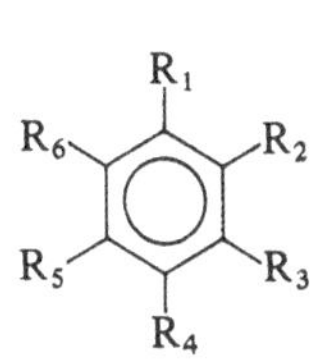

1. $R_1 = SO_2OC_2H_5$; $R_2 = R_3 = R_4 = R_5 = R_6 = H$
2. $R_1 = CH_2SO_2OC_2H_5$; $R_2 = R_3 = R_4 = R_5 = R_6 = H$
3. $R_1 = CH_3$; $R_4 = SO_2NHCH_2CH_2COOH$; $R_2 = R_3 = R_5 = R_6 = H$
4. $R_1 = CH_3$; $R_4 = SO_2NHCH_3$; $R_2 = R_3 = R_5 = R_6 = H$
5. $R_1 = COOH$; $R_2 = R_3 = R_4 = R_5 = R_6 = H$
6. $R_1 = R_2 = COOH$; $R_3 = R_4 = R_5 = R_6 = H$
7. $R_1 = R_3 = COOH$; $R_2 = R_4 = R_5 = R_6 = H$
8. $R_1 = R_4 = COOH$; $R_2 = R_3 = R_5 = R_6 = H$
9. $R_1 = R_2 = R_3 = COOH$; $R_4 = R_5 = R_6 = H$
10. $R_1 = R_2 = R_4 = COOH$; $R_3 = R_5 = R_6 = H$
11. $R_1 = R_3 = R_5 = COOH$; $R_2 = R_4 = R_6 = H$
12. $R_1 = R_2 = R_3 = R_4 = COOH$; $R_5 = R_6 = H$
13. $R_1 = R_2 = R_4 = R_5 = COOH$; $R_3 = R_6 = H$
14. $R_1 = R_2 = R_3 = R_5 = COOH$; $R_4 = R_6 = H$
15. $R_1 = R_2 = R_3 = R_4 = R_5 = COOH$; $R_6 = H$
16. $R_1 = R_2 = R_3 = R_4 = R_5 = R_6 = COOH$
17. $R_1 = R_2 = OH$; $R_3 = R_4 = R_5 = R_6 = H$
18. $R_1 = R_3 = OH$; $R_2 = R_4 = R_5 = R_6 = H$
19. $R_1 = R_3 = R_5 = OH$; $R_2 = R_4 = R_6 = H$
20. $R_1 = R_2 = R_3 = OH$; $R_4 = R_5 = R_6 = H$
21. $R_1 = CH_3$; $R_2 = R_4 = OH$; $R_3 = R_5 = R_6 = H$
22. $R_1 = CH_3$; $R_2 = R_6 = OH$; $R_3 = R_4 = R_5 = H$
23. $R_1 = CH_3$; $R_2 = R_4 = R_6 = OH$; $R_3 = R_5 = H$
24. $R_1 = CHO$; $R_4 = OH$; $R_2 = R_3 = R_5 = R_6 = H$
25. $R_1 = CHO$; $R_2 = OH$; $R_3 = R_4 = R_5 = R_6 = H$
26. $R_1 = CHO$; $R_3 = OCH_3$; $R_4 = OH$; $R_2 = R_5 = R_6 = H$
27. $R_1 = CHO$; $R_3 = R_5 = OCH_3$; $R_4 = OH$; $R_2 = R_6 = H$
28. $R_1 = COOH$; $R_2 = OH$; $R_3 = R_4 = R_5 = R_6 = H$
29. $R_1 = COOH$; $R_4 = OH$; $R_2 = R_3 = R_5 = R_6 = H$
30. $R_1 = COOH$; $R_3 = OCH_3$; $R_4 = OH$; $R_2 = R_5 = R_6 = H$
31. $R_1 = COOH$; $R_3 = R_5 = OCH_3$; $R_4 = OH$; $R_2 = R_6 = H$
32. $R_1 = COOH$; $R_3 = OH$; $R_2 = R_4 = R_5 = R_6 = H$
33. $R_1 = COOH$; $R_3 = R_4 = OH$; $R_2 = R_5 = R_6 = H$
34. $R_1 = R_5 = COOH$; $R_2 = OH$; $R_3 = R_4 = R_6 = H$
35. $R_1 = R_2 = COOH$; $R_3 = OH$; $R_4 = R_5 = R_6 = H$
36. $R_1 = R_2 = COOH$; $R_4 = OH$; $R_3 = R_5 = R_6 = H$
37. $R_1 = R_3 = COOH$; $R_2 = OH$; $R_4 = R_5 = R_6 = H$
38. $R_1 = R_3 = COOH$; $R_4 = OH$; $R_2 = R_5 = R_6 = H$
39. $R_1 = R_3 = COOH$; $R_5 = OH$; $R_2 = R_4 = R_6 = H$
39a. $R_1 = R_2 = R_3 = R_4 = R_5 = COOH$; $R_6 = OH$
40. $R_1 = R_3 = COOH$; $R_4 = OH$; $R_5 = OCH_3$; $R_2 = R_6 = H$
41. $R_1 = CH_2COOH$; $R_2 = COOH$; $R_3 = R_4 = R_5 = R_6 = H$
42. $R_1 = CH_2COOH$; $R_2 = R_3 = R_4 = COOH$; $R_5 = R_6 = H$
43. $R_1 = R_2 = R_3 = R_4 = COOH$; $R_5 = CH_2COOH$; $R_6 = H$
44. $R_1 = R_2 = R_3 = R_5 = COOH$; $R_4 = CH_2COOH$; $R_6 = H$
45. $R_1 = R_2 = R_3 = R_4 = R_5 = COOH$; $R_6 = CH_2COOH$
46. $R_1 = R_2 = R_3 = R_4 = COOH$; $R_5 = CH_2OH$; $R_6 = H$
47. $R_1 = R_2 = R_4 = R_5 = COOH$; $R_3 = CH_2OH$; $R_6 = H$
48. $R_1 = R_2 = R_3 = R_5 = COOH$; $R_4 = CH_2OH$; $R_6 = H$
49. $R_1 = CH_2CH_2COOH$; $R_3 = OCH_3$; $R_4 = OH$; $R_2 = R_5 = R_6 = H$
50. $R_1 = CH_2CH_2COOH$; $R_3 = R_5 = OCH_3$; $R_4 = OH$; $R_2 = R_6 = H$
51. $R_1 = R_2 = COO\text{-iso-}C_4H_9$; $R_3 = R_4 = R_5 = R_6 = H$
52. $R_1 = R_2 = COOC_8H_{17}$; $R_3 = R_4 = R_5 = R_6 = H$

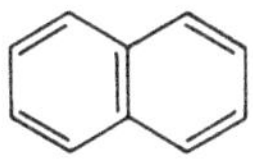
53
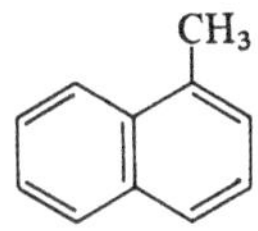
54
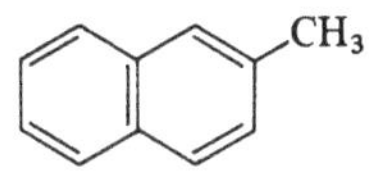
55
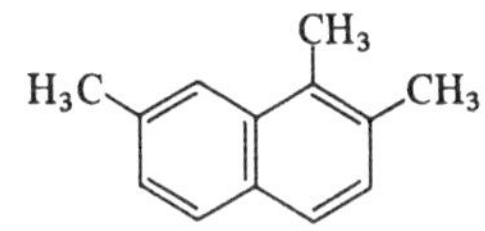
56

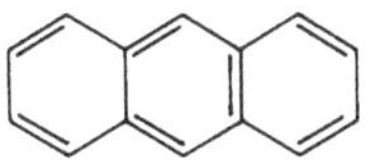
57
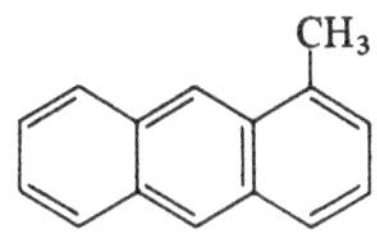
58
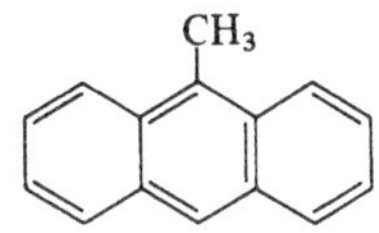
59

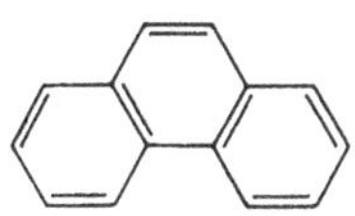
60
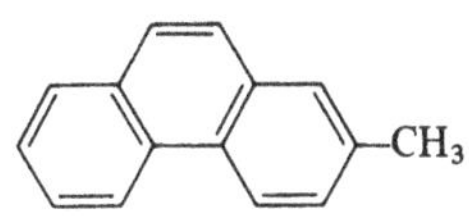
61
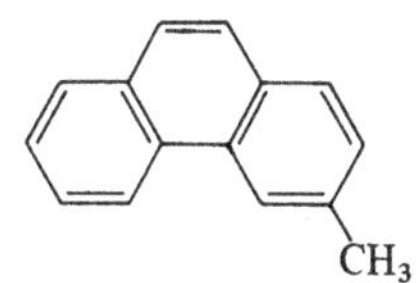
62

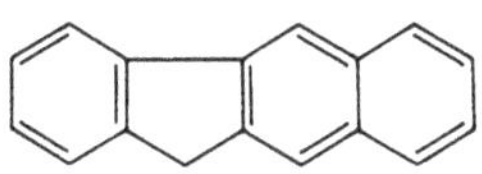
63
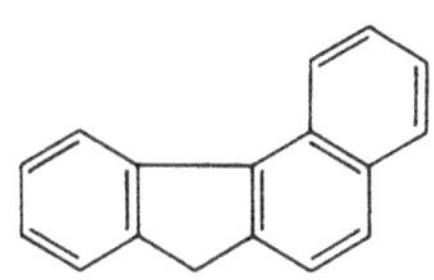
64
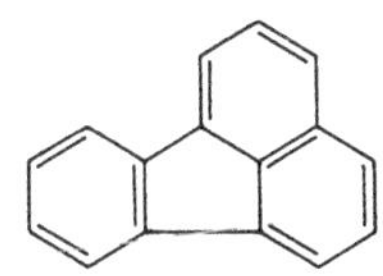
65

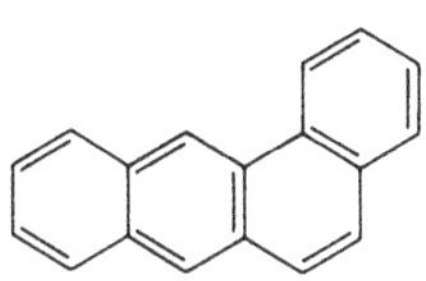
66
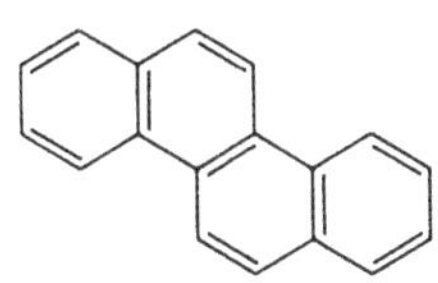
67
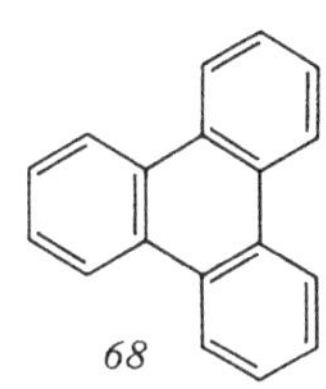
68

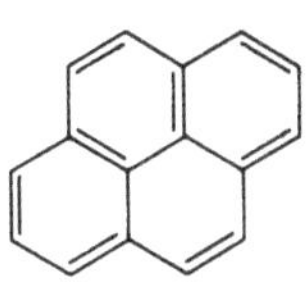
69
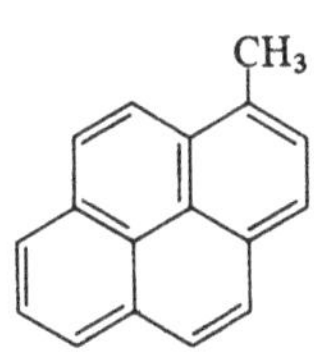
70
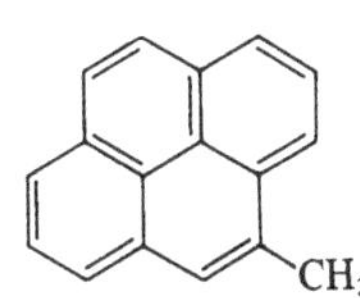
71

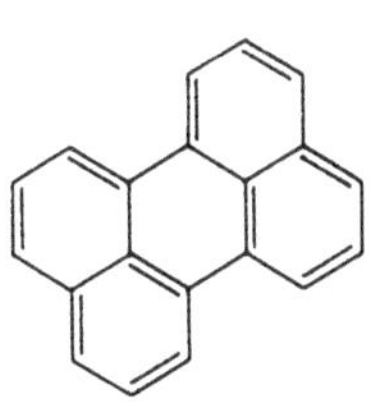
72
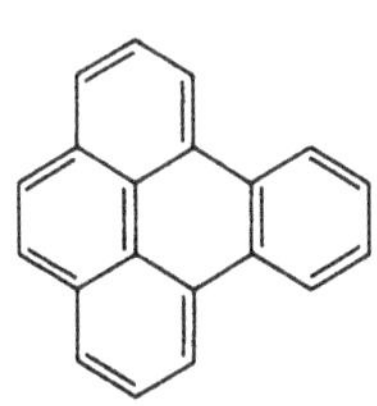
73
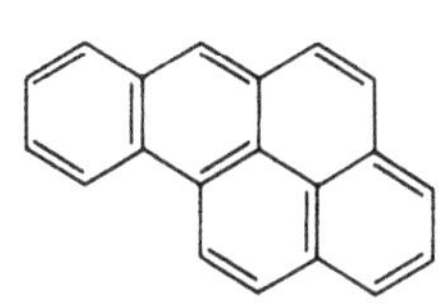
74

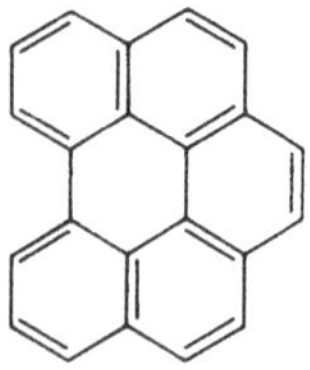

75

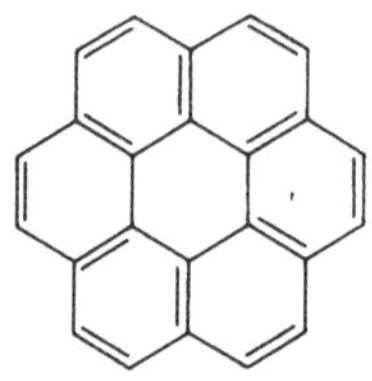

76

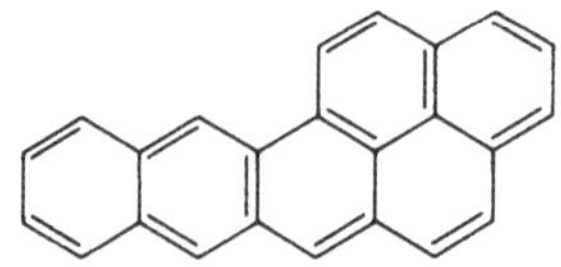

77

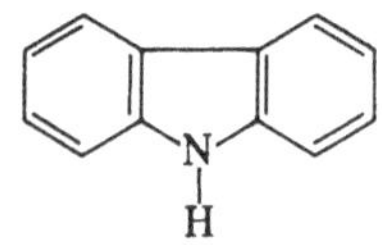

78

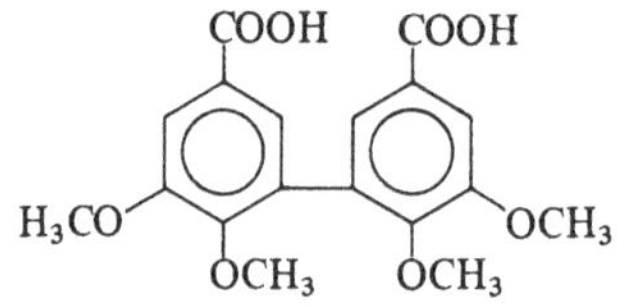

79

of alkaline $KMnO_4$ oxidation, some researchers *e.g.* Chen et al. [10], Maximov et al. [35] and Khan and Schnitzer [26], have used methylated HS's. Methylation prior to oxidation protects phenolic hydroxyl groups against attack by electrophilic $KMnO_4$ and thus permits the isolation of phenolic acids in addition to benzenecarboxylic and aliphatic acids. Degradation products like *p*-hydroxybenzoic acid (*29*), catechol (*17*), vanillin (*26*), vanillinic acid (*30*), syringaldehyde (*27*), syringic acid (*31*), guaiacylpropionic acid (*49*), syringlpropionic acid (*50*), protocatechuic acid (*33*) and 5-carbohydroxyvanillic acid (*40*) are well known lignin type compounds [45]. The majority of the aliphatic compounds and substituted benzenes were obtained through the oxidative degradation procedures (for further details see Choudhry [12]).

Some of the compounds, described above, appear as principal degradation products; while others as minor ones. For instance, Chen et al. [10] have reported that the most abundant aliphatic compounds resulting from the oxidation ($KMnO_4$) of HA's extracted from Italian and Israeli soils are di- and tricarboxylic acids and n-C_{16} and n-C_{18} fatty acids. Prominent phenolic acids are hydroxy benzene pentacarboxylic acid (*39*a) and to a lesser extent compound *35*, whereas the most abundant benzenecarboxylic acids are compounds *15, 13, 16, 10, 14*. Thus, tri-, tetra-, penta- and hexa-carboxylic acids are quantitatively the major HA degradation products. The yields of oxidation products from FA's are lower than those from HA's [10]. Furthermore, on $KMnO_4$ oxidation of methylated FA's and HA's, the maximum yields of all identified compounds amount to be 18% and 13%, respectively [10]; the contribution of aliphatic compounds to these yields is about less than 2% and the remainders aromatic compounds.

Some researchers [26, 37] have isolated and identified more than 100 organic compounds from soil FA's and HA's with the aid of the fractionation and identification scheme, developed by Barton and Schnitzer [7]; which does not involve any chemical degradation. For example, the amounts (in mg) of different groups of compounds gained from 100 g of methylated fulvic acids by these physical separation procedures are as follows: alkanes = 75.6; fatty acids = 47.2; phenolic acids = 297.9; benzene carboxylic acids = 194.4 and dialkyl phthalates = 350.0. Thus, the yield for each group of identified compounds ranges from 0.04% to 0.35%.

Working Hypotheses on Chemical Structure

Humus is obviously not a definable organic compound, and it is unlikely that the composition will be clarified in the foreseeable future. In spite of intensive studies for more than 100 years there does not seem to be any agreement regarding a unit for the humic substance molecules [20].

Recently, Anderson [2] has reported a hypothesis for humus formation and transformations, which is given in Fig. 1.

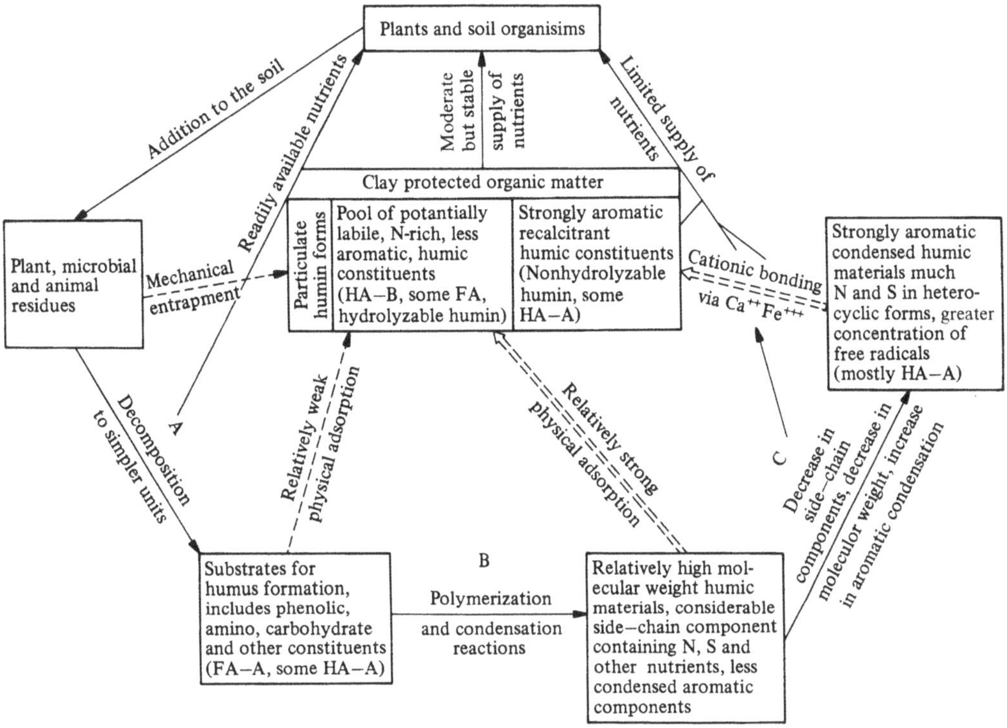

Fig. 1. Diagrammatic representation of processes of humus formation and transformation. (From Anderson [2])

According to Anderson's hypothesis: a) the decomposition of the added residues to the constituent parts such as lignin-derived phenolic units, carbohydrates, or amino compounds yields the building blocks or substrates for humus formation; b) the newly formed humic acids envisaged as having a relatively high aliphatic component, much of it in the form of side chains associated with a dominantly aromatic core; c) FA's, although primarily considered to be HA's precursors, may also be HA's degradation product; d) the humin fraction or non extractable humus is thought to include particulate organic matter entrapped within mineral aggregates, clay-adsorbed material like the HA-B (see Fig. 1) and resistant aromatic components.

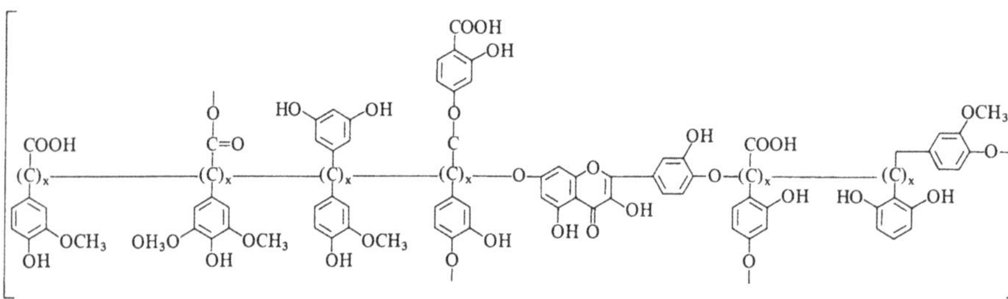

Fig. 2. A structure of the colored humus molecule proposed by Christman and Ghassemi [15]

In 1966, Christman and Ghassemi [15] proposed a structure for the colored humus molecule (Fig. 2). Based on the assumption that lignin plays an essential part in the humification processes, this suggested formula seems to be useful for theoretical purposes. Considering that the number of units may vary within a wide range, and that groups of organic and inorganic compounds may be attached to the unit and also substituted by others, it is easy to see the many existing possibilities which make a definite composition unlikely. The identification of lignin-type compounds e.g., p-hydroxybenzoic acid (*29*), catechol (*17*), vanillin (*26*), vanillinic acid (*30*), syringic acid (*31*), quaiacyl- and syringyl-propionic acid (*49, 50*; respectively) etc. among the degradation products of HS, may favor the idea that they are lignin-derived, but these lignin-derived components are not produced by a HA developed in a lignin-free environment in Antarctica [50]. According to Wilson et al. [65], there is increasing evidence that the origin of HS's are not lignin or coal.

Haworth [24] concluded that HA contains or readily gives rise to a complex aromatic core responsible for the electron spin resonance (ESR) signal and to which are attached chemically or physically: a) Polysaccharides; b) proteins; c) simple phenols; and d) metals as indicated in Fig. 3. The order of attachment of the groups

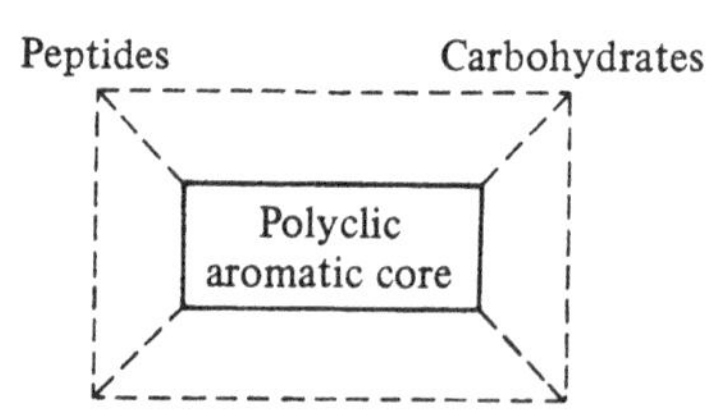

Fig. 3. Diagrammatic representation of humic acid [24]

is uncertain. The protein-core attachment appears to be quite stable against chemical and biological attack. He suggests that it is quite possible for some of the protein in HA to be hydrogen-bonded as in (formula *80*) or (formula *81*) and the remainder may be attached to the aromatic rings as in (formula *82*).

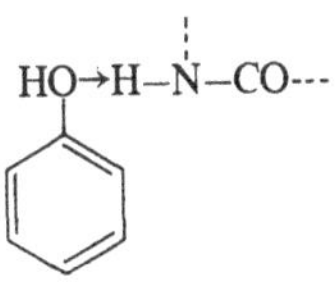

80

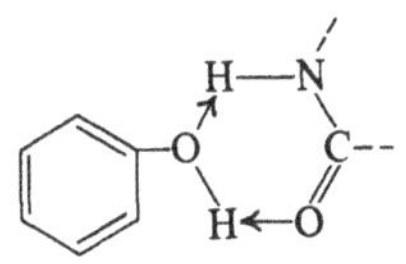

81

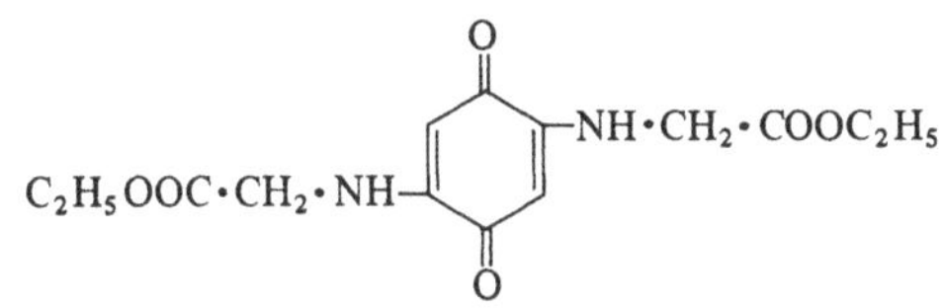

82

Majumdar and Rao [32] conclude from the studies on enzyme-degraded fulvic acid that the sugar and amino acid residues are present in very long chains attached to the aromatic core of the FA molecule. They suggest that the carbohydrate and amino acid portions are present as side chains attached to the aromatic core and not as bridge units between two blocks of aromatic core units. During the studies on the interactions between FA's, amino acids and carbohydrates in lake water, Haan and Boer [23] have suggested that the FA, amino acid and carbohydrate fractions that behave the same way on Sephadex G25 at both (3 and 7) pH value are associated with each other by strong (e. g. covalent) chemical bonds.

Recently, Eltantawy and Baverez [18] obtained X-ray diffraction traces for composts and soil HA's and observed three bands at 3.6, 2.1, and 1.2Å; which were ascribed to the aromatic structure or graphite-like layers. The presence of another band at 7.5 Å in the X-ray diffraction of the examined HA's indicated a nonaromatic fraction.

According to Schnitzer [49], upto 50% of the aliphatic structures in FA's and HA's consist of n-fatty acids esterified to phenolic OH groups of the following type (Fig. 4): where R_1 = COOH or $COCH_3$ or OH; R_2 = H or OH or COOH; R_3 = H or OH or OCH_3 or COOH; R_4 = OH esterified to fatty acid; R_5 = H or OH or OCH_3; R_6 = H or COOH and n equals mainly 14 and 16 for the HA and 14, 15, 16 and 18 for the FA. The remaining aliphatics are made up of more "loosely" held fatty acid and alkane that seem to be physically adsorbed on the humic materials and which are not structural humic components, and possibly of aliphatic chains joining aromatic rings. The appearance of phenolic and benzenecarbocylic acids as the major FA and HA degradation products is considered an evidence in favour of above mentioned view.

A proposed chemical structure for fulvic acids, which has appeared in some publications reported by Schnitzer and his associates [48, 49, 50], is given in Fig. 5.

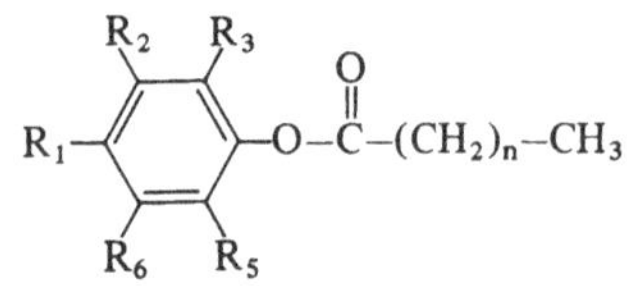

Fig. 4. Structure of phenol-fatty acid esters in HA's and FA's proposed by Schnitzer [49]

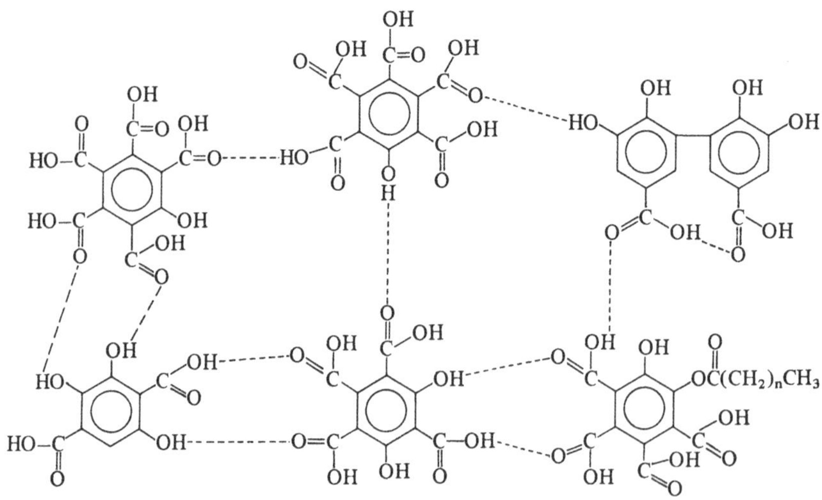

Fig. 5. A proposed partial chemical structure for fulvic acid. (Reproduced from Schnitzer [49])

Bonding between "building blocks" is by hydrogen-bonds, which makes the structure flexible, permits the "building blocks" to aggregate and disperse reversibly, depending on pH, ionic strength, etc., and also allows the FA to react with inorganic and organic soil constituents either via oxygen-containing functional groups on the large external surfaces, or by trapping them in internal voids.

Maximov et al. [35] have criticized Schnitzer's hypothesis on the structures of humic substances mentioned in the preceding paragraph. The energies of hydrogen bonds formed by carboxylic acids are not high and do not exceed 15 kcal/mol. Bond cleavage occurs under the action of even dilute alkalis with subsequent formation of salts. Breakdown of HA's by alkaline permanganate oxidation causes 50% of their carbon contents to be transformed into carbon dioxide and oxalic acid. If monomeric acids wehre the base of the HA structure, such drastic destruction would not be needed for liberation of these compounds. The substances formed through permanganate oxidation of HA's are essentially products of deep degradation of their macromolecules, not the "building blocks" retained in the initial molecules by a system of hydrogen bonds, as is assumed by Schnitzer [26, 53]. In this case, the appearing aliphatic and benzene carboxylic acids are just "echoes" or fingerprints of the initial structure [35]. Most of the yields in the literature for identified oxidation products of soil humic acids are likely too high [35]. Also, according to Majumdar and Rao [32], using strong acids and alkalis for hydrolysis, structural rearrangement may take place and characterization of the products may lead to misinterpretation of the structure of the parent molecule; and such rearrangement is less likely with enzymic hydrolysis.

Chen et al. [10] concluded that while there are some quantitative differences (i. e. low yields of phenolic acids) in the degradation products between the Mediterranean HA's and FA's and those from other climates, the general chemical structures must have been similar.

Finally it should be noted that soil fulvic acid and humic acid contain at least 20% and 19% aromatic protons, respectively [44].

Visible and Ultraviolet Light Absorbing Characteristics

In general, humic substances (HS's) yield uncharacteristic spectra in the visible (400–800 nm) and ultraviolet (200–400 nm) regions. Absorption spectra of alkaline and neutral aqueous solutions of HA's and FA's and of acidic, aqueous FA solutions are featureless, showing no maxima or minima: the optical density usually decreases as the wavelength increases [49]. Light absorption of HS's appears to increase with increase in (i) the degree of condensation of the aromatic rings which they contain; (ii) the ratio of carbon in aromatic "nuclei" to carbon in aliphatic or alicyclic side chains; (iii) total carbon content; and (iv) molecular weight (references cited in Choudhry [13]).

The ratio of optical densities (OD) or absorbances of dilute aqueous HA and FA solutions at 465 and 665 nm is often used for the characterization of HS's. This ratio, usually referred to as E_4/E_6, is independent of the concentration of the humic materials, but varies for HS's extracted from different soil types [27]. By some authors the E_4/E_6 ratios are termed as quotient values denoted by $Q_{4/6}$. Schnitzer [48] reported that these values for HA's and FA's extracted from soils formed under widely differing conditions are of the range 3.8–5.8 and 7.6–11.5, respectively. Kononova [27] believes that the magnitude of the E_4/E_6 ratio is related to the degree of condensation of the aromatic carbon network, with a low ratio indicative of a relatively high degree of condensation of aromatic humic substituents. Conversely a high E_4/E_6 ratio reflects a low degree of aromatic condensation and infers the presence of relatively large proportions of aliphatic structures. The E_4/E_6 ratio is independent of HA and FA concentrations at least in the 100–500 ppm-range and moreover, these ratios e.g. for FA (from Podzol) increase as the pH is raised from 1 to 6, attain maximum values between pH 6 and 8 and then begin to decrease gradually [9]. Recently, Banerjee [5] has reported that $Q_{4/6}$ values for parent HA's (extracted from Assam, Indian soil) and their fractions of molecular weight range $<4{,}000$; 4,000–10,00; and $>10{,}000$ are 5.0; 9.0; 7.2; and 3.6 respectively. This indicates that the fraction of lower molecular weight has higher $Q_{4/6}$ values. Thus the low molecular weight fractions show a relatively greater absorption in the blue region than in the red region of the spectrum.

A humic material that shows exceptional spectroscopic properties and that does exhibit a characteristic visible spectrum is the so-called "green HA" [13]. The green HA, dissolved in aqueous alkali, shows a λ_{max} near 450, 570, and 620 nm.

It has been suggested that the green HA is a metabolic product of fungal sclerotia [28].

Ultraviolet (UV) spectra of most humic substances are featureless, with the optical density decreasing as the wavelength increases. Occasionally an indication of a maximum can be discerned in the 260–300 nm region (references cited in Choudhry [13]). For instance, the UV spectrum of low-molecular-weight fraction of a Podzol Bh FA in alkaline solution exhibits a λ_{max} at 265 nm, which disappears on neutralization but shifts to 270 nm, when the solution is acidified [52]. Similarly, Sato and Kumada [46] reported a λ_{max} at 281 nm for the green HA dissolved in alkali. Schnitzer [47] has obtained a straight line by plotting molar extinction coefficient (ε) at 212 nm vs. number average molecular weights ($\bar{M}_n$) (from 274 to 915) of methylated FA fractions separated over Al_2O_3. He has interpreted this observation as an indication of the presence of repeating nonconjugated chromophoric units.

According to Swift et al. [62], the resemblance to lignin, as determined by IR, UV and UV ionization spectra, decreases as the molecular weight of the HA fractions decreases. Finally, it is noteworthy that UV spectra of HS's of diverse origins are very similar in spite of difference in elementary composition, sedimentation characteristics and other properties [67].

Fluorescent Characteristics

HA's and FA's fluoresce both under visible and UV light. Fluorometric methods have been used to observe the progress of fractionation by chromatographic methods as well as to determine the content of humic substances in the water bodies. Seal et al. [54] reported that FA's and HA's isolated from Indian black cotton soil exhibit similar green fluoresce with flat maxima in the 500–540 nm region. From these observations they conclude that FA's, hymatomelanic acids (HyA's) and HA's constitute a sequence of polymers. The low fluorescence yield was ascribed to fluorescence-inhibiting functional groups, molecular crowding and absorption of part of the emitted light by absorbing centers. SEAL et al. [54] conclude that the fluorescence is due either to the presence of an aromatic nucleus substituted by at least one electron-donating group, or to a conjugated unsaturated system, capable of a high degree of resonance.

It has been reported that pH affects the fluorescence of FA [50]. When dissolved in MeOH, FA shows a fluorescence maximum near 507 nm; in 0.01 *M* MeONa this maximum is lowered to 465 nm. Selective blocking of COOH and OH groups by methylation, acetylation and esterification does not shift the fluorescence maxima when the samples are dissolved in MeOH. When these same preparations are dissolved in MeONa, however, the fluorescence maxima for the untreated and for the esterified FA's are lowered to about 465 nm but those for the other preparations remain unchanged. This is interpreted to indicate that the fluorescence of FA is associated with the ionization of phenolic OH but not with that of COOH groups. According to Datta et al. [16] fluorescence spectra of Na-humate, -hymatomelanate, and -fulvate exhibit maxima at 470 nm in aqueous solutions. These maxima remain unaltered in the residues after hydrolysis with 6

NHCl. In ether, pyridine, acetone and DMF (dimethyl foramide) the maxima shift to 370 nm; while in alcohols it is at 400 nm. Thus, the fluorescence of HA's and FA's is affected by pH and the polarity of the solvent. Since the maximum appears only in very dilute solutions, the extinction coefficients of the fluorescing species appear to be very high.

Adhikari and Hazra [1] have reported the following fluorescent maxima (in nm) for the humus-metal complexes: Na-fulvate, 500; Ni-fulvate 487; Cu-fulvate, 450; Na-hymatomelanate, 434; Ni-hymatomelanate, 420, 550; Cu-hymatomelanate, 420, 550; Na-humate 489; Ni-humate, 475; and Cu-humate, 475. The fluorescence excitation spectra of all these fractions are more or less identical in shape. These observations lead to a conclusion that FA's, HyA's and HA's are identical groups of humic substances, representing some intermediate stages of a polymer series [1]. The difference in the intensity of observed fluorescence may be due to the concentration variation in fluorescing groups present in these substances. The fluorescence yield of different humus fraction-metal complexes are lower than those of humus fractions alone in all wavelenghts. The fluorescent excitation maxima of humus metal combinations are shifted in general towards the shorter wavelength region. In case of hymatomelanic acid-metal complexes besides the usual peak at 420 nm, a second peak is noticed in the longer wavelength region (550 nm). The latter peak is due to polymerization effect as fluorescence peak of polyphenols is shifted to longer wavelength region in synthetic HA's [1]. The fact that acid hydrolysis has no effect on fluorescent properties of HS's [6], certainly signifies that the fluorogens responsible for fluorescence properties have not been originated from amine acid moieties of humic substances [1].

Christman and Ghassemi [15] have found a linear relationship between color concentration of humus water and fluorescence. They found that humus water from different sources had the wavelength of maximum fluorescence-emission centered at 490 nm whith the combined excitation wavelength of 365, 404, and 436 nm. The wavelength of maximum emission is not shifted by acid or base, although the fluorescent intensity is pH-dependent.

Lévesque [31] has studied the fluorizing properties of various humus fractions, as for instance FA using Sephadex and he concludes that the high-molecular-weight fraction had little or no fluorescence. Low-molecular-weight fractions fluorized intensively and at higher wavelengths. They exhibit emission peaks of 520 nm, when excited at 468 nm.

Free Radical Characteristics

Humic substances are known to be rich in stable free radicals [13]. It is most likely that this free radical (FR) character of HS's play important roles in polymerization-depolymerization reactions, in reaction with other organic molecules, including pesticides and toxic pollutants, and in the physiological effects that these substances are known to exert [49].

Rex [42], was the first researcher, who discovered free radical characteristic of HS's and related compounds. He has reported that electron paramagnetic resonance (EPR) studies of wood, lignin, tanins, and humic acids, such as found in soils and coals, indicate an EPR line of 6 ± 2 gauss width (the peak to peak separation of the derivative signal being onsidered as line width) at g (spectroscopic splitting factor) $= 2.003 \pm 0.002$, exhibiting the presence of a free radical. According to Rex [42], FR's survive within the lignin micelles through the process of coal formation and persist apparently for geological time (10^8 years). He believes that the FR's are semiquinones formed by the dehydrogenation or oxidative removal of hydrogen from aromatic -OH, $-NH_2$, or -SH groups.

Tollin and his coworkers [64] observed that the unpaired spin concentrations of sodium salts of HA (extracted from a Podzol B forest soil) prepared by tratment of HA with i) Na in MeOH; ii) NaOMe in MeOH; and iii) Na in H_2O are about 25 times, 16 times and 7 times greater than that of the original (untreated) acid (1.4×10^{18} spins/g). Acidification of the reduced sodium humate leads to a product with lower radical content (0.6×10^{18} spins/g) than the original HA. Tollin et al. [64] accomodated these results to a humic acid model which contains quinone [83] moieties coexistent with semiquinone [84] and quinhydrone [85] moieties. The chemical transformations of reduction and basification of the proposed model HA moieties are illustrated below.

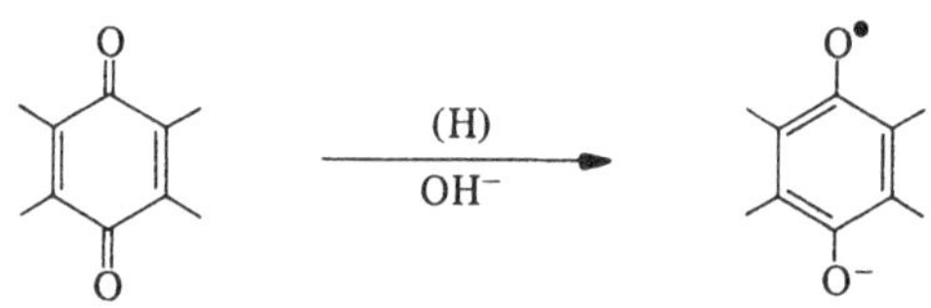

83 *84* (1)

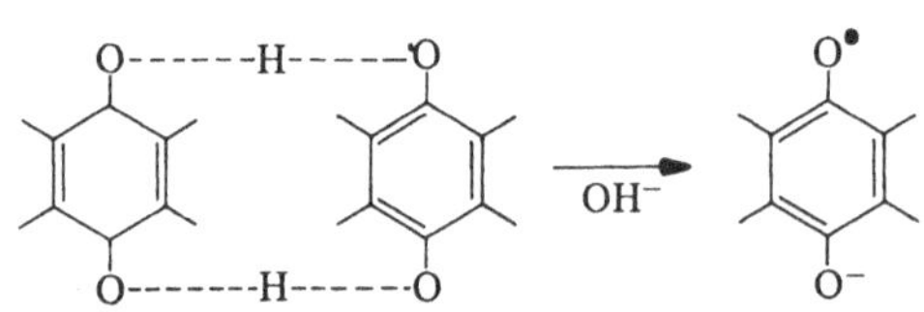

85 *84* (2)

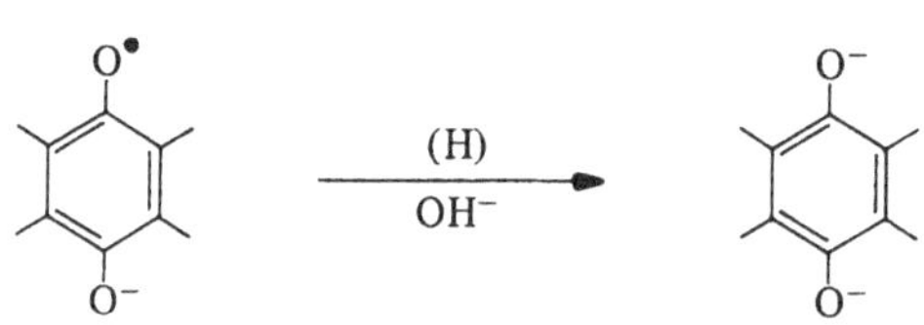

84 *86* (3)

Although Reactions 1 and 2 produce an increase in semiquinone-radical ions *84*, it is also possible that these ions (or semiquinones existing in HA prior to chemical treatment) may be reduced to the diamagnetic phenolate (*86*) as in Reaction 3. Thus, if all three reactions are possible, one would anticipate at least one (and probably more) maximum in spin concentrations as function of time of reduction or amount of reducing agent. Studies performed by Tollin et al. [64] for chemical reduction of HA confirms this expectation. The initial rapid increase in radical

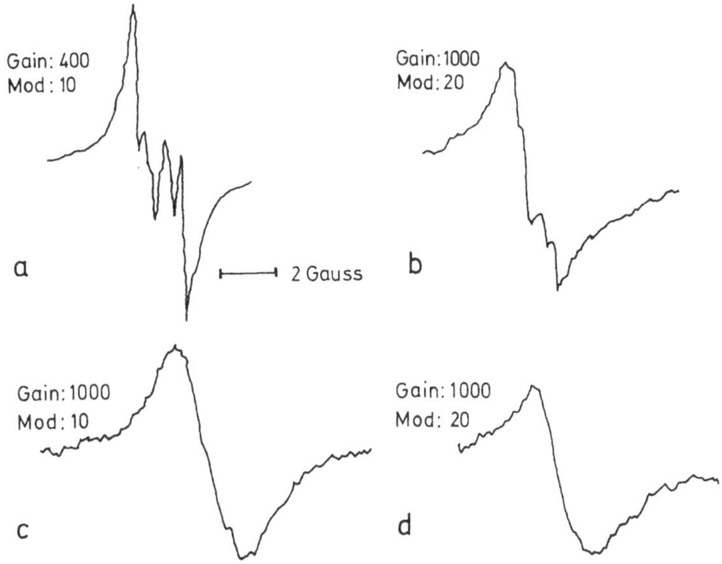

Fig. 6. ESR spectra: a and b represent Class I ABHA; while c and d represent Class II ABHA (reproduced from Haworth [24]

concentration (maximum attained after 30 min) is most likely a result of Reaction 2, whereas the subsequent variation in spin concentration could be caused by competition between Reactions 1 and 3. It is significant that reduction does not alter the general shape of the EPR spectrum, indicating that no marked change in radical species occurs [64].

Atherion et al. [3] have investigated the electron spin resonance (ESR) spectra of "acid-boiled humic acid" (ABHA) isolated from eighty different soils and peats. ABHA's were prepared by boiling HA's with 6N HCl. The solid ABHA indicates strong absorption at $g = 2.00$, which is some 4.1 gauss in breadth between points of maximum slope. The ESR spectra of ABHA in solution phase (solvent: 0.1 N NaOH) have been divided into two classes. Typical spectra are shown in Fig. 6 which is a plot of the gradients of the absorption curve against the magnetic field strength.

Class I, containing ABHA's from 51 sources, shows four-lined spectra with a breadth of 1.75–1.9 gauss, and with the lines in the same relative positions, although some spectra were better resolved than others; the range of Class I soils is illustrated by curves (a) and (b) in Fig. 6. The hyperfine structure (four lines) indicate that the unpaired electron is interacting with two non-equivalent protons with coupling constants 1.2 and 0.35 gauss. The unsymmetrical curve (a) in Fig. 6 suggests some heterogenity. Class II, containing 28 samples, exihibit ill-defined, structureless spectra, without clear peaks but with the same overall breadth of 1.75–1.9 gauss, as shown by curves (c) and (d) in Fig. 6. Later, Cheshire and Cranwell [11] reported, on the ESR spectra of ABHA's prepared from the humic acids extracted from the Scottish cultivated soils. They observed ESR spectra with a four-peak structure (Class I; pH <4.6) and almost structureless spectra (Class II; pH >4.6).

It was concluded by them that the type of HA (humic acid) in a soil is not related to the present pH of the soil but rather to the conditions of pH attending its formation. These observations and conclusions put forward by Cheshire and Cranwell [11] correspond to those of Haworth and his coworkers [3, 24].

From the ESR spectrometric studies of FA's, HA's and humins originating from soils of widely differing geographical origins and pedological histories, Riffaldi and Schnitzer [43] concluded that: (1) The free radical (FR) content decreases in the following order: humin > HA's > FA's, (ii) line widths decrease in the order: FA's > Humin > HA's and (iii) g-values of the humins are similar to those of HA's. Senesi et al. [56] have demonstrated an unsymmetrical triplet hyperfine splitting in ESR spectra of FA's of Podzol Bh origin, when treated at room temperature with dilute H_2O_2 (pH 7) and with silver oxide (pH 13). The three equally spaced lines of the triplet indicate interaction of the unpaired electron with two equivalent protons [57]. During the oxidation, reduction and irradiation experiments on fulvic acids (FA's) extracted from Podzol Bh soil, two types of FR's by means of ESR spectrometry could be detected; which are as follows: (a) permanent ones, having long life spans and (b) transient ones, not sufficiently long lived under oxidative conditions to be detected but sufficiently stable under reducing conditions and after irradiations [57]. Recently, Schnitzer and Levesque [51] have developed a novel method for determination of the degree of humification of peats by using ESR technique. According to these researchers, the FR's that are involved in humification are substituted semiquinones, whose concentrations increase as humification advances.

Irradiations

Commonly, three types of observations during irradiations of humic substances have been reported in the literature: (i) effects of irradiation on the FR content; (ii) photooxidation; and (iii) effects of irradiation on absorption, fluorescence and chemiluminescence spectra.

Tichy [63] reported that spin concentration and biological activity increase after irradiation of HA with UV-light. Sławinska et al. [58] studied the influence of visible light (395–780 nm) and O_2 on the ESR spectra of natural (from Podzol soil), commercial (Merck) and synthetic humic acids. All the materials investigated exhibit enhancement of the ESR signals by visible light. Parameters for ESR-signals of natural HA's are described in the Table 1. In an oxygen atmosphere, the g-value changes by 0.0010 Gauss (Gs). The relative changes of ESR-signal intensities induced by light, being characterized by the ratio $\chi = I_L I/_D$, in different atmospheres decrease in the following order: $O_2 > N_2 >$ air (see Table 1). From these studies, Sławinska et al. [58] concluded that dark HA absorb light energy very efficiently. Because of the HA-O_2 spins coupling, absorbed energy can probably be transfered to oxygen (Scheme I):

Scheme I

$$
\begin{array}{l}
{}^3HA\,(n-\pi)_{C=O} + {}^3O_2 \rightarrow {}^1HA + {}^1O_2^* \rightarrow (ODP)^* \\
\qquad\qquad\qquad\qquad\qquad\quad \downarrow \qquad\qquad\quad \downarrow \\
\qquad\qquad\qquad\qquad\quad {}^3O_2 + \text{light} \quad ODP + \text{light}
\end{array}
$$

Table 1. Parameters of ESR-signals of solutions of Podzol soil HA's[a] in darkness (D) and illumination (L) under N_2, O_2 and air – atmosphere, (Reproduced from Slawinska et al. [58])

Atmosphere	g-values, ±0.00005		Line widths (Gs)		$\chi^b = I_L/I_D$
	D	L	D	L	
N_2	2.0032	2.0035	≃4	≃5	5.47
O_2	2.0032	2.0042	≃5	≃5	5.50
air	2.0055	2.0056	≃5	5.1	5.13

[a] Solutions of HA in aqueous alkali 2.0 or 3.0 mg cm^{-3} were buffered with 0.02 *N* $NaHCO_3 - NaCO_3$ to pH 9.0–9.2 and saturated with air, O_2 or N_2 at 298 °K.

[b] I_L and I_D represent the intensities of the ESR-signals in illumination (L) and darkness (D), respectively

where ODP represents oxidative degradation products of HA. This assumption seems to be supported by the fact that light induced ultra-weak emission of light (chemiluminescence) arises under the same conditions as ESR-signals. According to the above scheme, chemiluminescence is a result of competitive, radiative decatication of singlet oxygen $^1O_2^*$ and/or electronically excited (*) ODP, formed in an exothermic ($\Delta H > 210$ kJ $mole^{-1}$) oxidative ring cleavage of quinone rearrangement in HA [58]. It should be mentioned that oxygen possesses two metastable singlet states with spectroscopic symmetry notations $^1\Sigma_g^+$ (37 kcal/mole) and $^1\Delta_g$ (22 kcal/mole). The $^1\Delta_g$ state survives longer (for $\sim 10^7$ collisions more) than the $^1\Sigma_g^+$ state. Now it is generally accepted that $^1\Delta_g\, O_2$ is involved in photo-oxidation [14]. Thus, HA would be expected to act as photosensitizers of some bonded substances. For example, herbicides (environmental chemicals) sorbed on HA might undergo detoxication stimulated by light and O_2 (air) [58].

It was reported by Senesi and Schnitzer [55] that the spin concentrations of FA samples irradiated by white light (500 W), in solid and solution (at pH 2.2) phases, are about 1.5 and 1.15–1.40 times greater than those of nonirradiated samples respectively; the increase being sustained for about 1 h. When the light is switched off, FR contents return to the same values that exhibited prior to irradiation. Contrary to Lagercrantz and Yhland [29], Senesi and Schnitzer [55] concluded that irradiation with white light produces reversible increases in the FR concentrations. According to the latter researchers, this suggests that when sunlight falls on surface soils high in humic substances (organic matter), large concentrations of free radicals (FR's) should be present in the materials.

Chen and his coworkers [10] have performed ultraviolet (253.7 nm) irradiation of dilute FA solutions, originated from the Bh horizon of the Armdale, a poorly drained Spodosol on Prince Edward Island, Canada. Dilute aqueous FA solutions of concentration levels (w/v) of 0.125 and 0.01% were exposed to UV irradiation for various lenghts of time at pH 3.5, 7.0 and 11.0. They concluded that: (i) The rate of photooxidation of aq. FA solutions by UV irradiation is pH-dependent, increasing as the pH rise e. g. it takes only 12 h of UV irradiation at pH 11.0 to produce the same weight as 23 h of irradiation at pH 3.5. (ii) Zero order reaction, which is common for photochemical reactions involving homogeneous systems, fit their experimental data during the first 4.5 h only. Regardless of pH, the photooxi-

Table 2. Compounds identified among products resulting from irradiation of 1.0 g of FA. (Chen et al. [10])

No.	Compounds identified[a, b]	0.01% aqueous FA solution (amounts in mg)		
		pH 3.5	pH 7.0	pH 11.0
–	n-C_{16} fatty acid	1.5		0.6
–	n-C_{18} fatty acid	1.3		0.7
6	1,2-benzene dicarboxylic	3.2		1.2
9	1,2,3-benzene tricarboxylic acid	14.4	17.0	1.9
11	1,3,5-benzene tricarboxylic acid	9.2	4.4	5.1
13	1,2,4,5-benzene tetracarboxylic acid			2.0
14	1,2,3,5-benzene tetracarboxylic acid	2.0	41.0	2.0
15	1,2,3,4,5-benzene pentacarboxylic acid	1.9	11.3	17.8
16	1,2,3,4,5,6-benzene hexacarboxylic acid	11.3	4.0	20.1
	Total identified	44.8	77.7	51.4

[a] The compounds were isolated and identified in fully methylated forms; diazomethane being used for methylation
[b] Chemical structures are described in the Section on degradation products

dation of 0.01% solution appears to be governed by phase boundary controlled reactions, that is, reactions controlled by the movement of an interface at a constant velocity involving a cylinder-shaped particle. (iii) In acid solutions, most of the organic matter appears to be oxidized to CO_2 and water. Organic S is converted to inorganic sulphate. Under neutral and alkaline conditions, organic C is oxidized to inorganic carbonate. (iv) Their data show that UV irradiation of dilute FA solutions destroys about 90% of the initial organic matter. Only the most resistant compounds, which are benzene carboxylic and aliphatic straight-chain fatty acids, resist decomposition. Phenolic acids, which are major FA "building blocks" and the chromophores responsible for the yellowish-brown color of FA, are destroyed. Compounds identified by Chen et al. [10] among products resistant to UV irradiation of the various pH's are listed in Table 2.

Major compounds identified are benzene-di-, -tri-, -tetra-, -penta-, and -hexacarboxylic acids. In addition, smaller amounts of n-C_{16} and n-C_{18} fatty are also identified. The compounds identified (Table 2) account for about 5–8% of the initial FA. Especially noteworthy is the absence of phenolic compounds among the organics identified since both benzene-carboxylic and phenolic acids are major FA degradation products. According to Chen et al. [10], their failure to detect phenolics amongst irradiation products may mean that these compounds are destroyed by UV irradiation; and in this regard the mechanism of photooxidation of FA resembles that of $KMnO_4$ oxidation of unmethylated fulvic acid. Finally, in the photooxidation of FA at low pH, decarboxylation of free COOH groups (diminuition and disappearance of 1720 cm^{-1} band) is a major reaction mechanism [10].

Recently, Sławinski and his associates [59] reported that absorption spectra of Fluka HA's measured after different times of photo-irradiation show a decrease of absorption in the whole measured spectral range (230 – 700 nm). They plotted $Q_{4/6}$ vs. τ_{ir} (time of irradiation). This plot indicated two phases of irradiation of the HA's. In the first phase of irradiation, the ratio $Q_{4/6}$ that is a conventional measure of degree or aromaticity of HA's core gradually increases. According to Sławinski et al. [59], it is undoubtedly caused by the photodegradation of complex aromatic structures of the core of HA's. The deflection of the curve $Q_{4/6} = f(\tau_{ir})$ observed in the second phase of photooxidation shows the decrease of the relative degradation rate of a highly condensed aromatic structure. It may be caused by a competitive cross-linking process involving photodimerization of complex aromatic structures [59].

Chemiluminescence spectra of 0.01% Fluka HA's (irradiated) measured with cut-off filters and a calibrated photomultiplier operating in the single photon counting mode reveal three main emission bands at 480–500, 570 and 615–650 nm [59, 60]. The relative intensities, which on the average are close to 1.0:0.6:0.7, depend on the τ_{ir} and the source of polymers. Intensity of the red band decreases with the increase of τ_{ir} and maximum intensity shifts toward the short wave lengths. The chemiluminescence emission bands at 480–500 nm match the fluorescence of the HA's as well as phosphorescence of the Fluka hymatomelanic acids (which emit at 78 K in the region 420–550 nm). According to Sławinski et al. [60], it seems that process of excitation resulting from the oxidation or an energy transfer from $^1O_2^*$ (singlet oxygen) contribute considerably to the emission. This is supported by the integral intensity of chemiluminescence decrases approx. 2 times for HA, when irradiation is performed in a N_2 atmosphere. The emission band at 570 nm corresponds to the long-wavelength band of HA's fluorescence. The red chemiluminescence at 615–650 nm corresponds neither to fluorescence nor phosphorescence. The red band is not sensitive to HA concentration which seems to eliminate the possibility of excimer or exciplex (see Choudhry et al. [14], for information of these species) formation. This band may be assigned to the emission from $^1O_2^*$ ($^1\Delta_g$) dimols [60] (Reaction 4):

$$2\,O_2\,(^1\Delta_g\,^1\Delta_g) \rightarrow 2\,O_2\,(^3\Sigma_g^-) + h\nu \;\; \lambda = 634\ \text{nm} \tag{4}$$

The emission in the region 480–570 nm may be due to the sensitized chemiluminescence from fluorescing part of HA's molecules (Reaction 5):

$$P_1^* + P \rightarrow P_1 + P^* \rightarrow P_1 + P + h\nu \tag{5}$$

where P_1^* is the primary excited product of photooxidation [60].

Finally, it should be mentioned that Zepp et al. [66] have presented evidence that rapid photochemical generation of singlet oxygen ($^1O_2^*$) occurs in inland and coastel water bodies of the southern-eastern United States. Steady state concentration of singlet oxygen varied from $2 \times 10^{-13} M$ to $22 \times 10^{-13} M$; that were photochemically generated by the action of sunlight. According to Zepp and his coworkers [66], these water bodies contain high concentrations of organics, most of which are FA's. Because most inland and coastel water bodies contain fulvic acids (FA's), it is reasonable to infer that photochemical generation of singlet oxygen ($^1O_2^*$) is a wide spread phenomenon in the aquatic environment.

References

1. Adhikari, M., Hazra, G.G: Humus-metal complex: Spectral studies. J. Indian Chem. Soc. *53*, 513 (1976)
2. Anderson, D.W.: Process of humus formation and transformation in soils of the Canadian great plains. J. Soil Sci. *30*, 77 (1979)
3. Atherton, N.M., Cranwell, P.A., Floyd, A.J., Haworth, R.D.: Humic acid-I. ESR spectra of humic acids. Tetrahedron *23*, 1653 (1967)
4. Bailly, J., Margulis, H.: Étude de quelques acides humiques sur gel de dextrane. Plant Soil *29*, 343 (1968)
5. Banerjee, S.K: Acidity, quotient values, and metal retention power of humic acids of varying molecular weight. J. Indian Soc. Soil Sci. *27*, 38 (1979)
6. Banerjee, S.K., Mukherjee, S.K.: Physicochemical studies of divalent transitional metal ions with humic and fulvic acids of Assam soil. J. Indian Soc. Soil Sci. *20*, 13 (1972)
7. Barton, D.H.R., Schnitzer, M.: A new experimental approach to the humic acid problem. Nature *198*, 217 (1963)
8. Chen, Y., Khan, S.U., Schnitzer, M.: Ultraviolet irradiation of dilute fulvic acid solutions. Soil Sci. Soc. Am. J. *42*, 292 (1978)
9. Chen A., Senesi, N., Schnitzer, M.: Information provided on humic substances by E_4/E_6 ratios. Soil Sci. Am. J. *41*, 352 (1977)
10. Chen, Y., Senesi, N., Schnitzer, M.: Chemical degradation of humic and fulvic acids extracted from Mediterranean soil. J. Soil Sci. *29*, 350 (1978)
11. Cheshire, M.V., Cranwell, P.A.: Electron-spin resonance of humic acids from cultivated soils. J. Soil Sci. *23*, 424 (1972)
12. Choudhry, G.G.: Humic substances. Part I: Structural aspects. Toxicol. Environ. Chem *4*, 209 (1981)
13. Choudhry, G.G.: Humic substances. Part II: Photophysical, photochemical and free radical characteristics. Toxicol. Environ. Chem. *4*, 261 (1981)
14. Choudhry, G.G., Roof, A.A.M., Hutzinger O.: Mechanisms in sensitized photochemistry of environmental chemicals. Toxicol. Environ. Chem. *2*, 259 (1979)
15. Christman, R.F., Ghassemi, M.: Chemical nature of organic color in water. J. Amer. Water Works Assoc. *58*, 723 (1966)
16. Datta, C., Ghosh, K., Mukherjee, S.K.: Flourescence excitation spectra of different fractions of humus. J. Indian Chem. Soc. *48*, 279 (1971)
17. Dubach, P., Metha, N.C., Jakab, T., Martin, F., Roulet, N.: Chemical investigations on soil humic substances. Geochim. Cosmochim. Acta *28*, 1567 (1964)
18. Eltantawy, I.M., Bavrez, M.: Structural study of humic acids by X-ray, electron spin resonance, and infrared spectroscopy. Soil Sci. Soc. Am. J. *42*, 903 (1978)
19. Flaig, W.: The chemistry of humic substances. Report of the FAO/IAEA Technical Meeting, 103 (1963)
20. Gjessing, E.T.: Physical and chemical characteristics of aquatic humus. pp. 3, 25, 27, 28, 83. Ann Arbor, MICH.: Ann Arbor Sci Publishers INC. (1976)
21. Grant, D.: Chemical structure of humic substances. Nature *270*, 709 (1977)
22. Griffith, S.M., Schnitzer, M.: The alkaline cupric oxide oxidation of humic and fulvic acids extracted from tropical volcanic soils. Soil Sci. *122*, 191 (1976)
23. Haan, H.de, Boer, T.de: A study of the possible interactions between fulvic acids, amino acids and carbohydrates from Tjeukemeer (a lake), based on gel filtration at pH 7.0. Water Res. *12*, 1035 (1978)
24. Haworth, R.D.: The chemical nature of humic acid. Soil Sci. *111*, 71 (1971)
25. Khan, S.U.: Distribution and characteristics of organic matter extracted from the black solonetzic and black chernozemic soils of Alberta: The humic acid fraction. Soil Sci. *112*, 401 (1971)
26. Khan, S.U., Schnitzer, M.: Permanganate oxidation of humic acids, fulvic acids and humins extracted from Ah horizons of black chernozem, a black solod, and black solonetz soil. Can. J. Soil Sci. *52*, 43 (1972)
27. Kononova, M.M.: Soil Organic Matter 2nd Ed. pp. 101, 104. Oxford: Pergamon Press (1966)
28. Kumada, K., Hurst, H.M.: Green humic acids and its possible origin as fungal metabolite. Nature *214*, 631 (1967)

29. Lagercrantz, C., Yhland, M.: Photo-induced free radical reactions in the solutions of some tars and humic acids. Acta Chem. Scan. *17*, 1299 (1963)
30. Lakatos, B., Meisel, J.: Biopolymer-metal complex systems V.PMR investigation of humic substances. Acta Agron. Acad. Sci. Hung. *27*, 313 (1978)
31. Lévesque, M.: Fluorescence and gel filtration of compounds. Soil Sci. *113*, 346 (1972)
32. Majumdar, S.K., Rao, C.V.N.: Physico-chemical studies of enzyme-degraded fulvic acid. J. Soil Sci. *29*, 489 (1978)
33. Mathur, S.P.: Infrared evidence of quinones in soil humus. Soil Sci. *113*, 136 (1972)
34. Maximov, O.B., Glebko, L.I.: Quinoid groups in humic acids. Georderma *11*, 17 (1974)
35. Maximov, O.B., Shvets, T.V., Elkiv, Yu.N.: On permanganate oxidation of humic acids. Georderma *19*, 63 (1977)
36. Odén, S.: Die huminsäuren. Kolloidchem. Beihefte *11*, 75 (1919)
37. Ogner, G., Schnitzer, M.: Chemistry of fulvic acid, a soil humic fraction and its relation to lignin. Can. J. Chem. *49*, 1053 (1971)
38. Oka, H., Sasaki, M., Itoh, M., Suzuki, A.: Studies on the structure of peat humic acid. I. Study on the chemical constitution of peat humic acids by high resolution nuclear magnetic resonance spectroscopy. Nenryo Kyokai-Shi 48/505, 295 (1969)
39. Posner, A.M.: Importance of electrolyte in the determination of molecular weights by "Sephadex" gel filtration with especial reference to humic acid. Nature *198*, 1161 (1963)
40. Rashid, M.A., King, L.H.: Molecular weight distribution measurements on humic and fulvic acid fraction from marine clays on the Scotian Shelf. Geochim Cosmochim. Acta *33*, 147 (1969)
41. Rashid, M.A., King, L.H.: Chemical characteristics of fractionated humic acids associated with marine sediments. Chem. Geol. 7, 37 (1971)
42. Rex, R.W.: Electron paramagnetic resonance studies of stable free radicals in lignins and humic acids. Nature *188*, 1185 (1960)
43. Riffaldi, R., Schnitzer, M.: Electron spin resonance spectrometry of humic substances. Soil Sci. Soc. Amer. Proc. *36*, 301 (1972)
44. Ruggiero, P., Interesse, F.S., Sciacovelli, O.: (^{1}H) and (^{13}C) NMR studies of the importance of aromatic structures in fulvic and humic acids. Geochim. Cosmochim Acta *43*, 1771 (1979)
45. Sarkanen, K.V., Ludwig, C.H. (eds.): Lignin: Occurrence, formation, structure and reactions. New York: Wiley-Interscience (1971)
46. Sato, O., Kumada, K.: The chemical nature of the green fraction of P type humic acid. Soil Sci. Plant Nutr. *13*, 121 (1967)
47. Schnitzer, M.: Characterization of humic constituents by spectroscopy. In. McLaren, A.D. (ed.): Soil biochemistry. Vol. 2, p. 60, New York: Marcel Dekker (1971)
48. Schnitzer, M.: Recent findings on the characterization of substances from soils from widely differing climatic zones. I.A.E.A., Vienna, 117 (1977)
49. Schnitzer, M.: Humic Substances: Chemistry and reactions. In Schnitzer, M. and Khan, S.U. (eds.): Soil organic matter. Vol. 8, p. 57. Amsterdam: Elsevier Sci. Publish Co. (1978)
50. Schnitzer, M., Khan, S.U.: Humic substances in the environment. pp. 2, 64, 67, 82, 99, 167. New York: Marcel Dekker (1972)
51. Schnitzer, M. Lévesque, M.: Electron spin resonance as a guide to the degree of humification of peats. Soil Sci. *127*, 140 (1979)
52. Schnitzer, M., Skinner, S.I.M.: Gel filtration of fulvic acid, a soil humic compound. I.A.E.A., Vienna, 41 (1968)
53. Schnitzer, M., Skinner, S.I.M.: The peracetic acid oxidation of humic substances. Soil Sci. *118*, 322 (1974)
54. Seal, B.K., Roy, K.B., Mukherjee, S.K.: Fluorescence emission spectra and structure of humic and fulvic acids. J. Indian Chem. Soc. *41*, 212 (1964)
55. Senesi, N., Schnitzer, M.: Effects of pH, reaction time, chemical reduction and irradiation on ESR spectra of fulvic acid. Soil Sci. *123*, 224 (1977)
56. Senesi, N., Chen, Y., Schnitzer, M.: Hyperfine splitting in electron spin resonance spectra of fulvic acid. Soil Biol. Biochem. *9*, 371 (1977)
57. Senesi, N., Chen, Y., Schnitzer, M.: The role of free radicals in the oxidation and reduction of fulvic acid. Soil Biol. Biochem. *9*, 397 (1977)
58. Sławinska, D., Sławinski, J., Sarna, T.: The effect of light on the ESR spectra of humic acids. J. Soil Sci. *26*, 93 (1975)

59. Sławinski, J., Puzyna, W., Sławinska, D.: Chemiluminescence in the photooxidation of humic acids. Photochem. Photobiol. *28*, 75 (1978)
60. Sławinski, J., Puzyna, W., Sławinska, D.: Chemiluminescence during photooxidation of melanins and soil humic acids arising from a singlet oxygen mechanism. Photochem. Photobiol. *28*, 459 (1978)
61. Stuermer, D.H., Payne, J.R.: Investigation of seawater and terrestrial humic substances with carbon-13 and proton nuclear magnetic resonance. Geochim. Cosmochim. Acta *40*, 1109 (1976)
62. Swift, R.S., Thornton, B.K., Posner, A.M.: Spectral characteristics of a humic acid fractionated with respect to molecular weight using an agar gel. Soil Sci. *110*, 93 (1970)
63. Tichy, V.: Biological activity of UV-irradiated humic acids. Trans. 5th Int. Symp. Studies about humus. Prague, *2*, 553 (1971)
64. Tollin, G., Reid, T., Steelink, C.: Structure of humic acid. IV. Electron-paramagnetic-resonance studies. Biochim. Biopolys. Acta *66*, 444 (1963)
65. Wilson, M.A., Jones, A.J., Williamson, B.: Nuclear magnetic resonance spectroscopy of humic material. Nature *276*, 487 (1978)
66. Zepp, R.G., Wolfe, N.L., Baughman, G.L., Hollis, R.C.: Singlet oxygen in natural waters. Nature. *267*, 421 (1977)
67. Ziechmann, W.: Spectroscopic investigations of lignin, humic substances and peat. Geochim. Cosmochim. Acta *28*, 1555 (1964)

Organic Material in Sea Water

P. J. Wangersky

Department of Oceanography, Dalhousie University
Halifax, N.S. B3H 4J1, Canada

Introduction

While the study of the composition of organic matter in sea water dates back to the *Challenger* expedition, and is therefore almost as old as the science of oceanography itself, the lack of really sensitive methods for the determination of the total amount of organic carbon present and for the multitude of individual compounds has hidden the importance of this field of research until fairly recently. With the advent of gas and high performance liquid chromatography, the linked gas chromatograph-mass spectrometer, and other physical methods for quantitative microanalysis has come the realization that the small amounts of organic carbon in sea water can act as regulators of many important processes in the oceans. The increased sophistication in sampling for minor and trace elements has emphasized the role of organic matter in controlling the distribution of these materials in the oceans. While we are still far from an understanding of all the ways in which the organic fraction influences the cycles of other components of the ocean, we can now at least point out what information is missing, and suggest ways of gathering that information.

Sampling Methods

While many of the quantities of interest to oceanographers, such as salinity and alkalinity, are operationally defined, this is never more the case than it is in the study of organic materials. Except for total organic carbon (TOC), all of the quantities normally measured are defined either by the method of sampling or the method of analysis. Furthermore, the problems of sample contamination and sample preservation are more difficult for organic materials than they are for any other component of sea water, with the possible exception of the trace metals. Therefore, it is particularly important that we understand what fraction of the universe the various methods of sampling are really isolating. As soon as a sampling method has been chosen, certain kinds of results have been excluded. No particle bigger than the mouth of a Niskin bottle will ever be caught by that bottle, and no particle bigger than the Niskin valve opening will ever be drawn from that bottle. For practical reasons we may wish to exclude certain categories of material from our samples; we may feel that the effort involved in their collection is not warranted by their frequency, for example. However, in making the decision to exclude certain categories of material we must be careful not to assume implicitly that they do not exist. We will discuss an example of such thinking later in this chapter.

We must be particularly careful in comparing measurements of what we assume to be the same materials, collected or analyzed by different methods. When our analyses give us values different from those commonly accepted, it is only human nature to feel that we have found the true path, and all of the others are wrong, perhaps even wrong-headed in not seeing the evident superiority of our techniques. A careful study of the methods involved will usually show that we have isolated or measured quite a different part of the universe in question, and that neither set of values can really be called "wrong".

The Surface Film

Of all of the components of the organic material in sea water, the surface film is the one most affected by the choice of sampling methods. This film should include only those materials, liquid and solid, resting on the sea surface, and should exclude any water. However, few samplers are able to remove only the surface layer, and each of the samplers taking a slice of the top layer takes a slice of a different thickness. Thus when the organic content of the surface film is reported in mg/l, the concentration reported will depend upon how much water was included in the slice. No quantitative measure means very much unless the thickness of the slice is also reported, and comparisons between different sampling methods is difficult.

Methods for collecting and concentrating this layer have been very ingenious. Most of them make use of the surface-active properties of the layer. One of the simplest and earliest, and still the most widely used, is the Garrett screen [1]. In this technique, a tray made of stainless steel screening is dipped vertically into the water, then turned and raised horizontally through the surface. The tray is then tipped on edge and drained. Variations on this method have included the use of a floating plastic or ceramic drum [2], a glass microscope slide [3], and plastic sheets and plates [4–6]. The use of the jet drops and bubble caps from bursting bubbles at the sea surface as a method for sampling a very thin layer of this surface was proposed by MacIntyre [7] and developed into a working method by Bezdek and Carlucci [8], Fasching et al. [9], and Blanchard and Syzdek [10]. A more unusual method employed a freezing plate to solidify and remove the upper 1,000 μm [11]. Some of these methods have been compared in intercalibration exercises [12, 13]; these tests have shown that not only do the different methods collect different amounts of the underlying water, they also select for different classes of organic compounds. Thus, to belabour the point a bit, as soon as the method of sampling has been chosen, the results of the analysis have been biased.

A standard method for sampling the surface layer is not likely to evolve from any of the methods currently in use, except perhaps by fiat. If some regulatory agency decides that the surface layer should be sampled for some pollutant, the method chosen by that agency will become the standard method. Because of the low cost and ready availability of the sampling gear, the most likely candidate for such standardization is the Garrett screen.

The Constituents of the Water Column

The greatest source of problems in the analysis of trace organic materials is contamination, either while the sample is being taken or in the course of sample handling on board ship. Only the larger oceanographic vessels are equipped with the clean room facilities which should be mandatory for all trace constituent analysis, organic or inorganic; therefore, the problem of contamination in sample handling must always be a major consideration in setting up a routine sampling protocol.

Contamination during the act of sampling can result from passage of a clean sampler through the contaminating surface layer, particularly contaminating after the ship has been on station for any length of time, or from the choice of an unsuitable sampler. The first problem, described by Gordon and Keizer [14] and by Zsolnay [15], can be avoided by the use of a sampler which opens only after it is safely below the surface layer. The second problem, the contaminating sampler, has been attacked by the elimination of organic materials, or by the substitution of more inert plastics for the usual PVC. Many such samplers have been described in the literature [16–19].

The changes taking place in the organic constitutents during sampling, sample handling, and storage have been discussed by Ahmed, et al. [20] and Otson et al. [21]. This is an area of research which needs much more work.

The Volatile Fraction

While most workers in this field have at least an intuitive understanding of what should be included as part of the volatile fraction, there is no definition of this fraction other than that which is implied by the method of collection. There are two major methods for separating and collecting this fraction: the head space method samples the vapor phase over the solution contained in a sealed bottle, and thus depends upon a knowledge of the distribution of a compound between liquid and gaseous phases [22–24]; the gas stripping method displaces the volatile materials from the aqueous phase, usually collecting and concentrating these materials by absorption by some solid phase [25–29].

It can easily be seen that both the kinds and amounts of compounds found will differ in the two methods of separation. When the aqueous phase is subjected to some treatment, such as heating or adding large quantities of some salt in order to decrease the solubility of the organic volatiles, the results of the head space analysis will more nearly approximate those of the gas stripping methods. At least at this time there is no general agreement on either the method of collection or the definition of "volatile material".

The Particulate Fraction

The definition of "particulate fraction" is based on the physical separation brought about by filtration through a device with pores of a known size, usually 0.45 or 0.50 μm. However, since most workers are interested in the determination of particulate organic carbon (POC), the filters actually used for the separation are felted glass fibre filters, such as the Whatman GF/C or GF/F. These filters have a nominal pore size; all particles larger than that size are retained, but smaller particles will not necessarily pass through. Indeed, it has been shown [30–33] that these filters retain an appreciable quantity of the smaller particles. The Selas "Flotronic®" silver filters have a much more uniform pore size, and would be expected to retain a somewhat different fraction of the particulate matter [34]. However, their cost generally makes their use impractical in any large-scale investigation. At this time, the working definition of particulate matter for most marine chemists and biologists is "the material caught on the filter I am using".

Methods other than filtration have been used for separation of particulate material. Centrifugation has been used by several investigators [35–37]. The separation would thus be made on the basis of specific gravity, and would yield a different fraction than would any of the filtration methods. All of the methods which involve capturing a sample of water in some sort of bottle or bag do not sample the complete spectrum of particles present in the water column. Even within the sample container, the particles present are not distributed homogeneously; the heavier particles sinking to the bottom of the sampler are in some cases below the level of the sampling valve, and are missed altogether [38, 39]. Those particles which are relatively rare would also be missed altogether if only a small volume of water were sampled [40].

The heavier particles, whose residence time in the water column is short, are not likely to be caught by any but the largest water samplers. After many years of examining thousands of filters under the microscope, I have yet to see my first copepod fecal pellet; yet these pellets are well accepted as the major carriers of diatom, coccolithophorid, and foraminiferal remains to the deep water. These and other larger, heavier particles were first captured quantitatively by high volume *in situ* filtration [41]. The high volume pump was not a simple design, and made special requirements on the capabilities of the research vessel. As a result, the older idea of the sediment trap was revived and improved [42–45]. Intercalibration experiments have shown, however, that the various trap designs catch different fractions of the particulate matter, and all of them catch less than does the high volume pump. We are still far from understanding exactly what we are sampling with our various collectors, but we do understand that the bottle samplers are measuring only material with a relatively long hang time. This is material which may be important for local regeneration, but which takes little part in the transport of carbon to deeper water.

The Dissolved Fraction

The dissolved fraction is what is left after we have removed all of the other fractions. Technically, we should only measure this fraction after we have blown off the volatile fraction and filtered off the particulate fraction. We always do remove some or most of the volatile fraction in the course of acidifying and sparging to remove the carbonate, but the act of filtration often adds more organic material as contaminant than it removes as particulates. Since the particulate fraction, except during phytoplankton blooms, is usually 10% or less of the dissolved organic carbon (DOC), the error introduced by including the particulates in the DOC is either ignored, or legitimized by reporting the analytical result as total organic carbon (TOC).

When our interest is in the TOC or DOC, the normal method of sampling is with water sampling bottles, either metallic, glass, or coated with a plastic resistant to leaching by sea water. Since we are dealing with quantities in the parts-per-million range, the usual precautions for sampling for trace analysis should be followed. Few oceanographic vessels are properly equipped for work of this kind; the sample handling room is only too often the winch room, almost awash in lubricants and hydraulic fluid for the winches. Research vessels are usually designed for the techniques and problems of yesterday, with only too little thought for the kinds of problems we will be investigating tomorrow. Sampling arrangements which were suitable for salinities and inorganic nutrient analyses may not be sufficient for trace metals and organic materials at the part per million level.

When we wish to investigate specific compounds or classes of compounds, we frequently find that some form of concentration of the organic fraction is necessary in order to bring our sample within the range of the method we wish to use. Individual compounds and classes of compounds are usually present at μmol or nanomol levels, and in a matrix of interfering organics and sea salts. There are two strategies we can employ in performing such separations and concentrations; we can bring up large quantities of water for processing on board ship, or we can send down our concentrator for *in situ* processing.

The first strategy has been by far the more popular, and much of the early work in this field has involved bringing on board fairly large water samples, out of which some particular organic fraction has been separated. Considerable ingenuity has been expended on methods of separation. One of the earlier methods involved co-precipitation, usually with a metal oxide or hydroxide [46–50]. On the whole, this method has proved to be unsatisfactory because of incompleteness of precipitation and because of the problems inherent in collecting large quantities of usually fluffy precipitate from large volumes of sea water.

An obvious next step was to contain the co-precipitant in a column and to run the sample through. This technique is still in common use, with a wide variety of adsorbents, each chosen to remove a particular fraction of the DOC. The adsorbent removing the greatest range of organic materials is probably activated charcoal [51–55]. In spite of its versatility, however, it has not been used at all widely because of its reluctance to let go of the material it has adsorbed; it adsorbs some 80% of the DOC in sea water, and releases some 80% of what it picks up, so that its overall efficiency is around 64%.

The macroporous resins, and in particular the XAD series, have been used to extract the marine equivalents of the humic and fulvic acids [56–61]. Humic and fulvic acids are terms derived from soil chemistry, where they refer to compounds isolated by particular chemical techniques. Much the same suite of compounds can be isolated from soils by adsorption on XAD-2. By analogy, those compounds isolated from sea water by adsorption on this resin have been termed marine humics and fulvics. The analogy does not extend to identity of composition; as we will discuss later, there are important differences between the terrestrial and marine compounds.

An important advantage of the column techniques is the ability to select an adsorbant wich will concentrate the component in which you are interested. As examples, the chlorinated hydrocarbons and the polynuclear aromatic hydrocarbons have been selectively concentrated on polyurethane foam [62–65], while the reversed-phase materials usually associated with high performance liquid chromatography have been used for the concentration of a variety of classes of compounds [66–68]. This technique offers a great deal of promise for future development, as the range of available adsorbants will increase with the increased use of HPLC.

The classic technique for the separation of organic materials from aqueous solution is liquid-liquid extraction, a technique which has also been applied with sea water, sometimes with a great deal of ingenuity [69–76]. A problem with these methods is the preparation of pure solvents; the advent of gas chromatography and HPLC have shown us the inadequacies of our traditional methods of reagent purification. Keeping the solvents pure in the atmosphere usually found on board ship is also no small problem. In general, extraction has proved to be a useful technique when a few special samples are to be run, but not when either a survey or a time series is needed.

Several unusual and interesting separation techniques have been suggested, but have not yet been used to any extent; these include a variety of foam and bubble fractionations [77–82], reverse osmosis [83], ultrafiltration [84, 85], freezing-out [86–89], steam distillation [90], and ligand-exchange chromatography [91]. Each of these methods has its advantages, but none is generally useful enough to have been adopted as a standard method.

The ideal method would separate and concentrate the organic materials before they encountered the contaminating environment of the research vessel. Several systems have been devised; these have utilized liquid-liquid extraction [92, 93] and adsorption on XAD resins [94]. This is obviously one direction in which future research must go. If systems for concentrating organic compounds from very large volumes of sea water *in situ* can be developed, the problems of contamination and of identification and measurement will be much less severe.

Analytical Methods

Any attempt to review analytical methods in this field is certain to be out-dated even as it is written; the impact of modern instrumental methods of analysis is just beginning to be felt in oceanography. As the physical methods of analysis, such as gas chromatography/mass spectrometry (GC/MS), high performance liquid chromatographhy (HPLC), photoacoustic spectrometry, Fourier transform infrared spectrophotometry (FT/IR), and the various forms of surface analysis by interaction with charged particles become more common in oceanography the questions we ask will become more sophisticated and perhaps more chemical than those we can now ask. Even now the literature is too vast to be reviewed at all thoroughly in a chapter of this length; I will only try to erect a framework of methodology, and will refer you to other sources for detailed discussion [95–97].

Organic Carbon

There is as yet no generally accepted method for the analysis of the carbon content of the volatile fraction, largely because there is no agreement on a method of separation defining this fraction. The methods proposed all involve some form of separation by bubbling with an inert gas, followed by concentration by adsorption, oxidation, and determination either directly as CO_2 [98, 99], or as CH_4 after reduction. The differences between the various methods involve details of sample treatment, such as temperature and length of bubbling time. One reason we know so little about the amount of organic carbon in the volatile fraction is, paradoxically, the ease of analyzing this fraction for specific compounds once it has been separated and concentrated. With GC and GC/MS we have gained some information about the specific, but at the expense of the general.

The determination of particulate organic carbon is straightforward classical microanalysis, once a proper sample has been taken. The only real difficulties encountered result from the low concentrations of the POC in deep water, usually 0–5 μg C/l, and the sometimes high and erratic carbon blank in the filters used. The low carbon concentrations make necessary the filtration of larger volumes of water, usually around 5 l, in order to bring samples from lower depth above background. If sufficient care is taken in the precombustion of the filters and in the handling of the sample, no insurmountable difficulties should be encountered. Although the early workers in this field usually built their own carbon analyzers, most laboratories now use commercially available units. A good description of a typical method is to be found in Ehrhardt [100]. The method can be adapted for shipboard analysis, a great advantage on a long voyage [101].

While there have been many ingenious methods devised for the analysis of dissolved organic carbon (DOC), a few methods have how been accepted as more or less standard, giving more or less the same results. The method employed by the greatest number of workers is the wet oxidation method employing potassium persulfate as the oxidant [102]. This is not a real-time method, and involves the storage of samples for later analysis in a shore laboratory. Attempts have been made to convert this to an auto-analyser technique, to be run on board ship. One successful conversion, using the change in conductivity in a 0.01 m NaOH solution upon absorption of CO_2 evolved from organic carbon as the detector, has been described [103, 104]. Other automatic analyzers have been built using ultraviolet light as the oxidizer [105, 106].

In the use of wet oxidation, there is always the worry that the oxidant we have chosen may not be strong enough to decompose all of the organic materials present in the sample. This worry has some basis in fact, since even such strong oxidants as various mixtures containing perchloric acid, and temperatures to 300 °C have been shown to give incomplete oxidation for some compounds [107]. A referee method, providing virtually complete oxidation, has been needed to ensure that the various wet oxidation methods are measuring all of the DOC. Such a method has been developed by MacKinnon [108], using evaporation of sea water under vacuum, followed by dry oxidation at 900 °C. A comparison of persulfate and ultraviolet wet oxidation with this dry oxidation method [109] showed that the ultraviolet oxidation found up to 10% more organic carbon than the persulfate, and the dry oxidation up to 10% more than the ultraviolet. The general outlines of the DOC distributions were not changed, however, and the dry oxidation is too cumbersome and slow a method to replace either of the wet oxidations as a standard method.

A direct injection, dry oxidation method which could be run on board ship in real time would still be the method of choice, particularly for long voyages. Such a method has been devised by Salonen [110]; it has not yet been tested or adopted by other workers. All of these methods are subject to constraints imposed by problems in blanks, in sampling errors, and in contamination which makes the interpretation of the data difficult [111].

Organic Nitrogen and Phosphorus

A knowledge of the nitrogen and phosphorus content of the organic matter is necessary to know the size of the reservoir of nutrients available for regeneration in the water mass being studied. In the past, these quantities were among the most difficult to obtain, the analyses being time-consuming and filled with possibilities for error. The standard method for organic nitrogen was the micro-Kjeldahl [112], while the organic phosphorus was determined after decomposition of the organic matter either by perchloric acid oxidation [113] or sulfuric acid at high pressure [114]. These methods were not easily adapted to shipboard operations, although they have been run at sea.

The use of high-intensity ultraviolet light as the oxidizing agent for organic materials simplified the analytical procedures, but in the early applications did not lead to any real decrease in time per analysis [115–118]. More recently, the techniques used for automatic analysis of DOC have been adapted to the analysis of total nitrogen and phosphorus in the sample after oxidation of the organic carbon [119–124]. This procedure still requires that inorganic nutrients be run, in order to find the organic N and P by difference.

Analysis by Compound Class

More specific information on the nature of the organic materials can be of three kinds; we can determine the classes of organic materials present, we can attempt to identify specific compounds in a particular compound class, or we can attempt to get both qualitative and quantitative measures of some compound or group of compounds. For many purposes, quantitative measures of compound class are all that are necessary. Detailed information on all of the component sugars in the polysaccharides present may tell us little more than we can learn from a simple determination of the amount of carbon bound up in polysaccharides. Many of the compound class determinations are spectrophotometric, depending upon chemical rather than physical properties.

Methods of this type have been devised for phenols [125, 126], lignins [127–129], amines [130, 131], carbohydrates, both simple [132] and polysaccharide [133–140], uronic acids [141, 142], hydrocarbons [143], amino acids [144], proteins [145], and surfactants [146–148]. Many of these determinations measure a fluorescent end product; this often permits the determination to be made directly on the sample of sea water, without a concentration step. The chances of contamination are thus lessened, and the method made more amenable to automatic analysis.

A somewhat different approach to analysis by compound class, one which uses the techniques of thin layer chromatography, has been developed with the use of the Iatroscan [149]. In this technique, a wire coated with powdered silica is used as a chromatographic column, and the distribution of compounds along the wire, after development in suitable solvents, is determined with a flame ionization detector. For use in sea water, some form of concentration of the organic fraction is necessary.

Another form of class analysis is the division of the organic fraction into molecular weight classes. This has usually been done either by the use of columns of molecular exclusion resins, such as Sephadex [150–152], or by dialysis through membranes of graduated pore size [153–156]. A difficulty with these techniques has been the necessity for preconcentration of the organic material, which usually has resulted in the loss of the low molecular weight material; this fraction has been estimated by difference, or by analysis for DOC in the solution after removal of the larger molecules. In either case, the estimate of the lowest molecular weight class is subject to considerable uncertainty.

Purely qualitative analysis, with no concern for either the efficiency of isolation or the total amount present, has usually been limited to polysaccharides and the humic and fulvic acids. The analysis of polysaccharides is straightforward, the only complicating factor being the possible presence of uronic acid residues. The humic and fulvic acids, on the other hand, are defined by the separation process, and no structure or composition is implied by the naming of the compounds. A considerable effort has been put into the use of chemical and physical methods for deducing possible structures for these compounds, in both sea water and fresh water [157–165], with results we will discuss later.

Analysis for Specific Compounds

There are two major approaches to the analysis for specific compounds in sea water. In one, analytical conditions are manipulated so that only the compound sought will be measured; in the other, a whole group of compounds is isolated and separated into components, usually by some physical method. The first approach was much favored early in the history of chemical oceanography, and often involved extremely ingenious methods of avoiding or masking interferences. Organic compounds determined in this manner included urea [166–169], monosaccharides [170], the B complex vitamins [171–174], glucose [175, 176], chlorophyll [177], glycolic acid [178], formic acid [179], and creatine [180]. While these methods had much to recommend them, including some very clever chemistry, they were often operating at or beyond the true limits of their sensitivity, and they tended to cozen the chemist into determining the compounds for which he had a method, rather than the compounds which were really important.

In recent years the tendency has been to separate a group of compounds, either physically or chemically, and then to separate this group into individual compounds, usually by some physical process. The early work of this kind used column and thin layer chromatography; most separations are now made by gas or high precision liquid chromatography (HPLC). As detectors have become more sensitive and more stable, the emphasis in gas chromatography has shifted from packed columns to capillary columns, and the coupled GC/MS is no longer an exotic in the marine chemistry laboratory. Gas chromatography is limited to those compounds which are naturally volatile, or which can be made into volatile derivatives. Many of the components of the organic materials in sea water can be made volatile only with considerable difficulty, or, as in the case of the polysaccharides and proteins, after breaking the original compounds down into their simplest components. Information on molecular size and arrangement must therefore be lost when this technique is used.

HPLC would seem to be the ideal companion technique for working with these larger molecules. The main difficulty in applying this technique to the analysis of sea water organics has been the relative insensitivity of the available detectors. Improvements in detector design have been helpful, but a sensitivity comparable to that of the GC has really only been achieved with fluorescence detectors. Another disadvantage has been the difficulty in coupling the HPLC to a mass spectrograph. Such matings have been announced many times in the literature, but no production line model has yet appeared. The utility of the technique is so obvious, however, that the appearance of suitable analytical machinery is only a matter of time.

Of the various organic components determined by such physical separations, by far the best known are the hydrocarbons, both natural and anthropogenic. The literature on methods for these compounds is far too great to summarize, or even to list in a chapter such as this. A good idea of the progress in this field can be obtained from the papers on intercalibration experiments; of the organic components, only the DOC and hydrocarbon methods have achieved enough universal acceptance to warrant such experiments [181–188]. While there is a great deal of overlap, as a general rule the aliphatics and the smaller aromatics are determined by GC or GC/MS, while the polyaromatic hydrocarbons (PAH) are determined by HPLC. The columns used and the conditions of analysis vary widely, and most researchers using HPLC for this analysis use some form of reversed-phase precolumn concentrator [67].

The chlorinated hydrocarbons and other pesticides have also received a great deal of attention from the analysts. Again, the methods of choice have been the various chromatographic separations, after some variety of preconcentration. A critical bibliographic review of the field is not possible in this article; a few papers will serve to illustrate the commoner methods, and to serve as an entry into the literature [189–193]. Any of the environmental chemistry journals will furnish many more examples.

The volatile organic compounds are obvious candidates for analysis by gas chromatography. The methods usually involve either head space analysis [22, 194–196] or concentration by gas stripping and adsorption, followed by desorption, usually thermal [197–199]. Most recent work has used capillary [200], rather than packed columns, and mass spectrometry for the identification of the components [201, 202]. Sterols [203, 204], fatty acids [205–211], and lipids in general [212, 213] are usually determined by GC or GC/MS. The early work on proteins, amino acids [214–217], and carbohydrates [218–221] achieved separations by column or thin layer chromatography (TLC) [222, 223]; later, volatile derivatives were discovered for amino acids [224–226] and monosaccharides [227, 228]. With the advent of HPLC and the discovery of fluorescent derivatives for these compounds, the earlier methods are slowly disappearing from the literature [229–231].

The list of compounds and methods given here is far from complete; at the present rate of creation of methods, any compendium claiming to be complete would have to be published in a loose-leaf binder, with monthly supplements. The great diversity of compounds and methods has been one of the problems in marine chemistry. Any two investigators working on the same group of compounds are likely to be using different methods of collection, different analytical methods, or different variations of the same method, which may or may not be measuring the same things. As a result, the data base in this field consists of scattered pieces of information whose comparability is likely to be doubtful at best.

Distributions

When we are studying the cycle of any component of the environment, the first facts to be established must be the distribution of this component in nature. While it is always a mistake to consider the standing crop to be the most important aspect of the cycle, it is equally a mistake to attempt to understand the cycle before we have reliable values for this standing crop, and some notion of the way it varies in time and space. To this end, then, the early work in the study of the cycle of organic carbon in nature concentrated on the determination of distributions, leaving the study of rates of change for the future. However, the oceans are vast, and oceanographers and oceanographic vessels are few; even now, our accumulated data on standing crops are insufficient, and the study of rates is a recent development. Following the scheme laid down in an earlier volume of this series (Vol. 1 A), we will look first at the distribution of organic materials in the different regions of the oceans.

Horizontal Distributions of Organic Materials

For the purposes of our study, the oceans can be divided into four regions: estuaries, coastal zones, boundary currents, and central gyres. Research in these regions has by no means been even-handed. In the earlier days of oceanography, the major emphasis was on understanding the large-scale processes in the oceans, and most of the research was concentrated in the open ocean, in the convergences, divergences, boundary currents, and central gyres. This was the period of the exploratory expeditions, assessing the current state of the oceans and outlining the problems to be solved by later expeditions, better equipped and more precisely directed. Unfortunately, this exploratory period had largely ended before the development of sensitive and reliable instruments and methods for the organic components of sea water.

Much of the recent research has been done closer to shore, in part because of the 200-mile economic zone and in part because of the increased importance of the fisheries. This is also the region with the most complicated current patterns, and with the greatest disturbance from man's activities. Thus we are in the position of attempting to solve the most difficult problems first, of examining the abnormal before we fully understand the normal.

Estuaries

Until relatively recently the estuaries and coastal zones were the *terra incognita* of oceanography; most oceanographic expeditions turned off their instrumentation and secured their laboratories while still a day out of their home ports. What knowledge we had of estuarine chemistry was limited to single expeditions to the estuaries of major rivers in far-away places, and to those estuaries in the vicinity of marine laboratories too small to support an ocean-going vessel. This situation has changed dramatically since 1970. While the major oceanographic institutions still concentrate on the major rivers, such as the Amazon [232–234], the St. Lawrence [235–238], and the Columbia [239], we now also have excellent studies of the organic content of such minor estuaries as the Delaware [240, 241], the Mackenzie [242], the Tjeukemeer [243], the Scheldt [244], the Wadden Sea [245, 246], the Ems-Dollart [247–249], and many others.

The rivers feeding these estuaries vary widely in particulate and dissolved organic content, but a few characteristics are held in common by most of the rivers. Many of them have a region of greatest sediment load at low salinity, usually characterized by high turbulence and resuspension of sediments. The particulate load carried by the river is normally coagulated into larger particles and deposited at very low salinities; thus little of the POC from terriginous sources actually enters the middle and lower estuary. Below this region of precipitation, the POC distribution is dependent upon the normal cycle of biological activity in the coastal sea water. On the other hand, the DOC normally behaves like a conservative component, with the end members the DOC of the river and that of the coastal ocean. There are certainly exceptions to this general rule, in specific rivers and in most rivers at specific times. Certainly, in times of flood the particulate load may well reach the open ocean, and during a bloom, either a normal spring bloom or a Red Tide, the excess production of easily utilized organic molecules may overwhelm the usual estuarine DOC distribution.

The estuarine contribution to the oceanic organic carbon cycle has been discussed on a world-wide basis [250] and in a purely North American context [251]. Perhaps the most complete and detailed synthesis is to be found in the report of a workshop held in Woods Hole in 1980 [252].

Coastal Zones

In many respects the coastal zones are by far the most important regions of the world oceans. It is in these zones that the major fisheries are to be found, the greatest proportion of marine transportation takes place, and the minerals and hydrocarbons of the sea floor can be reached using present-day technology; yet these are the regions whose oceanography is least known. In part, this is a result of the complexity of these regions. Both the shape of the shoreline and the bottom topography can influence the movements of water masses, and the effects of violent storms can be felt throughout the water column, even to the resuspension of bottom sediments [253, 254]. Local conditions thus serve to obscure the more general patterns, and each segment of coastline seems a special case. Oceanographers seeking to understand the major current systems thus concentrated on the areas of the oceans beyond the reach of coastal influences, and only since the adoption by most nations of the 200-mile economic zone have major efforts been expended on the oceanography of coastal regions.

In general, both the POC and DOC concentrations in coastal waters reflect the local productivity. Standing crops of all forms of organic matter tend to decrease with distance from land, but the differences are not startling [108], and tend to reflect local conditions, such as regions of upwelling, frontal zones [255], and intrusions of oceanic water on to the coastal shelf [256]. Seasonal and even daily variability in organic carbon concentrations may exceed the variability to be expected in going from coastal to oceanic waters [257–259]. In temperate and boreal regions, the POC in the mixed layer may consist largely of phytoplankton during the spring bloom; during the rest of the year, and all through the year in more tropical and less productive waters, the plankton in general seldom make up more than 10–20% of the POC, and correspondingly, somewhat less than 1% of the total organic carbon.

While the POC in coastal surface waters may vary over a range of 20–500 μg C l^{-1}, the DOC, as determined by either the persulfate or the UV wet oxidation methods, usually remains between 1–2 mg C l^{-1}. Higher values, up to 4 or 5 mg C l^{-1}, have been reported, but usually by some variant of a dry oxidation method more than commonly subject to contamination. These higher values may in fact be correct, but they are under suspicion unless confirmed either by a second TOC method or by equally high values for major components of the dissolved organic phase [260]. In the deeper coastal waters, both POC and DOC quickly drop to values within the range of those in the deeper ocean. There are certainly exceptions to this rule, the most notable being the values in those areas of extremely high productivity, such as the upwelling regions of Peru. When the waters below the mixed layer approach the anoxic, further heterotrophic utilization of organic material is slowed or perhaps even stopped. This can result in an elevated TOC level in these waters.

While the body of information on some specific coastal regions, such as the Baltic Sea and the Japan Sea, is growing rapidly, we are still not in a position to go beyond generalities in our discussion of the carbon cycle in these waters; we cannot yet make comparisons between regions, and we are just beginning to understand temporal and spatial differences within regions.

The Open Ocean

The distribution of dissolved and particulate organic matter in the open oceans reflects to some extent the productivity of the surface waters of the regions. The correlation with productivity is greater than with the DOC, at least in part because of the longer residence time of some fractions of the dissolved materials [261]. As we will discuss later, laboratory studies have shown that only about a third of the organic material put into solution in the decomposition of phytoplankton is used within days; the remaining fraction may last for weeks or months. The DOC thus integrates productivity over a longer period, smoothing out springtime peaks and winter troughs. The particulate material, on the other hand, seems to be removed relatively quickly, either by heterotrophic consumption or by collection into larger particles and subsequent sedimentation. As a result, the POC in the surface waters provides a reasonable estimate of recent productivity [263].

In general, surface POC values in the open ocean range from 20–200 μg C l^{-1}, depending upon the recent local productivity. Values can be still higher in patches of dinoflagellate or *Trichodesmium* blooms. The values decrease quickly with depth, reaching 8–10 μ C l^{-1} by 100 m. The decrease with depth appears to be exponential [262, 263], and rather similar in all of the oceanic areas studied. There are exceptions to this pattern; observers have found particle maxima associated with particular water masses, such as the Antarctic Intermediate Water [262], with boundaries between water masses, and hence at density discontinuities [264], and with an oxygen minimum layer [265]. High POC values are also associated with the resuspension of sediments in nepheloid layers in the bottom waters [266]. There is at least a suggestion in the data that the occasional high values found in deeper water, and often discarded as probable contaminations, are in fact real, reflecting very local collections of aggregated materials with an associated flora and fauna [262, 267].

It should be noted that these distributions are of particulate standing crop, and measure only that fraction of the particulate matter caught in water sampling bottles. The heavier particles with faster sinking rates are not included in these estimates, and will be discussed in a later section.

Generalizations concerning the distribution of DOC in the oceans are harder to make, in spite of the greater amount of data available. The data, while voluminous, have been collected using a variety of methods which are not strictly comparable. The disparity in methods is particularly obvious in the bimodality of the measurements made in deep water; one group of investigators finds values ranging from 0.5 to 0.7 mg C l^{-1}, while another group finds between 1.0 and 1.5 mg C l^{-1} [268–271]. This problem was treated in the section on methods.

Again, there seems to be at least a rough correlation between primary productivity and DOC, although this correlation is masked by the longer residence time of some fractions of the DOC. Since the kinds of compounds freed into solution by the phytoplankton depend upon the kinds of organisms present and also upon their nutritional state, it is not surprising that the correspondence between the present state of the plankton and the standing crop of DOC is not too good. However, the higher values for DOC are found in regions such as the oceanic convergences and divergences, the frontal zones, and along the edges of the current systems; in short, where turbulent mixing ensures a transport of regenerated nutrient from deeper water into the surface layers. The great central gyres, notorious as regions of low productivity, also are generally low in DOC [272]. The exceptions to this rule are the cold-core rings which traverse the gyres; these are separate universes, with only limited exchange with the waters of the gyres, and carry both the chemistry and biology of their source waters.

Vertical Distribution of Organic Materials

The differences between the vertical divisions of the oceans are somewhat more sharply marked than those of the horizontal divisions. The sharpest differences occur at the interfaces between phases: the surface microlayer and the benthic boundary layer. There is also some evidence for discontinuities, at least in POC, at the interfaces between water masses, but our present hit-or-miss methods of sampling the water column allow us to sample these regions only by accident. Much of our evidence, therefore, is anecdotal, and limited to the surface waters within the range of sampling by SCUBA divers. For the purposes of this discussion, the oceans will be divided vertically into five sections: the surface microlayer, the mixed layer, the deep water, the benthic boundary layer, and the sediment.

The Surface Microlayer

The presence of surface slicks on the ocean was probably recognized by the first primitive man with enough leisure to look up from the immediate task of paddling his log out to the fishing grounds. The effects of such slicks on the wave spectrum was certainly known, and used, by Biblical times. Modern interest in the phenomenon started with the discovery of the transport of organic phosphorus compounds into the surface layer on the surfaces of rising bubbles [273] and continued with the development of appropriate sampling gear [274]. A number of reviews have examined specific aspects of the chemistry of these films [275–278].

The most obvious manifestations of the surface microlayers are the windrow slicks seen on the ocean surface in near-calm weather, presumably the result of the piling up of hydrophobic organic materials in regions of downwelling of Langmuir convection cells in the surface layers. Such slicks consist of many molecular layers of organic matter. Even in those areas of the surface without visible organic films, a layer of hydrophobic materials is present [279]. This layer can be shown to be present by measurement of the surface potential of the sea water [13, 280–281]. If this surface film is removed, it is regenerated from organic materials from the main body of the surface water mass. This regeneration can be greatly speeded by the passage of bubbles through the water mass [280]; it is considered that the bubbles collect surface-active materials and deposit them in the surface layer.

Early work on the composition of this surface film suggested that the compounds were predominantly lipids, hydrocarbons, and naturally occurring surfactants [282–291]. More recent work has supported an earlier suggestion by Baier [292] that the greater part of the surface film is composed of polysaccharides, proteins, and humic materials [293–295]. The differences between the two groups of workers are probably a result of the methods of concentration and analysis used; the methods involving an extraction into non-polar solvents would tend to discriminate against the proteins and polysaccharides.

It is difficult to assign a definite concentration to the organic content of the surface film, since, as we mentioned in the section on sampling methods, the concentration found depends on how much of the underlying water is included by the sampler. A more sensible approach might be the measurement of the thickness of the surface layer, or the degree to which the sea surface was covered by such a layer, or possibly both. No method of making such measurements has yet been suggested, although the surface potential methods [281] show that the surface film is far from continuous, and offer some hope for future methods development.

The concentration of particulate organic matter in the surface layer is equally difficult to quantify, and for much the same reasons [12]. In general, the microlayer is enriched with particulate matter relative to the bulk water just below [296, 297]. The particulate matter present in a microlayer may be deposited from the atmosphere [298], or may be formed in situ through such processes as film compression and collapse due to wave action [299, 300].

In both the dissolved and particulate phases, the surface microlayer serves as a site for the concentration of heavy metals [301–309] and organic pollutants [310–318]. The concentration of toxic materials may thus reach dangerous proportions in the surface layer while the water column concentrations are barely detectable. It has been suggested that the accumulation of these anthropogenic materials might be responsible for a decrease in number of organisms such as *Sargassum*, which typically inhabit the surface, and are therefore exposed to the higher concentrations of the microlayer [319].

The Mixed Layer

The mixed layer as it is commonly described is a misnomer; the term was coined during the early days of oceanography, when temperature, salinity, and dissolved inorganic nutrients were determined at specific depths by the use of sampling bottles, and smooth curves were interpolated between the sampling points. With the advent of the continuous temperature, conductivity, and depth recorders it has become evident that the mixed layer could better be described as the layer which can be made homogeneous by the average storm, and that in between storms any number of discontinuities in temperature and salinity may exist [320]. In the case of the nonconservative properties, such as the organic materials, the standing crop depends upon a balance between production and use, and may vary considerably even during a single day [321, 322].

The production of organic carbon is tied, albeit loosely, to the distribution of photosynthesizing organisms in the sea, and thus to the upper layers. The uptake of the organic material is more widely spread; as a result, the standing crop does not strictly correlate with primary productivity, either in time or in space. Much of the organic material produced during the day is used practically in situ, and the inorganic nutrients, such as nitrogen and phosphorus, freed into the water column to support another cycle of primary production. However, a proportion of the nutrients fixed into organic matter is lost to the surface water on each cycle, as detritus sinks below the photic zone, or as zooplankton and fish feed at the surface and then release their excreta into deeper water. If regeneration were the only source of nutrients for primary productivity, we would expect the surface layers to become progressively impoverished, and finally almost sterile. Hoewever, a number of mechanisms, all related to some form of turbulent transport, serve to return nutrient-laden deeper water to the surface.

A mechanism mentioned in most oceanographic texts is eddy diffusivity. If two masses of water are moving in different directions, or in the same direction at different speeds, turbulent eddies will be set up the interface between the water masses, and will transport water which has properties somewhere between those of the two parent water masses. Such eddy currents can in fact be measured directly when the water masses are moving sufficiently rapidly; for most oceanographic purposes, the diffusivity is measured indirectly by estimating the diffusivity needed to produce a particular effect, such as the transport of nutrients required to sustain a particular level of primary productivity. That this must be a minor mechanism is shown by the rapidity with which discontinuities in the various water properties are established in an extended period of calm weather.

A more important source of regenerated nutrients is the mixing brought about by storms. Regeneration of inorganic nutrients exceeds re-use at relatively shallow depths; within the first few hundred meters a substantial proportion, perhaps 20–30% of the organically bound nutrients, have been released to the water column in a form usable by phytoplankton. If this deeper water is then mixed into the surface layers by a storm, in effect new nutrient has been added to the surface waters and primary productivity can increase. The coupling of productivity to storm activity has been observed, although no systematic studies of the effect have yet been undertaken.

In general, as we remarked earlier, the surface waters display a rapid decrease in all organic fractions; this decrease is virtually complete by 100 m in most waters, and marks the average depth of mixing by storm activity in the region. In areas of strong currents, such as the Gulf Stream, the higher POC and DOC values may extend down to 200 m. While the distribution of POC follows that of DOC at least roughly, one component of the POC, the larger aggregates sometimes called "marine snow", is much more patchy in its distribution, at least in the surface waters. In regions of higher productivity a major proportion of the POC may be found in these larger particles, with as little as 5% of the POC to be found in the spaces between aggregates [323]. Such large particles have been seen at all depths of the ocean, but are particularly common in the surface layers. Their presence in the water column has given rise to the concept of an ocean in which intense activity takes place within these "clouds" or "oases", with little happening in the intervening "deserts" (324). At this time, evidence for this concept is largely anecdotal, derived from observations from scientific submersibles.

The Deep Water

Below the thermocline, the concentration of all forms of organic carbon decreases with depth. The decrease in both POC and DOC is not greatly marked, largely because the concentrations are low to begin with. For a considerable period the existence of these decreases was a matter of dispute; the methods used for the measurements were not sufficiently precise and the variability between methods was too great, as was the natural variability of the components being measured. However, once a sufficient number of analyses collected and analyzed by the same techniques and by the same analyst had been accumulated, at least the POC could be shown to decrease exponentially with depth [325]. It is now generally accepted that the DOC also decreases with depth, although the form of the decrease is not specifically stated.

With the advent of the newer instrumental methods of analysis, such as high performance liquid chromatography and coupled gas chromatography-mass spectrometry, the fate of particular compounds can be followed through the water columns. The best known of such reaction sequences is that of the sterols [326, 327]. With compounds such as the simple sugars and free amino acids, depth in the water column, which can roughly be equated with time since release, correlates inversely with ease of use by heterotrophic organisms. Even at the very low concentrations typical of the central gyres of the oceans, there is a definite sequence in the disappearance of these easily-metabolized compounds. The sequence is not as clear-cut as might be hoped, since some monomers can be freed into the water column by the breakup of particulate matter and of the larger soluble proteins and polysaccharides, either by heterotrophy or by the action of extracellular enzymes adsorbed on particulate matter [328].

Since the chemical composition of a major fraction of the DOC is not known, we can say little about its fate in the oceans, other than that stable carbon isotope analysis suggests that most of it is utilized before it reaches the sea floor [329, 330]. A very few analyses suggest a shift in the distribution of molecular weight with depth [331]; much more must be done with these techniques, including the establishment of some sort of time scale, perhaps through links with the hydrography of the regions studied. A limitation on the kinds of analyses possible on the deep-water samples has been the very low concentrations; in order to obtain enough material for analysis, extractions and separations from very large volumes of water have been necessary. Such separations are difficult and easily contaminated. This limitation may be removed by the advent of in situ concentrating samplers.

Sediment-Water Interface

The importance of reacctions at the sediment-water interface was not recognized until fairly recently, in part because this interface is so difficult to sample. The older methods for sampling the sediment – the various grabs, the gravity and piston cores, and the bottom sleds – invariably missed the top few centimeters of the sediment, either by washing them away on the voyage back up to the research vessel, or by smearing them along the walls of the sampler. Our only evidence of activity in this layer was indirect, such as the heterotrophic uptake inferred from the difference between calculated rates of delivery of organic materials to the sea floor and actual rates of incorporation into the upper layers of the sediment. These estimates suggest that as much as 95% of what reaches the boundary layer is metabolized before burial in the sediments [329].

With the employment of scientific deep submersibles, it is now possible to obtain and protect samples of this interface. The box corer, particularly in the larger, square meter size, appears to bring up sediment with an undisturbed upper layer. These developments are still too new for any accumulation of information on the organic material in this benthic boundary layer to exist; it is to be hoped that once the more spectacular possibilities have been explored, some time and attention will be diverted to collections of the boundary layer for analyses of its organic content.

What information we do have on the composition of this boundary layer was acquired through serendipity. The deployment of sedimentation traps in deep water produced anomalous rates of sedimentation in some of the traps nearest the sea floor. These rates could only be explained by resuspension of the boundary layer during turbulent "events"; the materials in these traps were typical not of the particulates in the water column, but of the sediment, sometimes as much as a thousand meters below [332]. The sediment traps collect some as yet unknown fraction of whatever is present in the water column at the proper depth, making no distinction between material falling from the surface and that present through resuspension. Our inferences derived from this source concerning the nature of the boundary layer must therefore be approximate at best.

The Sediment

The incorporation of organic material into the sediments is a function of the balance between rate of supply of organic material to the sea floor and rate of supply of oxygen; once the sediment or the bottom water becomes anoxic, the rate of utilization of the deposited organic matter by bottom-living heterotrophs slows markedly, and the nature of the products of utilization changes. Over most of the ocean floor, the benthic boundary layer remains oxygenated, and it is only some distance down into the sediment that the redox potential becomes reducing enough to permit the re-solution of MnO_2.

Where the productivity of the surface waters is higher, and particularly in the nearshore regions, where the water column is shallow and a smaller proportion of the particulate material is metabolized before it reaches the bottom, the anoxic layer may reach the sediment-water interface. In some regions of especially high production or slow bottom currents the anoxic region may extend upward into the water column. This phenomenon may be permanent, as in the upwelling regions off Peru or in the Cariaco Trench, or it may be a temporary result of some particular combination of weather pattern and nutrient supply. In the latter case, one result may be a massive destruction of groundfish and bottom-living invertebrates.

The sequence of reactions is quite different in oxygenated and anoxic sediments; in the presence of oxygen, the end product of biological metabolism is CO_2, which can then take part in reactions with the inorganic components of the sediment. In anoxic sediments, the end product is methane, which is usually oxidized to CO_2 in the water column above [333]. In rare instances, usually in shallow, extremely productive lakes, the methane may escape into the atmosphere.

Even in regions where the bottom water is frequently renewed and the sediment is oxygenated for the upper meter, a small quantity of organic matter will persist in the sediment. Materials which commonly are completely destroyed before burial, such as the exoskeletons of copepods, can be preserved indefinitely if buried whole, perhaps by local slumping of the sediment. Coccolithophores and Foraminifera lay down their calcium carbonate on an organic matrix, which lasts certainly as long as the calcite is preserved, perhaps millions of years. Bits and pieces of organic material, unrecognizable as to origin but certainly organized by biological processes, can be found everywhere in the sedimentary record. However, in the open ocean the organic content of the sediment is very low, usually of the order of 1% or less of the dry weight.

Thus, most of the organic carbon fixed by photosynthesis in the surface layers of the ocean is metabolized and the inorganic nutrients returned to the water column. In inshore waters, where a larger proportion of the organic material may be buried in the sediment, and where anoxic conditions in that sediment may slow down nutrient regeneration, the surface layers of the sediment are susceptible to resuspension by storms. In this manner the organics in the sediment may be exposed to the possibility of biological oxidation several times before they are buried below the reach of turbulent resususpension. Without the return of inorganic nutrients to the water column by such processes, the productivity of the oceans, depending as it does on the balance between burial of nutrients in the sediments and re-supply from the continents, would be considerably lower than it is now.

Sources of Organic Materials

The organic materials in sea water were originally thought to have two major sources, the transport of terrestrial materials by rivers and streams and the death and decay of marine organisms. The cycle of organic materials in the oceans was considered to be relatively straightforward, the new fixed carbon from these sources being decomposed by bacterial activity or buried by sedimentation. As we have begun to be able to examine processes rather than to measure standing crops, we have found the cycle to be considerably more complicated and subject to more internal regulation.

Terrestrial Sources

Obvious sources of supply for any component of sea water are the great land masses. The intense productivity of the land results in an organic loading of the rivers and streams, visible at times as a reddish-brown coloration of the water. Even more visible are the large pieces of particulate matter contributed to estuaries and harbors by even medium-sized rivers. However, much of the particulate matter carried by fresh water never reaches even the estuary, but aggregates into larger particles and is deposited in the sediment in the early stages of mixing with salt water. This coagulation has two causes: whatever the original surface charge on the particles, the surface-active organic matter in sea water coats the particle surfaces, giving them all the same charge [334–338]; and the addition of ionic material to the water drives the charge closer to zero, thus increasing the probability of coagulation when particles collide. In most cases, the river-borne particulate matter sediments well up-stream, where the salinity is just becoming noticeable. In regions such as the Baltic, where the salinity is low and the contribution from the rivers very large, the particulate matter may be carried down into the coastal region before it precipitates [339]. In either case, the contribution of particulate organic matter to the oceanic water column is negligible.

We do not have nearly as much good data for the dissolved organic component, since reasonable methods of analysis for DOC are relatively recent. Furthermore, the organic carbon cycle in fresh water is not a static system; at least that portion contributed by biological activity is subject to the same heterotrophic metabolism as the material contributed to the oceans by phytoplankton. What eventually arrives at the mouth of the river are the products of recent photosynthesis and the accumulated more resistant products of land drainage and photosynthesis past. It is not too surprising that in many of the larger rivers the DOC appears to mix conservatively with sea water [340]. Contributions of organic matter to the oceans from the rivers do not appear to be as important as they were thought to be even ten years ago.

Perhaps the most important transporter of organic material from the land, in quality if not in quantity, is the wind [341]. Many of the man-made compounds which may be dangerous to marine life adhere to particulate matter and are sedimented and removed from the biological cycle when they enter fresh water systems. However, these compounds can be carried vast distances by the winds, and can then become a danger to those marine forms dependent on particulate organic matter either as a place to live or as a carbon source. The best example of this transport mechanism was the distribution of DDT over all of the world oceans.

Estuaries and Coastal Zones

While the estuaries and coastal zones are the recipients of organic matter carried from the land masses by rivers and streams, the most important contribution from the land is probably the inorganic nutrient load. There is now a considerable body of evidence which shows that the humic acids formed on land and carried by the rivers are precipitated in the estuaries; the humic acids of marine sediments, even a few kilometers from shore, are different both chemically and in their stable isotope composition from their terrestrial counterparts [342–346]. The contribution of terrestrial materials which is most visible in inshore waters stems from man's use of these waters as a roadway. The fact that our present economy is based on petroleum products is nowhere more evident than in the coastal shipping lanes and in the approaches to any considerable port city.

The most important source for organic materials in coastal waters is phytoplankton productivity, just as it is in the open ocean. However, the coastal waters have an additional source, the fixed algae. These algae have been shown to exude organic materials in considerable concentrations in the course of normal growth. Even more material is lost through damage due to desiccation during the normal tidal cycle, and most particularly through the breakup of plants during storms [347]. The material lost is probably primarily polysaccharides, although some species have been shown to exude polyphenols [348]. It has been postulated [349, 350] that these polyphenols, together with other large organic molecules, may form the yellow coloration, the "Gelbstoffe", characteristic of coastal waters. The methods used to collect and separate Gelbstoffe from the other organic materials in sea water have often been the same methods used to collect humic and fulvic acids. As a result, many workers have considered these substances to be identical, and there is now a considerable confusion in the literature between terrestrial humic and fulvic acids, the marine equivalents, and the Gelbstoffe.

While the presence of these yellow materials with a blue-white fluorescence demonstrates a difference in kind between coastal and open ocean dissolved organics, differences in amount are normally relatively small, no more than would be expected from the increased productivity of the inshore waters. There are, of course, exceptions to all general rules, and instances of particularly high POC or DOC can be found both inshore and offshore, but more often inshore. The regions of "milky water", areas where the sea surface is covered by particles of wax, apparently the result of massive kills of zooplankton [351], are examples, as are the fish-killing Red Tides and the *Trichodesmium* blooms. These anomalous happenings may actually be quite important in the regulation of the local ecosystems, and may contribute a considerable amount of organic carbon to the oceanic pool; however, oceanography at this time is not equipped to deal with anything other than the most general phenomena. At least for the present, such irregularities must be relegated to the status of footnotes and curiosities.

The Open Ocean

The primary source of dissolved and particulate organic materials in the open ocean, as in the coastal zone, is the photosynthesis carried out by phytoplankton. Dissolved organic matter is derived from the phytoplankton in several ways: some organisms secrete a mucopolysaccharide sheath which continually degrades, adding polysaccharides to both the dissolved and particulate pools [352]; those organisms which form blooms will often produce more organic matter than can be utilized immediately by the bacteria and zooplankton present; sloppy feeding by zooplankton will add to the organic pool [353]; some organics will result from metabolic activities of zooplankton and higher organism; and many species of phytoplankton, if not all, exude organic material into solution as a normal part of their growth cycles. There is considerable disagreement about the quantity of organic matter released to the ocean in this manner; some workers hold that such exudation occurs primarily with unhealthy or senescent cells [354], while others feel it to be a natural function of some species. The proportion of the photosynthetically fixed carbon released in this manner varies greatly, from a few percent to half or more of the production of organic matter.

The variability in the reported proportions released, both between and within species, is more easily understood if we consider exudation in relation to the nutritional state of the organism. When the diatom or dinoflagellate is supplied with enough nutrients, the photosynthetic cycle normally proceeds to the creation of a new individual. Under these conditions, we would not expect any great exudation of organic materials; the organic compounds synthesized would all go into the production of the new cell. However, if any of the necessary components were in short supply, the photosynthetic cycle could proceed only to a point, and the materials already synthesized either stored or discarded. Thus the proportion exuded by the plankton would be greatest in oligotrophic waters, as has been suggested by several workers, and would vary greatly in both quantity and quality, depending upon the nutritional state of the organism and just which nutrients were lacking.

Designing experiments to test this concept was difficult when the only method for raising phytoplankton was batch culture. In these cultures, the organisms undergo every condition of nutrient stress from overabundance, at the inoculation of the culture, to severe deprivation, at the point where "senescence" sets in. The organic materials released to the culture medium may thus change radically during the course of the experiment, and the medium, if sampled at the end of the experiment, would sum up the nutritional history of the culture. When samples were taken at several points in the course of growth of a number of green flagellate cultues, it was shown that carbohydrates were released in quantity only after active population growth had ceased [355]. In chemostat and turbidostat cultures, the nutritional state of the population can be regulated much more exactly; such cultures will eventually answer the question of what the phytoplankton can do.

The particulate organic matter must eventually derive from the activities of the phytoplankton, either directly or indirectly. Death and decay of all organisms result in the production of some particulate matter small enough or light enough to remain in suspension for some time. Many marine invertebrates produce structures, either for feeding or protection, which collect small particles [356, 357]. Bacteria can bind together particles of colloidal size, small enough to be classified as DOM by our usual methods, or can form particles simply by clumping together as they reproduce [358]. In recent years many physical and chemical mechanisms have been found for collecting together and binding large molecules or small particles into organic aggregates; these have included the formation and desiccation of jet drops and film caps from bursting bubbles at the sea surface [359], the aggregation of surface-active compounds and colloidal particles by bubble solution [360], and the collection and aggregation of small particles on density discontinuities and shear surfaces.

The biological activity of the open ocean is thus entirely dependent upon the productivity of the primary producers, the phytoplankton. We have, at best, only a rough idea of the extent of this productivity, since our normal methods of measurement usually ignore the dissolved organic material exuded by the phytoplankton, and in any case cannot take into account any exuded material metabolized by bacteria in the course of the productivity measurement. Our more advanced chemical techniques are beginning to furnish us with an inventory of the standing crop of organic materials in the ocean; however, at least to the present moment we know very little about those materials quickly utilized by bacteria, the compounds which should be present, but are not. This information can only be gained through experiments with axenic cultures.

Sinks

It is self-evident that the oceans must be close to a steady-state condition for organic carbon; indeed, as we examine the various mechanisms for the removal of organic carbon we will see that most such mechanisms show a strong component of negative feed-back. Any model purporting to describe the organic cycle in the oceans must incorporate such feed-back controls, as well as the proper time relationships, in order to be convincing. While the construction of models of this level of complexity is certainly difficult, it is a necessary exercise if we wish to predict the possible effects of man's activities, present and future, on the health of the oceans.

Photodecomposition

Of the various methods for the removal of organic materials from the oceans, the various photoreactions have been the least considered until very recently. While the extent of sea surface presented to sunlight is great, the absorption of the more energetic wavelengths of ultraviolet light was assumed to limit these reactions to the top few centimeters of the sea surface. Furthermore, the low concentration of organic material made direct reactions seem unlikely. However, we now know that such reactions can occur even at depths of several meters, and that many reactions involve short-lived intermediate compounds or radicals resulting from photooxidation of water or dissolved oxygen. We now have several examples of such reactions in the literature, and doubtless many more will be discovered [361–363]. The important point to be established at this time is quantitative; what proportion of the photosynthetically fixed organic carbon is removed or altered by this mechanism, and is such removal relatively uniform across the spectrum of compounds produced, or selective for only some classes of compounds?

It has also been established that some man-made organic materials, such as pesticides, and some petroleum products can be degraded by photoreactions into compounds more injurious to the environment than were the original materials [364, 365]. This possibility will force us to re-examine our laws for the regulation of pollutants.

Heterotrophy

The organic matter produced by the phytoplankton is used, in one way or another, by all of the other organisms in the ocean. In a sense, heterotrophy is how everything else lives; however, we shall use the term in its specialized meaning, that of making metabolic use of the dissolved organic matter. We will exclude from our necessarily limited discussion the function of organic compounds as signals [366]. Although qualitatively important in the regulation of such activities as breeding and the settlement of larvae, these compounds are quantitatively insignificant.

Larger Organisms

Much of the early interest in dissolved organic matter stemmed from Pütter's hypothesis [367] that larval forms of some marine organisms could use dissolved organic matter as a supplement or as a complete diet in the period immediately following hatching. While this hypothesis, based as it was on a possibly faulty analytical method, was discounted [368], it has reappeared at irregular intervals, usually at the introduction of improved analytical methods.

There seems little doubt at this time that most soft-bodied and many hard-bodied invertebrates can derive a considerable portion of their carbon requirement from compounds in solution at normal seawater concentrations [369]. There are even a number of organisms, including a bivalve [370] and all of the *Pogonophora* [371], which have no functional gut, and must therefore derive all of their organic carbon from this source. However, there is some question as to whether these organisms use the material directly or through an intermediate bacterial population [372]. At this point, Pütter's hypothesis is neither proven nor disproven; the larger organisms are able to take up dissolved organic materials, perhaps even to live on them, but the importance of this food source in the normal environment is not easy to assess.

Phytoplankton

Much effort has been expended on the study of heterotrophic uptake by phytoplankton, in part in the hope of explaining the seemingly inexplicable dumping of organic materials already discussed. If the material dumped during active photosynthesis could be re-absorbed by the phytoplankton after dark, the anomalous production would seem to make better evolutionary sense. Many species have been shown to take up at least the smaller organic molecules, and some species have actual transport mechanisms for specific compounds [373, 374]. Also, some species which do not normally grow heterotrophically have been shown to have enhanced growth when organic materials, particularly compounds containing nitrogen, are added to the medium.

However, whenever whole communities have been tested for heterotrophic growth, usually by the addition of ^{14}C-labelled organic substrates, the ^{14}C has invariably been found in the bacteria rather than in the phytoplankton [375–377]. We are forced to assume that the phytoplankton, or at least some species, can indeed live and grow heterotrophically, but that under normal circumstances the bacteria simply get there first.

Bacteria

Estimates on the proportion of the primary productivity used by bacteria without passage through some other organism have ranged from 20–80%. Few workers in the field would argue too strenuously against a figure of 50% as a rough estimate. If we add to this the bacterial usage of fecal materials, both particulate and dissolved, we can see that bacterial heterotrophy, with its attendent regeneration of inorganic nutrients, is by far the dominant catabolic process in the oceans.

The number of bacteria present in a given sample of sea water can now be counted with reasonable accuracy, using the acridine orange direct count method [378]. This method renders the bacteria fluorescent, and therefore easy to count, except in those cases where the bacterial population of a particle or aggregate is so dense that the whole particle fluoresces. The method does not distinguish between actively growing bacteria and those simply present. Several methods have been proposed for making this distinction. These include the direct counting of cells in the process of division [379], the uptake of ^{14}C-labelled substrates in the dark [380], sometimes after size fractionation [381], autoradiography [382], and the uptake of (^{3}H)thymidine, a DNA precursor [383, 384].

When the various radioactive uptake methods have been compared with carbon balances or O_2 utilization methods for the same waters, the uptake methods appear to underestimate the true rates of heterotrophy [385–388]. This is not surprising, since we know that no single substrate is used with equal efficiency by all bacteria; the choice of substrate determines the uptake rate measured [389–392]. This whole area of research has recently been reviewed [393].

The important measurements which must be made are all measurements of rates, much more difficult to make than measurements of standing crops. The measurements are particularly difficult to interpret when we consider that the substrate actually being metabolized is complex, composed of materials of vastly different digestibility. When phytoplankton were allowed to decompose aerobically in sea water in the dark, about a third of the organic material was used up in a few weeks, another third disappeared after a few months, and the last third was disappearing over a period of years [394]. The regeneration rate must obviously depend upon which fraction of the organic matter is under consideration.

Removal to the Sediment

As we have said, the ultimate sink for organic materials must be the sediment. The routes taken on the way to that sink are more complex than we had originally imagined.

Incorporation of DOM into POM

The early concept of the particulate organic matter as largely detrital, a combination of materials carried into the oceans from land and the debris left over from the death and decomposition of marine organisms, turned out to be unsatisfactory, since little material from land reaches the open ocean, and the temporal distribution of particulates cannot be reconciled with the annual cycles of plankton production. In particular, the POC load is too high during the winter months, when both phytoplankton and zooplankton populations are at their annual low points [395].

The dilemma was at least partially resolved when the process of particle formation by the collection of organic materials on the surfaces of bubbles was discovered [396]. This process made available to particle feeders and to those bacteria preferring the attached mode of existence some portion of the pool of dissolved organic matter, a pool 10–100 times greater than the particulate pool. It also provided a process for removing organic matter from the surface layers of the ocean, since the organic aggregates incorporated diatom frustules, clay particles, and other denser materials, and thus increased their sinking rates.

Another process leading to the removal of organic matter from the surface waters is adsorption on both organic and inorganic surfaces. Any surface immersed in sea water quickly acquires an organic coating [397–399]. To some extent, this coating protects the underlying material from reaction with sea water; it may also serve as the first step in bacterial colonization. If the particle is large, or considerably denser than sea water, the organic material may be removed to deeper water before the organic coating is removed by bacterial activity.

Particulate matter in the surface waters often exists in the form of marine snow, loosely connected collections of aggregrates, debris, phytoplankton, and microzooplankton [400]. These are effectively little self-contained universes, with primary and secondary production closely coupled in space and time [401]. If we consider the microzooplankton as producers of denser fecal material, easily lost to deeper water, these aggregates could be considered as small factories for the transformation and translocation of carbon in the ocean. While the majority of the carbon fixed by photosynthesis would be regenerated on the spot by the metabolic processes of the whole community, some small fraction would be continuously transported out of the community. At this time we do not know what proportion of the primary productivity of the ocean occurs in such universes; our methods of collection provide averages over a large volume of water. Some method of determining particulate size distributions, gentle enough to preserve the integrity of these universes, is necessary before we can even estimate what proportion of the phytoplankton community is associated with the aggregates.

The Large Particle Problem

When the water column is sampled in the usual manner, using sampling bottles of whatever volume, we actually sample only the standing crop of particles, those small enough or light enough to remain suspended in sea water. Those larger, heavier particles, such as intact *Foraminifera* or fecal pellets of larger organisms, are too rare and fall too rapidly to be caught very often even in 20 l samplers. However, recent work with high-volume in situ pumps [402–404] and sediment traps has indicated that these large particles can account for a major portion of the transport of organic matter out of the surface waters.

While much of our information on fluxes is derived from recent sediment trap investigations, these estimates are open to considerable question. Intercalibration tests between different trap designs have shown fairly large differences in calculated fluxes [405], and experiments devised to test the absolute reliability of some trap designs have shown a dependence of the efficiency on local hydrodynamics [406]. When tested against the in situ pumping systems, the traps have invariably produced lower flux rates than the pump. At least for the time being, we must accept the sediment trap data as minimum estimates of flux, perhaps more valuable for their qualitative than for quantitative information.

Sinking of POM

We must consider the particulate organic matter as a system of two major components, a standing crop of suspended material and a continuous rain of larger, heavier particles. The two components are certainly interdependent; the suspended aggregates may produce heavier particles through incorporation into fecal pellets, through collection on the webs of various marine invertebrates, or through collection and aggregation on the surfaces of breaking or dissolving bubbles. Similarly, the heavier particles may disintegrate on their way to the sea floor. Most fecal pellets are covered by a skin of organic material, which preserves their integrity as a particle. This skin is attacked by bacteria, and may be completely destroyed within the first 2,000 m of free fall [407]. Once the skin has disintegrated, the particle begins to break up into smaller particles and clouds of particles, all of which may fall less rapidly. If the particles are small enough, they may become part of the suspended load; some fraction of the organic materials may even go back into solution.

Our picture of the organic carbon cycle thus includes three major organic pools, the dissolved, the particulate suspended, and the particulate falling through the water column, with relatively simple passage from one form to another, but with a continual decrease in total organic content with depth.

Heterotrophy at the Sea Floor

The amount of organic carbon actually reaching the sea floor is a function of water depth and surface productivity. It is certainly not a linear function; there is a definite discontinuity if the lower part of the water column becomes anoxic, since anaerobic decomposition is so much slower than biological oxidation. However, even with aerobic conditions all the way to the bottom and a reasonable depth of water column, much more organic material reaches the sea floor than is ever incorporated into the sediments [329]; this is clearly shown by the existence of a bottom fauna even in the deepest trenches. Since a fair proportion of this bottom fauna is fixed in place, it can further be inferred that the rain of organic materials is reasonably constant. The existence of an abovebottom travelling fauna, as well, suggests that mechanisms exist to take care of occasional bonanzas [408, 409]. If we knew more about the basal metabolism and grazing ranges of these benthic organisms, we could set minimum limits to the amount of carbon arriving at the sea floor, limits which might be more accurate than our current sediment trap data.

Early workers assumed that any organic material missed by benthic invertebrates would soon be metabolized by the numerous benthic bacteria. This concept was put into question by the "Alvin" experiment [410], where soup, fruit, and sandwiches left on board the Alvin when the deep submersible sank in 2,000 m of water were still undecomposed when the vessel was recovered a year later. Subsequent work on the effects of temperature and pressure on the metabolic rates of bacteria isolated from the sea floor have produced rates orders of magnitude slower than those of bacteria from surface water [411–413]. If these metabolic rates were the general rule in benthic bacteria, we must infer that little of the material reaching the sea floor escaped the larger benthic organisms.

A further complication was the difficulty in finding bacteria adapted to high pressures. Almost all of those isolated from sediments in deep water displayed a tolerance, at best, to the pressures from which they had been taken, and actually grew much better at lesser pressures. This should not have been surprising; every particle settling to the sea floor carries with it a burden of bacteria, collected primarily from the depth of water in which the particle originated. The majority of bacteria at the sea floor may thus be intruders from the waters above, adapted to lesser pressures and higher temperatures, and barely surviving in the conditions under which they were collected [414].

If we assume that a bacterial fauna adapted to sea floor pressures and temperatures must have evolved, we must consider where it should be found, and what it should look like. Food falls in the deep sea must be relatively rare. Even the most mobile of benthic bacteria is limited by its size to a small search radius, and must spend most of its lifetime in some resting form, waiting for cues to signal the arrival of food in the immediate vicinity. Alternatively, the microbe could evolve to take advantage of the superior food gathering ability of a larger organisms, and exist as a passenger in the digestive tract of a benthic invertebrate. Bacteria isolated from such sources have been shown to be truly barophilic, with metabolic rates under pressure comparable to those of surface-living bacteria grown under surface conditions [411, 413–415]. These may be the true benthic bacteria, responsible for the final decomposition of organic matter not used by the larger benthic organisms.

This distribution of benthic bacteria would imply that almost all of the oxidation of organic matter in the deep sea sediment occurs in the layers available to bioturbation, with only a very slow utilization below these depths. The distribution of organic matter in deep sea sediment cores certainly does not contradict this idea, since any organic matter, even materials normally subject to quick disintegration, which is somehow buried deeper than a few hundred centimeters in the sediment may remain virtually intact.

Our concept, then, of heterotrophy at the sea floor is one of intense activity by the larger benthic organisms and the bacteria living within them, sometimes commensally and sometimes symbiotically, as in the *Pogonophora*, and a very much slower "mopping up" by free-living bacteria, largely misplaced from the surface waters. The proportioning of the activity will depend upon the depth of the water column, since the shallower and warmer the water, the more intense will be the activity of the free-living bacteria.

The Cycle of Organic Carbon in the Sea

The primary source of organic carbon to the oceans is the productivity of the phytoplankton; the contribution from terrestrial sources is minor, although potentially harmful. A considerable portion of this productivity must come from microflagellates and cyanobacteria, although we do not yet know how much. Although we have evidence of chemosynthesis by bacteria at the hot water vents in the sea floor spreading zones, this productivity seems to have little effect outside the regions of the vents, with their specialized ecosystems.

A large portion of the primary productivity is used immediately in the surface layers, and the inorganic nutrients recycled. The size of this portion has variously been estimated to be as much as 80%. The chief users are bacteria; their competitors are all of the rest of the oceanic food web. At each exchange in the food web, the bacteria collect their portion of the proceeds. Some of the organic carbon fixed photosynthetically is freed into the water column in dissolved form. Those portions of this material which survive immediate bacterial attack take part in a variety of chemical and physical reactions, some of which, such as the formation of organic aggregates or adsorption on inorganic particles, serve to remove them from the surface layers.

Decomposition of both dissolved and particulate organic material continues during the descent through the water column, with only 5–10% of the organic carbon originally leaving the surface layers actually arriving at the sea floor. This small amount is still further decomposed by bottom organism, until only in regions with high surface productivity and shallow water columns is any great amount of organic carbon actually buried in the sediment.

In quantitative terms, we can consider the oceanic organic carbon cycle to have only two important members, the phytoplankton as primary producers and the heterotrophic bacteria as decomposers. All of the rest of the users of the oceanic production, including man and his fisheries, are relatively minor side-loops on the system.

Acknowledgements

The writing of this chapter was supported by grants from the Natural Sciences and Engineering Research Council of Canada.

References

1. Garrett, W.D.: Limnol. Oceanogr. *10*, 602 (1965)
2. Harvey, G.W.: Limnol. Oceanogr. *11*, 608 (1966)
3. Harvey, G.W., Burzell, L.A.: Limnol. Oceanogr. *17*, 156 (1972)
4. Larsson, K., Odham, G., Sodergren, A.: Mar. Chem. *2*, 49 (1974)
5. Miget, R., Kator, H., Oppenheimer, C., Laseter, J.L., Ledet, E.J.: Anal. Chem. *46*, 1154 (1974)
6. Kjelleberg, S., Stenstrom, T.A., Odham, G.: Mar. Biol. *53*, 21 (1979)
7. MacIntyre, F.: J. Phys. Chem. *72*, 589 (1968)
8. Bezdek, H.F., Carlucci, A.F.: Limnol. Oceanogr. *19*, 126 (1974)
9. Fasching, J.L., Courant, R.A., Duce, R.A., Piotrowicz, S.R.: J. Rech. Atmos. *8*, 649 (1974)
10. Blanchard, D.C., Syzdek, L.D.: Limnol. Oceanogr. *20*, 762 (1975)
11. Hamilton, E.I., Clifton, R.J.: Limnol. Oceanogr. *24*, 188 (1979)
12. Daumas, R.A., Laborde, P.L., Marty, J.C., Saliot, A.: Limnol. Oceanogr. *21*, 319 (1976)
13. Van Vleet, E.S., Williams, P.M.: Limnol. Oceanogr. *25*, 764 (1980)
14. Gordon, D.C. Jr., Keizer, P.D.: Fish. Mar. Service Tech. Rept. No. 481 (1974) 24 pp.
15. Zsolnay, A.: Mar. Pollut. Bull. *9*, 23 (1978)
16. Clark, R.C. Jr., Blumer, M., Raymond, S.O.: Deep-Sea Res. *14*, 125 (1967)
17. Bertoni, R., Melchiorri-Santolini, U.: Mem. Ist. Ital. Idrobiol. *29*, 97 (1972)
18. Gump, B.H., Herz, H.A., May, W.E., Chesler, S.N., Dyzel, S.M., Enagonio, D.P.: Anal. Chem. *47*, 1223 (1975)
19. Gagosian, R.B., Dean, J.P. Jr., Hamblin, R., Zafiriou, O.C.: Limnol. Oceanogr. *24*, 583 (1979)
20. Ahmed, S.M., Beasley, M.D., Efromson, A.C., Hites, R.A.: Anal. Chem. *46*,1858 (1974)
21. Otson, R., Williams, D.T., Bothwell, P.D., McCullough, R.S., Tate, R.A.: Bull. Environ. Contam. Toxicol. *23*, 311 (1979)
22. Corwin, J.F.: Organic Matter in Natural Waters, Hood, D.W., ed., Inst. Mar. Sci., Univ. Alaska Publ. No. 1, (1970) pp. 169–180
23. Otson, R., Williams, D.T., Bothwell, P.D.: Environ. Sci. Technol. *13*, 936 (1979)
24. Otson, R., Williams, D.T.: Anal. Chem. *54*, 942 (1982)
25. Mieure, J.P., Dietrich, M.W.: J. Chromatog. Sci. *11*, 559 (1973)
26. Zlatkis, A., Lichtenstein, H.A., Tichbee, A.: Chromatographia *6*, 67 (1973)
27. Grob, K., Zuercher, F.: J. Chromatog. *117*, 285 (1976)
28. Kuo, P.P.K., Chian, E.S.K., DeWalle, F.B., Kim, J.H.: Anal. Chem. *49*, 1023 (1977)
29. MacKinnon, M.D.: Mar. Chem. *8*, 143 (1979)
30. Sheldon, R.W., Sutcliffe, W.H. Jr.: Limnol. Oceanogr. *14*, 441 (1969)

31. Cranston, R.E., Buckley, D.E.: Bedford Inst. Oceanogr. Rept. Ser., BI-R-72-7 (1972) 14 pp
32. Sheldon, R.W.: Limnol. Oceanogr. *17*, 494 (1972)
33. Postel, J. R., Rao, V.N.R.: Proc. Indian Acad. Sci. *82 B*, 221 (1975)
34. Wangersky, P.J., Hincks, A.V.: Analytical Techniques in Environmental Chemistry, Albaiges, J. (Ed.) Pergamon Press (1980), pp. 53–62
35. Lammers, W.T.: Environ. Sci. Technol. *1*, 52 (1967)
36. Jacobs, M.B., Ewing, M.: Science *163*, 380 (1969)
37. Etcheber, H., Jouanneau, J.-M.: Estuar. Coast. Mar. Sci. 11:701 (1980)
38. Burns, N.M., Pashley, A.E.: J. Fish. Res. Bd. Can. *31*, 291 (1974)
39. Calvert, S.E., McCartney, M.J.: Limnol. Oceanogr. *24*, 532 (1979)
40. Delmas, D.: Tethys *9*, 279 (1980)
41. Bishop, J.K.B., Edmond, J.M.: J. Mar. Res. *34*, 181 (1976)
42. Wiebe, P.H., Boyd, S.H., Winget, C.: J. Mar. Res. *34*, 341 (1976)
43. Honjo, S.: J. Mar. Res. *36*, 469 (1978)
44. Blomqvist, S., Hakanson, L.: Arch. Hydrobiol. *91*, 101 (1981)
45. Staresinic, N., von Brockel, K., Smodlaka, N., Clifford, C.H.: J. Mar. Res. *40*, 273 (1982)
46. Park, K., Williams, W.T., Prescott, J.M., Hood, D.W.: Science *138*, 531 (1962)
47. Wiliams, P.M., Zirino, A.: Nature, Lond. *204*, 462 (1964)
48. Chapman, G., Rae, A.C.: Nature, Lond. *214*, 627 (1967)
49. Meyers, P.A., Quinn, J.G.: Geochim. Cosmochim. Acta *35*, 628 (1971)
50. Sridharan, R.G. Jr., Lee, G.F.: Environ. Sci. Technol. *6*, 1031 (1972)
51. Jeffrey, L.M., Hood, D.W.: J. Mar. Res. *17*, 247 (1958)
52. DeWalle, F.B., Chian, E.S.K.: J. Environ. Eng. Div., ASCE *100*, 1089 (1974)
53. Grob, K., Grob, K. Jr., Grob, G.: J. Chromatog. *106*, 299 (1975)
54. Kerr, R.A., Quinn, J.G.: Deep-Sea Res. *22*, 107 (1975)
55. Conti, F., Goretti, G., Lagana, A., Petronio, B.M.: Ann. Chim. *68*, 783 (1978)
56. Stuermer, D.H., Harvey, G.R.: Deep-Sea Res. *24*, 303 (1977)
57. Ryan, J.P., Fritz, J.S.: J. Chromatog. Sci. *16*, 488 (1978)
58. Aiken, G.R., Thurman, E.M., Malcolm, R.L., Walton, H.F.: Anal. Chem. *51*, 1799 (1979)
59. Dressler, M.: Chromatog. Rev. *23*, 167 (1979)
60. Picer, N., Picer, M.: J. Chromatog. *193*, 357 (1980)
61. Ram, N.H., Morris, J.C.: Environ. Sci. Technol. *16*, 170 (1982)
62. Lewis, R.G., Brown, A.R., Jackson, M.D.: Anal. Chem. *49*, 1668 (1977)
63. Navratil, J.D., Sievers, R.E., Walton, H.F.: Anal. Chem. *49*, 2260 (1977)
64. Saxena, J., Kozuchowski, J., Basu, D.K.: Environ. Sci. Technol. *11*, 682 (1977)
65. Basu, D.K., Saxena, J.: Environ. Sci. Technol. *12*, 791 (1978)
66. Derenbach, J.B., Ehrhardt, M., Osterroht, C., Petrick, G.: Mar. Chem. *6*, 351 (1978)
67. Saner, W.A., Jadamec, R., Sager, R.W., Killeen, T.J.: Anal. Chem. *51*, 2180 (1979)
68. Mills, G.L., Quinn, J.G.: Mar. Chem. *10*, 93 (1981)
69. Goldberg, M.C., DeLong, L., Kahn, L.: Environ. Sci. Technol. *5*, 161 (1971)
70. Osterroht, C.: Kiel. Meeresforsch. *28*, 48 (1972)
71. Hughes, D.R., Belcher, R.S., O'Brien, E.J.: Bull. Environ. Contam. Toxicol. *10*, 170 (1973)
72. Acheson, M.A., Harrison, R.M., Perry, R., Wellings, R.A.: Wat. Res. *10*, 207 (1976)
73. Wu, C., Suffet, I.H.: Anal. Chem. *49*, 231 (1977)
74. Yasuhara, A., Fuwa, K.: Chemosphere *6*, 179 (1977)
75. Natusch, D.F.S., Tomkins, B.A.: Anal. Chem. *50*, 1429 (1978)
76. Wun, C.K., Rho, J., Walker, R.W., Litsky, W.: Hydrobiologia *71*, 289 (1980)
77. Karger, B.L., Rogers, L.B.: Anal. Chem. *33*, 1165 (1961)
78. Lemlich, R., Lavi, E.: Science *134*, 191 (1961)
79. Dorman, D.C., Lemlich, R.: Nature, Lond. *207*, 145 (1965)
80. Karger, B.L., DeVivo, D.G.: Sep. Sci. *3*, 393 (1968)
81. Wallace, G.T. Jr., Wilson, D.F.: U.S. Naval Res. Lab. Rept. *6958*, 1 (1969)
82. Wallace, G.T. Jr., Duce, R.A.: Mar. Chem. *3*, 157 (1975)
83. Fang, H.H.P., Chian, E.S.K.: Environ. Sci. Technol. *10*, 364 (1976)
84. Gjessing, E.T.: Environ. Sci. Technol. *4*, 437 (1970)
85. Kwak, J.C.T., Nelson, W.P., Gamble, D.S.: Geochim. Cosmochim. Acta *41*, 993 (1977)
86. Shapiro, J.: Science *133*, 2063 (1961)

87. Shapiro, J.: Anal. Chem. *39*, 280 (1967)
88. Habermann, H.M.: Science *140*, 292 (1963)
89. Kammerer, P.J. Jr., Lee, G.F.: Environ. Sci. Technol. *3*, 276 (1969)
90. Peters, T.L.: Anal. Chem. *52*, 211 (1980)
91. Siegal, A., Degens, E.T.: Science *151*, 1098 (1966)
92. Ahnoff, M., Josefsson, B.: Anal. Chem. *48*, 1268 (1976)
93. Stepan, S.F., Smith, J.F., Flego, U., Reukers, J.: Wat. Res. *12*, 447 (1978)
94. Ehrhardt, M.: Deep-Sea Res. *25*, 119 (1978)
95. Grasshoff, K.: Methods of Seawater Analysis, Verlag Chemie, N.Y. (1976)
96. Wangersky, P.J., Zika, R.G.: Mar. Anal. Chem. Stds. Progr., Rept. No. 3, NRCC 16566, 1 (1978)
97. Dawson, R., Liebezeit, G.: Marine Organic Chemistry, Duursma, E.K., Dawson, R., eds., Elsevier Sci. Publ., N.Y., pp. 445–496 (1981)
98. Games, L.M., Hayes, J.M.: Anal. Chem. *48*, 130 (1976)
99. MacKinnon, M.D.: Mar. Chem. *8*, 143 (1979)
100. Ehrhardt, M.: Methods of Seawater Analysis, Grasshoff, K., ed., Verlag Chemie, N.Y., pp. 215–220 (1976)
101. Hagell, G.T., Pocklington, R.: Bedford Oceanogr. Rept. Ser./BI-R-73-14, 1 (1973)
102. Menzel, D.W., Vaccaro, R.F.: Limnol. Oceanogr. *9*, 138 (1964)
103. Ehrhardt, M.: Deep-Sea Res. *16*, 393 (1969)
104. Ehrhardt, M.: Methods of Seawater Analysis, Grasshoff, K., ed., Verlag Chemie, N.Y., pp. 289–297 (1976)
105. Collins, K.J., Wiliams, P.J. le B.: Mar. Chem. *5*, 123 (1977)
106. Van Steenderen, R.A., Lin, J.-S.: Anal. Chem. *53*, 2157 (1981)
107. Martinie, G.D., Schilt, A.A.: Anal. Chem. *48*, 70 (1976)
108. MacKinnon, M.D.: Mar. Chem. *7*, 17 (1978)
109. Gershey, R.M., MacKinnon, M.D., Williams, P.J. le B., Moore, R.M.: Mar. Chem. *7*, 289 (1979)
110. Salonen, K.: Limnol. Oceanogr. *24*, 177 (1979)
111. Reijnders, H.F.R., Romer, F.G., Griepink, B.: Z. Anal. Chem. *285*, 21 (1977)
112. Adamski, J.M.: Anal. Chem. *48*, 1194 (1976)
113. Hansen, A.L., Robinson, R.J.: J. Mar. Res. *12*, 31 (1953)
114. Harvey, H.W.: J. Mar. Biol. Ass. U.K. *27*, 337 (1948)
115. Armstrong, F.A.J., Tibbitts, S.: J. Mar. Biol. Ass. U.K. *48*, 143 (1968)
116. Lyutsarev, S.V., Sapozhnikov, V.V., Selifonova, E.P.: Okeanologiia *13*, 903 (1973)
117. Bruegmann, L., Wilde, A.: Acta Hydrochim. Hydrobiol. *3*, 203 (1975)
118. Goosen, J.T.H., Klooserboer, J.G.: Anal. Chem. *50*, 707 (1978)
119. Grasshoff, K.: A. Anal. Chem. *220*, 89 (1966)
120. Canelli, E., Mitchell, D.G.: Wat. Res. *9*, 1093 (1975)
121. D'Elia, C.F.: Monogr. Oceanogr. Methodol., U.N. *5*, 485 (1978)
122. Solorzano, L., Sharp, J.H.: Limnol. Oceanogr. *25*, 751 (1980)
123. Solorzano, L., Sharp, J.H.: Limnol. Oceanogr. *25*, 754 (1980)
124. Valderrama, J.C.: Mar. Chem. *10*, 109 (1981)
125. Aleeksova, L.M.: Tr. Gos. Oceanogr. Inst. *1972*, 60 (1972)
126. Afghan, B.K., Belliveau, P.E., Larose, R.H., Ryan, J.F.: Anal. Chim. Acta, *71*, 355 (1974)
127. Almgren, T., Josefsson, B.: Sv. Papperstidn. *76*, 19 (1973)
128. Pocklington, R., MacGregor, C.D.: Intern. J. Environ. Anal. Chem. *3*, 81 (1973)
129. Wilander, A., Kvarnas, H., Lindell, T.: Tech. Rept. Fish. Mar. Serv. Res. Div., No. 649, 1 (1974)
130. Batley, G.E., Florence, T.M., Kennedy, J.R.: Talanta *20*, 987 (1973)
131. Benson, J.R., Hare, P.E.: Proc. Nat. Acad. Sci. USA *72*, 619 (1975)
132. Johnson, K.M., Sieburth, J.McN.: Mar. Chem. *5*, 1 (1977)
133. Lewis, G.J., Rakestraw, N.W.: J. Mar. Res. *14*, 153 (1955)
134. Zein-Eldin, Z.B., May, B.Z.: Anal. Chem. *30*, 1935 (1958)
135. Handa, N.: J. Oceanogr. Soc. Japan *22*, 1 (1966)
136. Marvin, K.T., Proctor, R.R.: Fish. Bull. *65*, 683 (1966)
137. Josefsson, B., Uppstrom, L., Ostling, G.: Deep-Sea Res. *19*, 385 (1972)
138. Hirayama, H.: Anal. Chim. Acta *70*, 141 (1974)
139. Burney, C.M., Sieburth, J.McN.: Mar. Chem. *5*, 15 (1977)
140. Eberlein, K., Hammer, K.D.: Fresenius Z. Analyt. Chem. *301*, 17 (1980)

141. Williams, P.M., Craigie, J.S.: Organic Matter in Natural Waters, Hood, D.W., (Ed.) Inst. Mar. Sci. Alaska Occ. Publ. No. 1, pp. 509–519 (1970)
142. Castagne, A.E., Siddiqui, I.R.: Carbohyd. Res. *42*, 382 (1975)
143. Gordon, D.C. Jr., Keizer, P.D.: Fish. Mar. Service Tech. Rept. *481*, 1 (1974)
144. Josefsson, B., Lindroth, P., Ostling, G.: Anal. Chim. Acta *89*, 21 (1977)
145. Setchell, F.W.: Mar. Chem. *10*, 301 (1981)
146. Le Bihan, A., Courtot-Coupez, J.: Bull. Soc. Chim. France *1970*, 406 (1970)
147. Kozarac, Z., Cosovic, B., Branica, M.: Mar. Sci. Commun. *1*, 147 (1975)
148. Gagnon, M.: Wat. Res. *13*, 53 (1979)
149. Ackman, R.G., Woyewoda, A.D.: J. Chromatog. Sci. *17*, 514 (1979)
150. Gjessing, E., Lee, G.F.: Environ. Sci. Technol. *1*, 631 (1967)
151. Sirutkina, I.S., Varshal, G.M., Lure, Y.Y., Stepanova, N.P.: Zh. Anal. Khim. *29*, 1626 81974)
152. Starikova, N.D., Yablokova, O.G., Korzhikova, L.I.: Okeanologiia *16*, 1007 (1976)
153. Baturnina, M.V., Lyutsarev, S.V., Romankevich, E.: Trans. P. P. Shirov Inst. Oceanogr. *102*, 76 (1975)
154. Smith, R.G. Jr.: Anal. Chem. *48*, 74 (1976)
155. Wheeler, J.R.: Limol. Oceanogr. *21*, 846 (1976)
156. Hoffman, M.R., Yost, E.C., Eisenreich, S.J., Maler, W.J.: Environ. Sci. Technol. *15*, 655 (1981)
157. Salfeld, J.C.: Humic Substances, Povoledo, D., Golterman, H.L., (Eds.) Center for Agric. Publ. Docum., Wageningen, pp. 269–280 (1972)
158. Alberts, J.J., Schindler, J.E., Nutter, D.E. Jr., Davis, E.: Geochim. Cosmochim. Acta *40*, 369 (1976)
159. MacFarlane, R.B.: Geochim. Cosmochim. Acta *42*, 1579 (1978)
160. De Haan, H., De Boer, T., Halma, G.: Freshw. Biol. *9*, 315 (1979)
161. Hatcher, P.G., VanderHart, D.L., Earl, W.L.: Org. Geochem. *2*, 87 (1980)
162. Saito, Y., Hayano, S.: J. Oceanogr. Soc. Japan *36*, 59 (1980)
163. Saito, Y., Hayano, S.: J. Oceanogr. Soc. Japan *36*, 286 (1981)
164. Reuter, J.H., Perdue, E.M.: Geochim. Cosmochim. Acta *45*, 2017 (1981)
165. Harvey, G.R., Boran, D.A., Chesal, L.A., Tokar, J.M.: Mar. Chem. *12*, 119 (1983)
166. Newell, B.S., Morgan, B., Cundy, J.: J. Mar. Res. *25*, 201 (1967)
167. DeManche, J.M., Curl, H. Jr., Coughenower, D.D.: Limnol. Oceanogr. *18*, 686 (1973)
168. Nakas, J.P., Litchfield, C.D.: Estuar. Coast. Mar. Sci. *5*, 143 (1977)
169. Aminot, A., Kerouel, R.: Can. J. Fish. Aquat. Sci. *39*, 174 (1982)
170. Antia, N.J., Lee, C.Y.: Fish. Res. Bd. Can. MS Rept. Ser., No. 168 (1963)
171. Gold, K.: Limnol. Oceanogr. *9*, 343 (1964)
172. Litchfield, C.D., Hood, D.W.: Appl. Microbiol. *13*, 886 (1965)
173. Carlucci, A.F., Silbernagel, S.B.: Can. J. Microbiol. *13*, 979 (1967)
174. Ohwada, K.: Mar. Biol. *14*, 10 (1972)
175. Hicks, S.E., Carey, F.G.: Limnol. Oceanogr. *13*, 361 (1968)
176. Cavari, B.Z., Phelps, G.: Appl. Environ. Microbiol. *33*, 1237 (1977)
177. Ryding, S.-O.: Vatten *31*, 327 (1975)
178. Shah, N.M., Fogg, G.E.: J. Mar. Biol. Ass. U.K. *53*, 321 (1973)
179. Stradomskaya, A.G., Goncharova, I.A.: Gidrokhim. Mater. *43*, 57 (1967)
180. Whitledge, T.E., Dugdale, R.C.: Limnol. Oceanogr. *17*, 309 (1972)
181. Farrington, J.W., Teal, J.M., Quinn, J.G., Wade, T., Burns, K.: Bull. Environ. Contam. Toxicol. *10*, 129 (1973)
182. Farrington, J.W., Teal, J.M., Medeiros, G.C., Burns, K.A., Robinson, E.A. Jr., Quinn, J.G., Wade, T.L.: Anal. Chem. *48*, 1711 (1976)
183. Medeiros, G.C., Farrington, J.W.: NBS Spec. Publ. 409, Mar. Pollut. Monit. (Petrol.), 167 (1974)
184. Traxler, P., Pierce, R.H. Jr.: Natl. Bur. Stds. (U.S.) Spec. Publ. *409*, 161 (1974)
185. Carlberg, S.R.: Ambio Spec. Rept. *5*, 269 (1975/77)
186. Chesler, S.N., Hertz, H.S., May, W.E., Wise, S.A.: NOAA/ERL-AR-77-13, pp. 1–14 (1977)
187. Hilpert, L.R., May, W.E., Wise, S.A., Chester, S.N., Hertz, H.S.: Anal. Chem. *50*, 458 (1978)
188. MacLeod, W.D. Jr., Prohaska, P.G., Gennaro, D.D., Brown, D.W.: Anal. Chem. *54*, 386 (1982)
189. Renberg, L.: Anal. Chem. *50*, 1836 (1978)
190. Dietz, E.A. Jr., Singley, K.F.: Anal. Chem. *51*, 1809 (1979)
191. Duinker, J.C., Hillebrand, M.T.J.: Neth. J. Sea Res. *13*, 256 (1979)

192. Ofstad, E.B., Lunde, G., Drangsholt, H.: Intern. J. Environ. Anal. Chem. *6*, 119 (1979)
193. Biggs, D.C., Powers, C.D., Rowland, R.G., O'Connors, H.B., Wurster, C.F.: Environ. Pollut. Ser. A *22*, 101 (1980)
194. Bassette, R., Ozeris, S., Whitnah, C.H.: Anal. Chem. *34*, 1540 (1962)
195. Friant, S.L., Suffet, I.H.: Anal. Chem. *51*, 2167 (1979)
196. Otson, R., Williams, D.T.: Anal. Chem. *54*, 942 (1982)
197. Swinnerton, J.W., Linnenbom, V.J.: Science *156*, 1119 (1967)
198. Kuo, P.P.K., Chian, E.S.K., DeWalle, F.B., Kim, J.H.: Anal. Chem. *49*, 1023 (1977)
199. Pankow, J.F., Isabelle, L.M., Kristensen, T.J.: Anal. Chem. *54*. 1815 (1982)
200. Bertsch, W., Anderson, E., Holzer, G.: J. Chromatog. *112*, 701 (1975)
201. Schwartzenbach, R.P., Bromund, R.H., Gschwend, P.M., Zafiriou, O.C.: Org. Geochem. *1*, 93 (1979)
202. Gschwend, P., Zafiriou, O.C., Gagosian, R.B.: Limnol. Oceanogr. *25*, 1044 (1980)
203. Saliot, A., Barbier, M.: Deep-Sea Res. *20*, 1077 (1973)
204. Artaud, J., Iatrides, M.C., Tisse, C., Zahra, J.P., Estienne, J.: Analusis *8*, 277 (1980)
205. Slowey, J.F., Jeffrey, L.M., Hood, D.W.: Geochim. Cosmochim. Acta *26*, 607 (1962)
206. Ackman, R.G., Jangaard, P.M.: J. Am. Oil Chem. Soc. *40*, 744 (1963)
207. Ackmann, R.G., Sipos, J.C., Tocher, C.S.: J. Fish. Res. Bd. Can. *24*, 635 (1967)
208. Mahadevan, Y., Stenroos, L.: Anal. Chem. *39*, 1652 (1967)
209. Kanazawa, A., Teshima, S.-I.: J. Oceanogr. Soc. Japan *27*, 207 (1971)
210. Barcelona, M.J., Liljestrand, H.M., Morgan, J.J.: Anal. Chem. *52*, 321 (1980)
211. Ehrhardt, M., Osterroht, C., Petrick, G.: Mar. Chem. *10*, 67 (1980)
212. Kennicutt, M.C. II, Jeffrey, L.M.: Mar. Chem. *10*, 367 (1981)
213. Kennicutt, M.C. II, Jeffrey, L.M.: Mar. Chem. *10*, 389 (1981)
214. Coughenower, D.D., Curl, H.C. Jr.: Limnol. Oceanogr. *20*, 128 (1975)
215. Dawson, R., Pritchard, R.G.: Mar. Chem. *6*, 27 (1978)
216. Gardner, W.S.: Mar. Chem. *6*, 15 (1978)
217. Garrasi, C., Degens, E.T., Mopper, K.: Mar. Chem. *8*, 71 (1979)
218. Larsson, L.I., Samuelson, O.: Mikrochem. Acta *2*, 328 (1967)
219. Josefsson, B.O.: Anal. Chim. Acta *52*, 65 (1970)
220. Mopper, K., Degens, E.T.: Anal. Biochem. *45*, 147 (1972)
221. Mopper, K.: Mar. Chem. *5*, 585 (1977)
222. Armstrong, D.W., McNeely, M.: Anal. Letts. *12*, 1285 (1979)
223. Macko, S.A., Green, E.J.: Mar. Chem. *8*, 181 (1979)
224. Hosaku, K., Maita, Y.: J. Oceanogr. Soc. Japan *27*, 27 (1971)
225. Pocklington, R.: Anal. Biochem. *45*, 403 (1972)
226. Gardner, W.S., Lee, G.F.: Environ. Sci. Technol. *7*, 719 (1973)
227. Modzeleski, J.E., Laurie, W.A., Nagy, B.: Geochim. Cosmochim. Acta *35*, 825 (1971)
228. Tesarik, K.: J. Chromatog. *65*, 295 (1972)
229. Hullett, D.A., Eisenreich, S.J.: Anal. Chem. *51*, 1953 (1979)
230. Lindroth, P., Mopper, K.: Anal. Chem. *51*, 1667 (1979)
231. Lin, J.-K., Lai, C.C.: Anal. Chem. *52*, 630 (1980)
232. Williams, P.M.: Nature, Lond. *208*, 937 (1968)
233. Richey, J.E., Brock, J.T., Naiman, R.J., Wissmar, R.C., Stallard, R.F.: Science *207*, 1348 (1980)
234. Sholkovitz, E.R., Price, N.B.: Geochim. Cosmochim. Acta *44*, 163 (1980)
235. Kranck, K.: Natur. Can. *106*, 163 (1979)
236. Pocklington, R., Leonard, J.D.: J. Fish. Res. Bd. Can. *36*, 1250 (1979)
237. Tan, F.C., Strain, P.M.: J. Fish. Res. Bd. Can. *36*, 678 (1979)
238. Tan, F.C., Strain, P.M.: Geochim. Cosmochim. Acta *47*, 125 (1983)
239. Dahm, C.N., Gregory, S.V., Park, P.K.: Estuar. Coast. Shelf Sci. *13*, 645 (1981)
240. Sheldon, L.S., Hites, R.A.: Environ. Sci. Technol. *12*, 1188 (1978)
241. Sharp, J.H., Culberson, C.H., Church, T.M.: Limnol. Oceanogr. *27*, 1015 (1982)
242. Peake, E., Baker, B.L., Hodgson, G.W.: Geochim. Cosmochim. Acta *36*, 867 (1972)
243. De Haan, H.: Verh. Internat. Verein. Limnol. *18*, 685 (1972)
244. Joiris, C., Billen, G., Wijnant, J.: Estuar. Coast. Shelf Sci. *11*, 279 (1980)
245. Manuels, M.W., Postma, H.: Neth. J. Sea Res. *8*, 292 (1974)
246. Cadée, G.C.: Neth. J. Sea Res. *15*, 228 (1982)

247. Laane, R.W.P.M.: Estuar. Coast. Mar. Sci. *10*, 589 (1980)
248. Laane, R.W.P.M.: Neth. J. Sea Res. *14*, 192 (1980)
249. Laane, R.W.P.M.: Neth. J. Sea Res. *15*, 88 (1981)
250. Schlesinger, W.H., Melack, J.M.: Tellus *33*, 172 (1981)
251. Mulholland, P.J., Watts, J.A.: Tellus *34*, 176 (1982)
252. Likens, G.E., ed.: Flux of Organic Carbon by Rivers to the Oceans, U.S. Dept. Energy Conference Rept. CONF-8009140, pp. 397 (1981)
253. Bothner, M.H., Parmenter, C.M., Milliman, J.D.: Estuar. coast. Shelf Sci. *13*, 213 (1981)
254. Fanning, K.A., Carder, K.L., Betzer, P.R.: Deep Sea Res. *29*, 953 (1982)
255. Zaneveld, J.R.V., Pak, H.J.: J. Geophys. Res. *84*, 7781 (1979)
256. Paffenhofer, G.A., Deibel, D., Atkinson, L.P., Dunstan, W.M.: Deep-Sea Res. *27*, 435 (1980)
257. Kamat, S.B., Indian J. Mar. Sci. *5*, 232 (1976)
258. Yanada, M., Maita, Y.: Bull. Fac. Fish. Hokkaido Univ. *27*, 152 (1976)
259. Harvey, G.R.: Mar. Chem. *12*, 333 (1983)
260. Harvey, G.R., Requejo, A.G., McGillivary, P.A., Tokar, J.M.: Science *205*, 999 (1979)
261. Ogura, N.: Deep sea Res. *17*, 221 (1970)
262. Wangersky, P.J.: Deep-Sea Res. *23*, 457 (1976)
263. Nakajima, K., Nishizawa, S.: Biological Oceanography of the Northern North Pacific Ocean, A.Y. Takenouti, ed., Idemitsu Shoten, Tokyo, p. 495 (1972)
264. Costin, J.M.: J. Geophys. Res. *75*, 4144 (1970)
265. Pak, H., Codispotti, L.A., Zaneveld, J.R.V.: Deep-Sea Res. *27*, 783 (1980)
266. Feely, R.A., Sullivan, L., Sackett, W.M.: Suspended Solids in Water, R.J. Gibbs, ed., Plenum Publ. Corp., N.Y., p. 281 (1974)
267. Wangersky, P.J.: Helgol. Wiss. Meeresunters. 20, 546 (1977)
268. Ogura, N.: Nature, Lond. *227*, 1335 (1970)
269. Kinney, P.J., Loder, T.C., Groves, J.: Limnol. Oceanogr. *16*, 132 (1982)
270. Ljutsarev, S.V., Mirkina, S.D., Romankevich, E.A., Smetankin, A.V.: Pol. Archs. Hydrobiol. *24* (Suppl.), 91 (1977)
271. Bordovsky, O.K., Mirkina, S.D., Korzhikova, L.I.: Okeanologiia *19*, 59 (1979)
272. Karabashev, G.S., Solovyev, A.N.: Pol. Archs. Hydrobiol. *24* (Suppl.), 115 (1977)
273. Baylor, E.R., Sutcliffe, W.H. Jr., Hirschfeld, D.S.: Deep Sea Res. *9*, 120 (1962)
274. Garrett, W.D.: Deep-Sea Res. *14*, 221 (1967)
275. Blanchard, D.C.: In: Applied Chemistry at Protein Interfaces, Adv. in Chem. Ser. 145, R.E. Baier, ed., Amer. Chem. Soc., Washington, D.C., 360 (1975)
276. Wangersky, P.J.: Ann. Rev. Ecol. Syst. *7*, 161 (1976)
277. Lion, L. W., Leckie, J.O.: A. Rev. Earth Planet. Sci. *9*, 449 (1981)
278. Hardy, J.T.: Prog. Oceanogr. *11*, 307 (1982)
279. Barger, W.R., Daniel, W.H., Garrett, W.D.: Deep Sea Res. *21*, 83 (1974)
280. Jarvis, N.L.: Limnol. Oceanogr. *12*, 213 (1967)
281. Williams, P.M., Van Vleet, E.S., Booth, C.R.: J. Mar. Res. *38*, 193 (1980)
282. Garrett, W.D., Timmons, C.O., Jarvis, N.L., Kagarise, R.E.: NRL Rept. *5925*, 1 (1963)
283. Jarvis, N.L., Garrett, W.D., Scheiman, M.A., Timmons, C.O.: Limnol. Oceanogr. *12*, 88 (1967)
284. Barger, W.R., Garrett, W.D.: J. Geophys. Res. *75*, 4561 (1970)
285. Barger, W.R., Garrett, W.D.: J. Geophys. Res. *81*, 3151 (1976)
286. Duce, R.A., Quinn, J.G., Olney, C.E., Piotrowicz, S.R., Ray, B.J., Wade, T.L.: Science *176*, 161 (1972)
287. Marty, J.C., Saliot, A.: J. rech. Atmos. *8*, 563 (1974)
288. Marty, J.C., Saliot, A.: Deep-Sea Res. *23*, 863 (1976)
289. Marty, J.C., Saliot, A., Tissier, M.J.: C. r. Hebd. Sèanc. Acad. Sci., Paris, (D) *286*, 833 (1978)
290. Marty, J.C., Choinitère, A.: Naturaliste Can. *106*, 141 (1979)
291. Meyers, P.A., Kawka, O.E.: J. Gt. Lakes Res. *8*, 288 (1982)
292. Baier, R.E.: J. Geophys. Res. *77*, 5062 (1972)
293. Gucinski, H., Goupil, D.W.: Ocean Sci. Engng. *6*, 351 (1981)
294. Carlson, D.J.: Nature, Lond. *296*, 426 (1982)
295. Van Vleet, E.S., Wiliams, P.M.: Limnol. Oceanogr. 28, 401 (1983)
296. Nishizawa, S., Nakajima, K.: Bull. Plankt. Soc. Japan *18*, 12 (1971)
297. Nakajima, K.: Mem. Fac. Fish. Hokkaido Univ. *20*, 1 (1973)

298. Hunter, K.A.: Mar. Chem. *9*, 49 (1980)
299. Healy, T.W., La Mer, V.K.: J. Phys. Chem. *68*, 3535 (1964)
300. Wheeler, J.R.: Limnol. Oceanogr. *20*, 338 (1975)
301. Fitzgerald, W.F., Hunt, C.D.: J. Rech. Atmos. *8*, 629 (1974)
302. Eisenreich, S.J., Elzerman, A.W., Armstrong, D.E.: Environ. Sci. Technol. *12*, 413 (1978)
303. Pojasek, R.B., Zajicek, O.T.: Wat. Res. *12*, 7 (1978)
304. Elzerman, A.W., Armstrong, D.E.: Limnol. Oceanogr. *24*, 133 (1979)
305. Elzerman, A.W., Armstrong, D.E., Andren, A.W.: Environ. Sci. Technol. *13*, 720 (1979)
306. Owen, R.H., Meyers, P.A., Mackin, J.E.: Geophys. Res. Letts. *6*, 147 (1979)
307. Bacon, M.P., Elzerman, A.W.: Nature, Lond. *284*, 332 (1980)
308. Lion, L.W., Leckie, J.O.: Environ. Geol. *3*, 293 (1981)
309. Armstrong, D.E., Elzerman, A.W.: J. Gt. Lakes Res. *8*, 282 (1982)
310. Garrett, W.D.: J. Rech. Atmos. *8*, 553 (1974)
311. Hardy, R., Mackie, P.R., Whittle, K.J., McIntyre, A.D., Blackman, R.A.A.: Rapp. P.-v. Run. Cons. Perm. Int. Explor. Mer *171*, 61 (1975/77)
312. Anufrieva, N.M., Gornitsky, A.B., Nesterova, M.P., Nemirovskaya, I.A.: Okeanologiia *16*, 255 (1976)
313. Benzhitskii, A.G., Polikarpov, G.G.: Sov. J. Mar. Biol. *3*, 157 (1977)
314. Stadler, D.F.: Dtsch. Hydrogr. Z. *30*, 189 (1977)
315. Mikhailov, V.I.: Okeanologiia *18*, 841 (1978)
316. Mikhailov, V.I.: Okeanologiia *19*, 819 (1979)
317. Wu, T.L., Lambert, L., Hastings, D., Banning, D.: Bull. Environ. Contam. Toxicol. *24*, 411 (1980)
318. Gearing, P.J., Gearing, J.N.: Mar. Environ. Res. *6*, 133 (1982)
319. Stoner, A.W.: Deep-Sea Res. *30*, 469 (1983)
320. Soloviev, A.V., Vershinsky, N.V.: Deep-Sea Res. *29*, 1437 (1982)
321. Burney, C.M., Davis, P.G., Johnson, K.M., Sieburth, J.McN.: Mar. Biol. *67*, 311 (1982)
322. Walsh, G.E.: Limnol. Oceanogr. *10*, 577 (1965)
323. Shanks, A.L., Trent, J.D.: Deep-Sea Res. *27*, 137 (1980)
324. Wangersky, P.J.: Helgoländer Wiss. Meeresunters. *30*, 546 (1977)
325. Wangersky, P.J.: Deep-Sea Res. *23*, 457 (1976)
326. Gagosian, R.B.: Geochim. Cosmochim. Acta *39*, 1443 (1975)
327. Gagosian, R.B.: Limnol. Oceanogr. *21*, 702 (1976)
328. Aizatulin, T.A., Khailov, K.M.: Okeanologiia *12*, 809 (1972)
329. Eadie, B.J., Jeffrey, L.M.: Mar. Chem. *1*, 199 (1973)
330. Eadie, B.J., Jeffrey, L.M., Sackett, W.M.: Geochim. Cosmochim. Acta *42*, 1265 (1978)
331. Baturnina, M.V., Ljutsarev, S.V., Romankevich, E.: Pol. Archs. Hydrobiol. *24* (Suppl.), 99 (1977)
332. Honjo, S.: J. Mar. Res. *36*, 469 (1978)
333. Scranton, M.I., Brewer, P.G.: Limnol. Oceanogr. *23*, 1207 (1970)
334. Neihof, R.A., Loeb, G.I.: Limnol. Oceanogr. *17*, 7 (1972)
335. Neihof, R.A., Loeb, G.I.: J. Mar. Res. *32*, 5 (1974)
336. Hunter, K.A., Liss, P.S.: Nature, Lond. *282*, 823 (1979)
337. Hunter, K.A.: Limnol. Oceanogr. *25*, 807 (1980)
338. Hunter, K.A.: Geochim. Cosmochim. Acta *47*, 467 (1983)
339. Brown, M.: Mar. Chem. *3*, 253 (1975)
340. Sharp, J.H., Culberson, C.H., Church, T.M.: Limnol. Oceanogr. *27*, 1015 (1982)
341. Bidleman, T.F., Leonard, R.: Atmos. Environ. *16*, 1099 (1982)
342. Nissenbaum, A., Kaplan, I.R.: Limnol. Oceanogr. *17*, 570 (1972)
343. Hair, M.E., Bassett, C.R.: Estuar. Coast. Mar. Sci. *1*, 107 (1973)
344. Nissenbaum, A.: Mar. Chem. *2*, 59 (1974)
345. Brown, M.: Estuar. Coast. Mar. Sci. *5*, 309 (1977)
346. Stuermer, D.H., Peters, K.E., Kaplan, I.R.: Geochim. Cosmochim. Acta *42*, 989 (1978)
347. Sieburth, J. McN.: J. Exp. Mar. Biol. Ecol. **2**, 290 (1969)
348. Ragan, M.A., Jensen, A.: J. Exp. Mar. Biol. Ecol. *36*, 91 (1979)
349. Sieburth, J.McN., Jensen, A.: J. Exp. Mar. Biol. Ecol. *2*, 174 (1968)
350. Sieburth, J.McN., Jensen, A.: J. Exp. Mar. Biol. Ecol. *3*, 275 (1969)
351. Sargent, J.R.: Sci. Prog. *65*, 437 (1978)
352. Ramus, J.: J. Phycol. *8*, 97 (1972)

353. Lampert, W.: Limnol. Oceanogr. *23*, 831 (1978)
354. Sharp, J.H.: Limnol. Oceanogr. *22*, 381 (1977)
355. Guillard, R.R.L., Wangersky, P.J.: Limnol. Oceanogr. *3*, 449 (1958)
356. Alldredge, A.L.: Science *177*, 885 (1972)
357. Alldredge, A.L.: Limnol. Oceanogr. *21*, 14 (1976)
358. Seki, H.: Mar. Biol. *9*, 4 (1971)
359. Baylor, E.R., Sutcliffe, W.H. Jr.: Limnol. Oceanogr. *8*, 213 (1963)
360. Johnson, B.D.: Limnol. Oceanogr. *21*, 444 (1976)
361. Zepp, R.G., Cline, D.M.: Environ. Sci. Technol. *11*, 359 (1977)
362. Zika, R.G.: Unpubl. Ph. D. thesis, Dalhousie Univ., 346 pp (1977)
363. Cooper, W.J., Zika, R.G.: Science *220*, 711 (1983)
364. Zepp, R.G., Wolfe, N.L., Gordon, J.A., Baughman, G.L.: Environ. Sci. Technol. *9*, 1144 (1975)
365. Zepp, R.G., Baughman, G.L., Schlotzhauer, P.F.: Chemosphere, *10*, 109 (1981)
366. Kittredge, J.S., Takahashi, F.T., Lindsey, J., Lasker, R.: Fish. Bull. *72*, 1 (1974)
367. Pütter, A.: Die Ernährung der Wassertiere und der Stoffhaushalt der Gewässer, Fisher, Jena (1909)
368. Keys, A., Christensen, E.H., Krogh, A.: J. Mar. Biol. Ass. U.K. *20*, 181 (1935)
369. Stewart, M.G.: Oceanogr. Mar. Biol. a Rev. *17*, 163 (1979)
370. Reid, R.G.B., Bernard, F.R.: Science *208*, 609 (1980)
371. Southward, A.J., Southward, E.C.: Sarsia *50*, 29 (1972)
372. Southward, A.J., Southward, E.C.: Kieler Meeresforsch. *5*, 445 (1981)
373. Hellebust, J.A.: Limnol. Oceanogr. *10*, 192 (1965)
374. Hellebust, J.A.: J. Phycol. *3*, 132 (1967)
375. Hollibaugh, J.T.: Mar. Biol. *36*, 303 (1976)
376. Hollibaugh, J.T.: Mar. Biol. *45*, 191 (1978)
377. Sepers, A.B.J.: Hydrobiologia *52*, 39 (1977)
378. Daley, R.J., Hobbie, J.E.: Limnol. Oceanogr. *20*, 875 (1975)
379. Hagstrom, A., Larsson, U., Horstedt, P., Normark, S.: Appl. Environ. Microbiol. *37*, 805 (1979)
380. Wright, R.T., Hobbie, J.T.: Ecology *47*, 447 (1966)
381. Williams, P.J. le B.: J. Mar. Biol. Ass. U.K. *50*, 859 (1970)
382. Wright, R.T., Hobbie, J.E.: Limnol. Oceanogr. *10*, 22 (1965)
383. Karl, D.M.: Appl. Environ. Microbiol. *38*, 850 (1979)
384. Fuhrman, J.A., Azam, F.: Mar. Biol. *66*, 109 (1982)
385. Gocke, K.: Mar. Biol. *33*, 49 (1975)
386. Gocke, K.: Mar. Biol. *42*, 131 (1977)
387. Joiris, C.: Helgoländer Wiss. Meeresunters. *30*, 611 (1977)
388. Sepers, A.B.J., Cahet, G., Goossens, H.: Mar. Biol. *66*, 237 (1982)
389. Hamilton, R.D., Austin, K.E.: Can. J. Microbiol. *13*, 1165 (1967)
390. Hamilton, R.D., Preslan, J.E.: Limnol. Oceanogr. *15*, 395 (1970)
391. Bertoni, R., Callieri, C., Torelli, R., Zola, O.: Mem. Ist. Ital. Idrobiol. *40*, 129 (1982)
392. Douglas, D.J.: Unpub. Ph. D. thesis, Dalhousie Univ. (1983)
393. Van Es, F.B., Meyer-Reil, L.-A.: Adv. Microb. Ecol. *6*, 111 (1982)
394. Otsuki, A., Hanya, T.: Limnol. Oceanogr. *17*, 248 (1972)
395. Riley, G.A.: Limnol. Oceanogr. *8*, 372 (1963)
396. Baylor, E.R., Sutcliffe, W.H. Jr., Hirschfeld, D.S.: Deep-Sea Res. *9*, 120 (1962)
397. Loeb, G.I., Neihof, R.A.: Adv. in Chem. *145*, 319 (1975)
398. Loeb, G.I., Neihof, R.A.: J. Mar. Res. *35*, 283 (1977)
399. Chave, K.E.: Science *148*, 1723 (1965)
400. Tent, J.D., Shanks, A.L., Silver, M.W.: Limnol. Oceanogr. *23*, 626 (1978)
401. Alldredge, A.L., Cox, J.L.: J. Mar. Res. *40*, 517 (1982)
402. Bishop, J.K.B., Edmond, J.M., Ketten, D.R., Bacon, M.P., Silker, W.B.: Deep-Sea Res. *24*, 511 (1977)
403. Bishop, J.K.B., Ketten, D.R., Edmond, J.M.: Deep-Sea Res. *25*, 1121 (1978)
404. Bishop, J.K.B., Collier, R.W., Kettens, D.R., Edmond, J.M.: Deep-Sea Res. *27*, 615 (1980)
405. Bruland, K.W., Franks, R.P., Landing, W.M., Soutar, A.: Earth Planet. Sci. Letts. *53*, 400 (1968)
406. Blomqvist, S., Hakanson, L.: Arch. Hydrobiol. *91*, 101 (1981)
407. Small, L.F., Fowler, S.W., Unlu, M.Y.: Mar. Biol. *51*, 233 (1979)

408. Barnes, A.T., Quetin, L.B., Childress, J.J., Pawson, D.L.: Science *194*, 1083 (1976)
409. Barham, E.G.: Science *205*, 1129 (1979)
410. Jannasch, H.W., Eimhjellen, K., Wieser, C.O., Farman Jarmaian, A.: Science *171*, 672 (1971)
411. Jannasch, H.W., Wirsen, C.O.: Science *180*, 641 (1973)
412. Albright, L.J.: Can. J. Microbiol. *21*, 1406 (1975)
413. Wirsen, C.O., Jannasch, H.W.: Mar. Biol. *31*, 201 (1975)
414. Jannasch, H.W., Wirsen, C.O., Taylor, C.D.: Science *216*, 1315 (1982)
415. Tabor, P.S., Deming, J.W., Ohwada, K., Colwell, R.R.: Appl. Environ. Microciol. *44*, 413 (1982)

Marine Gelbstoff

M. Ehrhardt

Institut für Meereskunde an der Universität Kiel
Düsternbrooker Weg 20, D-2300 Kiel 1, Federal Republic of Germany

Early Observations – the Origin of the Concept

The extremely variable coloration of seawater, ranging from the deep, luminous blue of the eastern Mediterranean over a cold, glassy green to the muddy yellow of certain estuaries and the ever changing hues of copper red, grey, and verdigris seen sometimes under a low sun is a phenomenon that has been observed since ancient times and inspired great works of art.

It was only during the last century, however, that first attempts were made to explain the causes of this coloration scientifically. Kalle [1] concluded that dissolved yellow substances contribute substantially to any deviation from the blue colour of pure water into the longer wavelengths region of the visible spectrum. Preliminary investigations indicated that yellow substances in seawater were closely related to terrestrial humic material. However, it passed through collodium membranes suggesting that it was not a colloid but a substance or combination of closely related substances in true molecular solution. Chemically the material defied unequivocal characterization, a situation which has remained essentially unchanged. Thus, for want of a systematic designation Kalle [2] called the material "Gelbstoff", the German word for "yellow material".

In modern and contemporary scientific literature the terms Gelbstoff, yellow material, marine humus, humic material, fulvic acids, etc. are often used synonymously, and this discussion based on and quoting the work of various investigators is no exception. A short glossary thus appears to be useful in order to avoid confusion concerning origin and properties:

Humic acids are operationally defined as being soluble in aqueous base, but insoluble in aqueous acids; *fulvic acids* are soluble in both acids and bases.

Marine Gelbstoff is a material formed in the sea with certain spectral characteristics. It is not derived from terrestrial sources and appears to be composed of fulvic and humic components both of which are structurally different from their terrestrial counterparts.

Humic material is a rather indistinct term embracing humic, fulvic, and other complex substances.

In near-shore or land-locked sea areas, terrestrial humic and fulvic material, mainly derived from the decay of lignin containing land plants, add a yellow component to the colour of seawater. Although many of its chemical and physical properties are quite similar to those of autochthonous marine Gelbstoff, only the latter is the principal subject of this chapter.

Already in his early work Kalle suggested that Gelbstoff was a relatively stable product of phytoplankton metabolism and that it was formed in the ocean continuously, because it could be observed in sea areas remote from any terrestrial influence. Even in the German Bight near the mouth of the river Elbe Kalle [3] consistently found higher concentrations of Gelbstoff than would result from mixing between seawater and river water, containing terrestrial humics. This was a strong indication that there is indeed on autochthonous Gelbstoff truly of marine origin (see also [4, 5]).

While investigating the colour of seawater and its causes, Kalle [6, 7] observed that UV irradiated seawater emitted a blue fluorescence light. Because with increasing salinity the ratio of the intensities of the blue fluorescence to Gelbstoff concentrations increased, he concluded that he was dealing with at least two different compounds. It is quite certain, however, that marine Gelbstoff as well as terrestrial humic material is not composed a definable number of specific chemical compounds. Evidence is accumulating that under suitable conditions numerous substances, from either terrestrial sources or from marine flora, form an assemblage of structurally related compounds. These compounds and to a small degree possibly even compounds of abiotic origin appear to undergo processes of maturation triggered by light, chemical agents, adsorption onto catalytically active surfaces, etc. The results are increase of molecular weight, decrease in solubility and reduced biological availability with, ultimately, the formation of a kerogen-like substance [8–15].

Thus, the term Gelbstoff, marine when the progenitors originated from marine flora or terrestrial when land plants provided the molecular building blocks, would designate a transitory, statistical structure with a certain, more or less well defined range of physical and chemical properties.

Isolation of Marine Gelbstoff

Any attempt to characterize the molecular nature of marine Gelbstoff ultimately depends on the isolation of a sufficient quantity of material for chemical and spectroscopic investigations. Because of the low concentrations of Gelbstoff especially in open ocean waters (< 1 mg dm^{-3}), an efficient method of isolation must be capable of processing large volumes of seawater. Also, the technique must be delicate enough to avoid alterations in the chemical composition of a complex and probably rather labile substance. Sorption appears to be the method of choice, and a variety of solid phases have been tested for their ability to concentrate Gelbstoff from seawater.

An early method of selectively adsorbing Gelbstoff onto Nylon [8] was later shown to suffer from organic contamination originating from the Nylon fabric [16].

Initial attempts with *activated charcoal as sorbant* have met with limited success [17, 18], but Kerr and Quinn [19] using a special charcoal preparation obtained 100 mg samples of organic matter from acidified seawater (pH = 2) with quantitative adsorption of the coloured material in the fulvic acid fraction, the fraction of adsorbed dissolved organic carbon (DOC) ranging from 55 to 86%. Aqueous base and methanol desorbed 39 to 99% of the organic carbon from the activated charcoal. The desorbed material was applied to alumina columns from which 0.05 molar NaOH eluted the major portion as a yellow to orange band.

The use of aqueous ammonia, methanol, and methanolic ammonia to desorb three different fractions of organic matter collected from seawater by sorption onto *activated charcoal* is described in a later publication [20]. This method offers the obvious advantage that the base can be removed quantitatively by evaporation leaving behind a material free from inorganic ions. Overall recovery of dissolved organic matter was similar, ranging from 37 to 118% depending upon the location and depth of sample collection. Mantoura and Riley [21, 22], Stuermer and Harvey [16], Stuermer and Payne [23], and Gagosian and Stuermer [12] employed the technique, first described by Riley and Taylor [24], of sorbing organic material dissolved in seawater onto the macroreticular resin Amberlite XAD-2, a styrene-divinyl benzene copolymer. The unfiltered water was acidified to pH 2 with hydrochloric acid and passed through the resin at an experimentally determined optimum flow rate of less than 2 bed volumes per minute [22]. The charged resin was rinsed with pure water or 0.01 molar hydrochloric acid to remove salts, and the sorbed material was eluted with aqueous ammonia.

Stuermer and Harvey [22] separated the base eluate into humic and fulvic acids, because the ratio of these components provides some structural information, fulvic acids being less consolidated and having a lower molecular weight than marine humic acids. After removal of lipophilic substances by extracting the concentrated base eluate with methylene chloride the humic and fulvic acids were recovered by lyophilizing the aqueous phase. Fulvic acids were then dissolved in 0.01 molar HCl, the insoluble humic fraction being separted by centrifugation, dissolution in ammonia and lyophilization in order to remove excess ammonia. Harvey et al. [25] using a gas lift system [26] to collect large volumes of seawater employ essentially the same sorption technique, but they mention explicitely that exposure to light and air was avoided throughout the isolation procedure apparently in anticipation of chemical alterations likely to be induced by these agents. The marine fulvic acids thus obtained after lyophilization were pale yellow powders, while the marine humic acids were yellow-brown.

Sources of Marine Gelbstoff

According to Kalle's original suggestion Gelbstoff not only originates from land and the littoral zone, but it is also produced in the open sea. For many years the only evidence for the latter hypothesis was circumstantial. It was based on the occurrence of Gelbstoff in the open sea, where terrestrial influences should be minimal, and on the fact that the 7×10^8 metric tons of organic matter contributed annually by rivers represent less than 1% of the organic matter contained in the oceans.

The hypothesis could be effectively tested, however, by comparing the ratio of the stable carbon isotopes ^{13}C and ^{12}C in marine Gelbstoff with that of phytoplankton on the one hand and that of land plants on the other, both of which could produce the source material. In seawater the carbon isotope ^{13}C is enriched relative to the atmosphere so that the $\delta^{13}C$ value, defined as

$$\delta^{13}C = \frac{(^{13}C/^{12}C)\text{sample} - (^{13}C/^{12}C)\text{standard}}{(^{13}C/^{12}C)\text{standard}} \times 1{,}000$$

of phytoplankton is higher than in land plants, because phytoplankton depends on dissolved HCO_3^- as its source of carbon [27]. If marine Gelbstoff is genetically related to marine phytoplankton, it should exhibit a similar $\delta^{13}C$ value.

Gelbstoff concentrations in open ocean waters rarely exceed a few hundred microgram per dm^3 so that until quite recently it was extremely difficult, if not practically impossible, to obtain sufficient quantities of pelagic Gelbstoff for carbon isotope determinations or any other detailed chemical analysis. Nissenbaum and Kaplan [10] determined the chemical and isotopic composition of humic acids from marine sediments sampled off southern California, in the Gulf of California, in the Pacific Ocean off Oregon and near Hawaii and in the Atlantic Ocean 100 km offshore north of Paramaribo and also from a number of terrestrial sediment samples. Most of the $\delta^{13}C$ values of the marine samples were in the range of -20 to -22‰ which differs significantly from the $\delta^{13}C$ values of -25 to -26‰ usually found in soil humic material derived from land plants or in sedimentary humic acids from coastal and littoral environments [10, 28].

The results of elemental analyses of carbon, hydrogen, and nitrogen were less revealing, but surprisingly high concentrations of sulfur were found in some of the marine humic acids. Since the samples rich in sulfur originated from areas with a very active biological sulfur cycle the authors suggested that the sulfur might not have been present in the original dissolved organic matter (Gelbstoff), but was possibly introduced during diagenesis in the sediments. Together, these findings strongly support the theory that marine sedimentary humic material and, hence, marine Gelbstoff is ultimately derived from marine phytoplankton.

In a later publication Nissenbaum [29] reported deuterium concentrations (δD values analogous to $\delta^{13}C$ values) in marine sedimentary and soil humic material. The δD values of marine samples showed a rather small scatter of $\delta D = (-105 \pm 5‰)$ suggesting a common precursor for the marine humates regardless of the sampling location, but different from the source material of terrestrial humics, whose δD values covered a much broader range (-57 to -97). $\delta^{13}C$ values of dissolved marine humic substances are reported by Stuermer and Harvey [21, 20]. Even in coastal seawater the $\delta^{13}C$ values (-22.7 to -23.8) were close to those obtained for Sargasso Sea and equatorial Atlantic surface water (-22.8) which agrees well with the $\delta^{13}C$ values measured by Nissenbaum and Kaplan for sedimentary marine humics. In the years 1968, 1969 and 1970, Sieburth and Jensen published four studies on Algal Substances in the Sea [8, 9, 30, 31]. Two dimensional paper chromatography showed that terrestrial Gelbstoff from different locations could be distinguished from each other and from marine Gelbstoff. Not only did the compositions of Gelbstoff from different localities vary, but there was also a marked seasonal change in the concentration of marine Gelbstoff. During the winter months from January through March it was barely detectable in samples of inshore waters. These are more indications of an autochthonous origin of marine Gelbstoff supported by the interesting observation that Gelbstoff from a bog water precipitated within a week at room temperature when an equal amount of seawater was added to its solution. In distilled water no precipitate was formed. This suggests that a least some terrestrial Gelbstoff, once introduced into seawater, will not remain in solution for any length of time and can thus not contribute to Gelbstoff found in the open sea.

The second paper in the series [30] aims at supporting the hypothesis that littoral brown algae such as *Fucus vesiculosus* exude phenolic substances and carbohydrates into the surrounding seawater where they react abiotically to form Gelbstoff. The experiments indeed showed that appreciable quantities of phenolic compounds (3.8% and 0.89% of the dry matter of *Fucus vesiculosus* and *Ascophyllum nodosum*, respectively) and similar quantities of carbohydrates were lost to the water of the storage tanks. Excretion or loss of these substances may well be different under natural conditions, but it seems unlikely that they were mere artefacts of the experiment.

After removal of the algae a considerable production of Gelbstoff took place from 60 mg × kg^{-1} dry weight to roughly ten times the initial concentrations after 14 days at 6 °C, while phenol concentrations decreased by a factor of approximately 5 from the beginning to the end of the storage period. Analogous experiments with *Laminaria hyperborea* gave similar results. After prolonged storage two dimensional paper chromatograms of the products were indistinguishable from those of coastal seawater extracts. Taking into account the rather limited efficiency of paper chromtographic separations this result has to be interpreted with caution, but by no means can it be considered as irrelevant.

Exudates of *Ascophyllum nodosum* were separated with Sephadex G-15 into colourless fractions containing phenolic, carbohydrate, and proteinaceous material. After reaction a yellow, Gelbstoff-like material was formed. This could also be synthesized when phloroglucinol was used instead of the algal phenolic fraction. The use of specific reagents revealed that the algae excreted simple *phenols* which, in the slightly alkaline medium of seawater, are slowly transformed into *polyphenols* which then enter the reaction sequence leading to Gelbstoff. The authors suggest that the blue fluoresceing compound(s) first described by Kalle may be a product of the polyphenols reacting among themselves. Because phenols, proteins, and carbohydrates, although chromatographically different from those excreted by the algae were also found in bog waters, it was proposed that the invoked mechanism of formation may be common to both freshwater and marine Gelbstoff.

Gillbricht [32] measured concentrations of Gelbstoff off West Africa between approximately 30° and 35 °N by its UV absorption at 230, 240, and 250 nm. Because the highest concentrations coincided with the near-shore temperature and salinity minima and phosphate maximum – all being indicators of upwelling water –, and because Gelbstoff concentrations decreased with increasing distance from the shore he concluded that the phytoplankton of the upwelling water was the source of the Gelbstoff. The observation that Gelbstoff concentrations varied with changing velocities of the wind inducing the upwelling was taken as further evidence. The phase difference between the periodically changing wind velocities and the alteration of the Gelbstoff concentrations was explained by the retarded reaction of the water to varying atmospheric conditions and the slow response of biological systems to changes in the environment. The latter contribution to the time lag could also be interpreted, however, by assuming that phytoplankton does not – as Sieburth and Jensen found for benthic algae – excrete Gelbstoff directly, but that it is slowly formed in the water from precursors not absorbing at the wavelengths used by Gillbricht for measuring Gelbstoff concentrations.

Furthermore, contributions of benthic algae, known to occur in the area [33] to the Gelbstoff observed by Gillbricht cannot be excluded. If the mechanisms discussed by Sieburth and Jensen were operative and benthic algae provided the source material, the mass excreted per unit time would be less diluted during periods of slow water movement, and its concentration should increase. This is precisely what Gillbricht described, but he assumed that the *plankton*, sinking during periods of slack water, is thus damaged and induced to excrete Gelbstoff. Since both explanations are possible, Gillbricht's observations do not provide unequivocal evidence that phytoplankton is the source of marine Gelbstoff.

An interesting method to distinguish between terrestrial- and marine-derived humic material in marine sediments is based on the determination of phenolic aldehydes. Products of the decomposition of lignin which is biosynthesized by higher vascular plants, but not by phytoplankton [34], appear to be incorporated into terrestrial humic material [35, 36]. By mild oxydation with nitrobenzene both lignin and terrestrial humic matter yield the phenolic aldehydes vanillin (3-methoxy-4-hydroxybenzaldehyde) and syringic aldehyde (3,5-dimethoxy-4-hydroxybenzaldehyde) which can thus be used as *chemical tracers to study the origin of sedimentary humic material* [37]. If organic substances in continental shelf sediments are mainly derived from terrestrial sources, the authors argue, and if the rates of decomposition of the source material for the phenolic aldehydes and of the bulk of the complex organic matter in the sediments are approximately equal, the concentration ratios should remain constant with increasing distance from the shore, while the absolute concentrations would be expected to decrease. The authors, however, found the ratio of phenolic aldehydes to total organic material to decrease from approximately 1% in sediment samples from inland rivers along the coast of Georgia to less than 0.1% in samples taken on the continental shelf more than 10 nautical miles off the coasts of Georgia and Texas. Because phenolic aldehydes were also obtained in small concentrations (1–3% p-hydroxybenzaldehyde per dry weight) by oxydation of samples of salt marsh grasses (*Spartina alterniflora* and *Juncus roemeriana*) these plants thriving in the littoral zone also seem to contribute to the humic material in near-shore waters, but their influence does not extend into the open sea. Thus, it appears that there are at least three different sources of Gelbstoff in the sea: *land plants and littoral flora predominating as sources of Gelbstoff in inshore waters and phytoplankton providing the precursors of pelagic Gelbstoff* (see also [38]).

Keeping in mind the potential precursors of littoral Gelbstoff which have been characterized tentatively by Sieburth and Jensen and leaving out a detailed discussion of terrestrial humic material we should now approach the questions: what are the compounds that could be visualized as potential source material for marine Gelbstoff, and what reaction sequence(s) could lead to its formation?

In the open sea, the greater part of living organic matter occurs in the form of phyto- and zooplankton; 35–55% of its dry weight consists of proteins in the form of structural polymers, enzymes, and transport systems [39]. Carbohydrates and triglycerides are the next most important constituents. It is not surprising therefore that free and combined amino acids and mono- and oligomeric sugars make up the bulk of the low molecular weight organic material dissolved in seawater [40, 41, 27]. Minor components of the dissolved organic material are fatty acids, hydrocarbons, amino sugars, phenols, pigments (chlorophyll and products of its degradation), ketones [42], aldehydes and dimethyl polysulfides [43], sterols [12], and probably many more compound types whose presence in seawater has so far remained unknown due to analytical limitations. There are no compelling reasons to exclude any of these classes of compounds as precursors of marine Gelbstoff, but reactivity and concentration certainly influence the degree to which they participate in its formation. As a possible mechanism for the formation of marine Gelbstoff from relatively simple molecular subunits the Maillard reaction has been invoked occasionally [1, 29]. It has also been rejected as a feasible pathway [14], because the conditions under which it has been studied (elevated temperature, high concentrations of reactants) are unlike those prevailing in seawater. It should not be overlooked, however, that although dissolved organic compounds are highly diluted in bulk seawater, their concentrations in the surface microlayer or on the surfaces of air bubbles may increase by orders of magnitude [45–47]. Thermal energy for the induction of chemical reactions is on a relatively low level in seawater, but radiative energy from sunlight is plentiful, and although most of the compounds suggested as starting material for the formation of Gelbstoff do not themselves absorb in the wavelengths region of solar radiation at sea level, the recent characterization of photocatalysts in seawater [42, 48] renders their participation in photochemically induced reactions more likely.

The *Maillard reaction* requires thermal activation energy and relatively high concentrations of reactants. Moreover, as Pocklington [14] writes: "Until someone actually demonstrates a Maillard-type ‚browning‘ reaction between likely precursors at reasonable concentrations under seawater conditions, this remains a possibility, not a chemical explanation". A short discussion may still be useful if it were only to enable the reader to form his or her own conclusions as to its importance for the formation of marine Gelbstoff.

The Maillard reaction may be subdivided into four sections [49]

a) the formation of glycosyl amines
b) Amadori or Heyns rearrangement of the glycosyl amines
c) conversion of the rearrangement products into α-dicarbonyl compounds
d) transformation of the α-dicarbonyl compounds into related reactive species.

A reducing sugar can react in the open-chain form with nucleophiles, e.g. the NH_2-group of amino acids

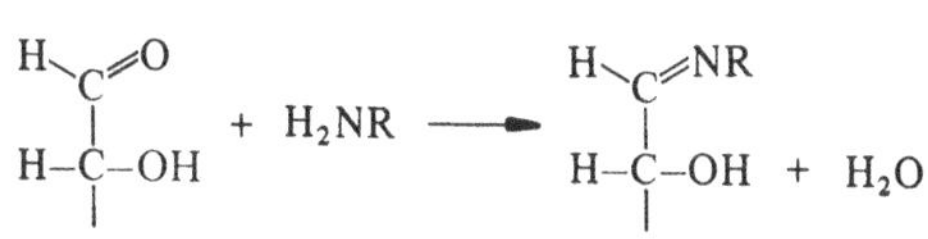

The condensation products which can be isolated as metal complexes are unstable in their free form, since the reaction may reverse. However, an Amadori (aldose

amines) or Heyns (ketose amines) rearrangement converts the condensation products into relatively stable compounds

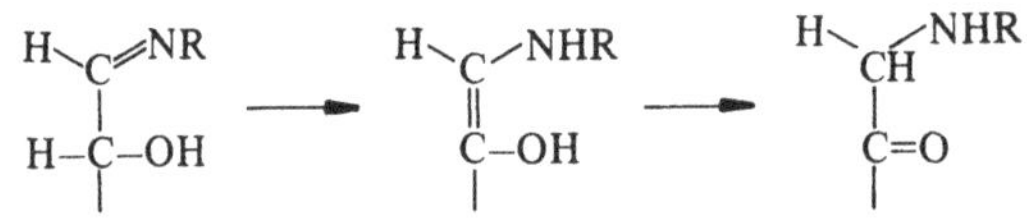

According to Angrick and Rewicki [49], the rearrangement products are only partially resorbed in the intestinal tract of animals, they cannot be digested and are excreted virtually unchanged. This observation is interesting in the light of the well-known considerable biological stability of more highly condensed dissolved organic matter in seawater [50, 51, 11].

The Amadori or Heyns rearrangement products are enolized and hydrolyzed after migration of the OH-group adjacent to the enol double bond. The original reducing sugars are thus converted into highly reactive, hydroxylated 1,2-dicarbonyl compounds

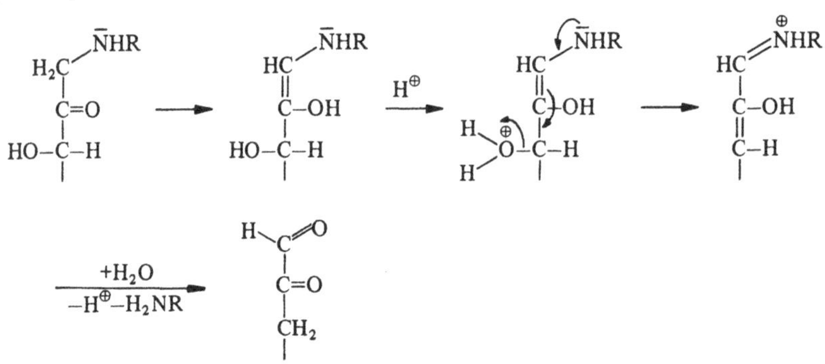

The 1,2-dicarbonyl compounds may then condense with the NH_2-group of amino acids and convert them, via Strecker degradation, into aldehydes minus one carbon atom which is removed as carbon dioxide:

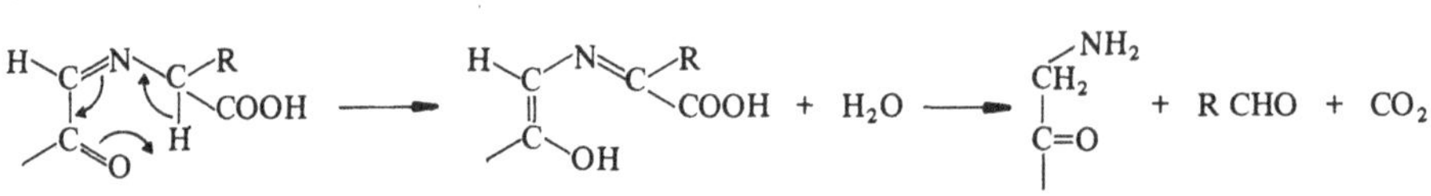

The 1,2-dicarbonyl compounds together with α-aminocarbonyl compounds and aldehydes which are known to occur in seawater [43, 52] represent an enormous synthetic potential. Even if the rates of the individual reactions in this sequence were exceedingly slow due to the special conditions prevailing in seawater, they could play a role in the formation of marine Gelbstoff.

Photochemical C-C bond formation already suggested by Sieburth and Jensen [9] is indicated in a more specific manner by Harvey et al. [25] who present experimental evidence that triglycerides of polyunsaturated fatty acids could thus be transformed into marine Gelbstoff. The observation of Wheeler [53] that a yellow, dissolved material with a UV spectrum nearly identical to that of marine fulvic acid [54] could be produced by UV irradition of linolenic acid in sterilized artificial seawater is quoted in support of this theory as well as [55] findings that hydroxylation and oxydative crosslinking occur in polyunsaturated fatty acids, when their aqueous emulsions are irradiated with UV light in the presence of air.

A reaction sequence for the photochemical formation of marine Gelbstoff analogous to the Maillard reaction discussed above has not yet been proposed. The reader interested in the potentialities of marine photochemistry is referred to the preview article by Zafiriou [56] and references cited therein.

Differentiation Between Terrestrial and Marine Gelbstoff Based on Spectroscopic Properties

The term "Gelbstoff", one should remember, was coined in reference to the common color of a group of closely related organic substances in seawater, but in addition to the optical qualities in the visible range there are more spectroscopic properties which characterize these compounds. Many of them can be used to differentiate between terrestrial and marine Gelbstoffe.

The visible and UV spectra of Gelbstoffe give little information on their chemical structures. The absorbances increase steadily from roughly 700 nm to the end of the easily accessible UV range at 200 nm. However, there are relative differences in the absorbances of Gelbstoffe due to their derivation from different sources and their degree of maturation, i.e. increased crosslinking and aromatization on their way from relatively simple, low molecular weight condensates to highly complex, kerogen-like substances.

Based on this observation, the ratio of absorbances at 465 and 665 nm, the E_4/E_6 ratio, is used as a parameter for typifying soil humic substances [57].

The E_4/E_6 ratio of soil humic substances usually decreases with increasing molecular complexity, but the specific absorbance in the visible range normalized to the carbon content of the samples is less intense for the soil fulvic than for the soil humic acids [58]. With open ocean Gelbstoff, however, a lower E_4/E_6 ratio is associated with a lighter color [19] indicating different source materials and/or different diagenetic processes in the marine enviroment. Also, equal absorbances of samples with equal organic carbon concentrations clearly shift towards shorter wavelenghts as one proceeds from an estuarine sediment extract to estuarine and open ocean Gelbstoff dissolved in surface waters [19].

Kerr and Quinn [20] collected dissolved organic matter from coastal and open sea water by adsorption onto activated charcoal. They eluted three fractions with 7M ammonium hydroxide, distilled methanol, and 7M ammonium hydroxide in methanol/water (50/50), and determined extinction ratios for 350 and 465 nm. Higher rations were found for more productive sea areas, but based on these ratios no clear distinction appeared to be possible between coastal and open ocean sea areas. Absorptivities at 465 nm, however, of the marine samples, though increasing from fraction 1 to 3, were considerably lower (appr. 0.05 $cm^{-1}g^{-1}dm^3$) than those reported for dissolved organic matter isolated from ponds, rivers, and sediments (appr. 1.0–2.0 $cm^{-1}g^{-1}dm^3$; [59]. The absorptivities of oceanic dissolved organic matter are even lower than those of dissolved organic matter generated under laboratory conditions from decomposing phytoplankton indicating that the natural mechanisms of Gelbstoff formation may not be simulated easily in vitro [20].

Gagosian and Stuermer [12] using UV absorptions at 420 and 665 nm for differentiating between coastal and oceanic fulvic acids found an $E_{420/665}$ ratio of 8.2 for coastal water fulvic acids and 4.8 for Sargasso Sea fulvic acids. They also investigated UV excitation of fulvic acids from different sea areas in water as solvent. Coastal water fulvic acid was excited most strongly at 325 nm, but fulvic acid from the Sargasso Sea had the excitation maximum at 332 nm. The corresponding fluorescence maxima were observed at 415 (primary), 437 (secondary), and, in the case of Sargasso Sea fulvic acid at 405 nm. The fluorescence intensity per unit weight was lower for the latter than for fulvic acids from coastal waters.

IR spectra of humic material in general and marine Gelbstoff in particular are rather uncharacteristic, broad and diffuse absorption bands indicating a multitude of similar structural elements. Absorption near 1700 cm^{-1} is characteristic for carbonyl groups, C-H stretching vibrations cause absorption near 3000 cm^{-1}, and a wide band extending beyond 3500 cm^{-1} signifies the presence of OH and NH groups [23].

Considering their indistinct nature it is not surprising that no significant differences were found in the IR spectra of open sea fulvic acids and those from other environments. More informative are ^{13}C and ^{1}H NMR spectra. An order of decreasing aromaticity is found from terrestrial to marine humic acids with even fewer aromatically bound carbon and hydrogen atoms in marine fulvic acids [23]. The decreasing aromaticity is consistent with an increasing H/C ratio calculated from elemental analyses. A plausible explanation for the lower aromaticity of marine humic and fulvic matter is the lack of aromatic structures in the source material. The highly aromatic lignin, an important precursor of terrestrial humics [61], although being a minor component of marsh grasses and some littoral plants [37] is not found in phytoplankton which appears to provide the source material for Gelbstoff in the open sea.

Attempts at Structure Elucidation

The molecular complexity of marine humic and fulvic acids which are components of Gelbstoff is a tantalizing challenge to the analyst's skill. Several attempts have been made to elucidate chemical structures in order to derive information on source materials, mechanisms of formation, and environmental significance of these substances.

The blurred UV, IR, NMR, ESR spectra, the results of direct probe distillation mass spectrometry, and the wide range of molecular weights from a few hundred to approximately 5000 as determined by ultrafiltration [62] and gel permeation chromatography [12] prove that Gelbstoff is a multi-component mixture of structurally related compounds of which even the fulvic acid fraction defies gas chromatographic separation [12, 63]. Thus, any endeavour to characterize structural elements would be more likely to succeed, if first its complexity was reduced either by oxidation or by reduction. Both methods have been applied.

Gagosian and Stuermer [12] and Stuermer and Harvey [63] investigated structural characteristics of *the fulvic acid fraction* of marine humus after careful reductive elimination of functional groups. The fulvic acids isolated from Sargasso Sea surface water and coastal waters from the northeastern United States were methylated with diazomethane. The methyl esters were then hydrogenated with H_2 at 5200 psi and 80 °C for 36 hours over copper-barium chromite resulting in the reduction of ester, acid, aldehyde, ketone, and olefin moieties while aromatic rings and amide groups were left unchanged. After reaction of the alcoholic and phenolic hydroxyl groups thus generated and cleavage of ether bonds with dibromo-diphenylphosphorane the bromides were again hydrogenated under normal pressure and at slightly elevated temperature with $Pd/BaCO_3$ as catalyst. After column chromatographic pre-separation GC and GC-MS analyses characterized a series of normal alkanes with 12 to 26 carbon atoms and a high even to odd carbon preference, a minor contribution of branched alkanes, and mono-substituted alkyl benzenes with 10, 11 and 12 carbon atoms in the aliphatic chain as products of the reduction sequence.

The distribution of the aliphatic hydrocarbons and their relative concentrations were in agreement with the pattern of fatty acids of marine plankton and bacteria suggesting that these were their sources. The ratio of aliphatic to aromatic hydrocarbons in the reduction products was 20 to 1, together they represented 3% of the organic carbon in the fulvic acid fraction. In addition to the hydrocarbons a series of fatty acid methyl esters was found in a more polar column chromatographic fraction whose chain length corresponded with the number of carbon atoms of the n-alkanes.

From these results and from the stoichiometric formula $C_8H_{15}O_5N$ derived from elemental analyses a hypothetical structure was developed which is shown in Fig. 1. (Taken from [63].)

One recognizes the more important compound classes of marine dissolved organic matter such as carbohydrates, amino acids, and fatty acids as subunits.

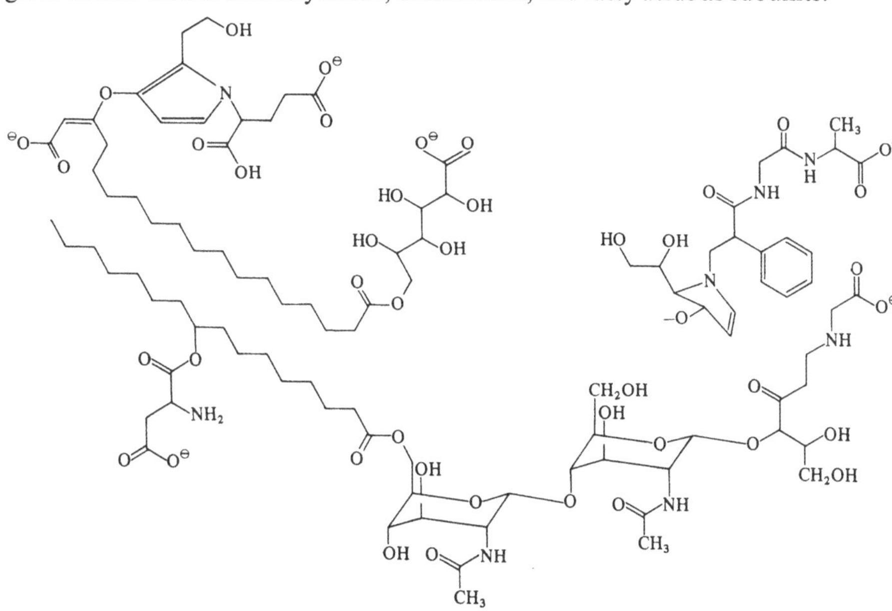

Fig. 1. Hypothetical structure of fulvic acids from Marine Gelbstoff proposed by Stuermer and Harvey [63]

Harvey et al. [25] oxidized fulvic acid components of marine Gelbstoff with permanganate in alkaline aqueous solution and esterified the reaction products with boron trifluoride/methanol. Capillary gas chromatography in combination with mass spectrometry characterized the most abundant oxidation products as dibasic succinic (C_4), glutaric (C_5), adipic (C_6), pimelic (C_7), suberic (C_8), and azelaic (C_9) acids.

Both fulvic and humic components were esterified easily with boron trichloride/methanol suggesting that the carboxyl groups were positioned in the molecules in such a way that they could readily react.

Proton NMR spectra revealed a practically complete lack of aromatic protons in the fulvic and merely 3–9% aromatic protons in the fulvic acids. The position of the centre of the aromatic proton resonance at 7.0 (δ) ppm is interesting as it indicates that the aromatic protons whose resonance in benzene appears at 7.2 ppm are shielded to some extend by electron-donating substituents such as alkyl or hydroxyl groups. The strongest resonances were observed in the methylene (1.2–1.8 ppm) and methyl (appr. 0.8 ppm) region followed by methylene groups adjacent to functional groups (1.9–2.5 ppm). Together they accounted for more than 72% of all carbon bound hydrogen atoms. A weak and variable resonance centered at approximately 3.5 ppm is ascribed to protons on a carbon atom bearing an ether or hydroxyl group.

These data and the known occurrence of polyunsaturated fatty acids in marine plankton which have been shown to undergo oxidative polymerization under the influence of light and oxygen [55, 53] lead the authors to propose a structure resulting from a light-assisted cross-linking of the unsaturated fatty acid moieties in polyunsaturated triglycerides with hydrolytic removal of the glycerol. Triglycerides were suggested as source material because the close spatial proximity of the fatty acid residues would aid in bond formation.

The observation of Stuermer [54] and Gagosian and Stuermer [12] that the ratio of humic to fulvic acid components increases with increasing depth is interpreted as time dependent progressive aromatization. The structures proposed by Harvey et al. [25] are shown in Fig. 2. (Taken from [25].)

The molecules do not contain nitrogen. Its incorporation as indicated by elemental analyses is believed to be incidental and not a fundamental requirement which contrasts with the hypothesis of a Maillard type condensation reactions being responsible for Gelbstoff formation.

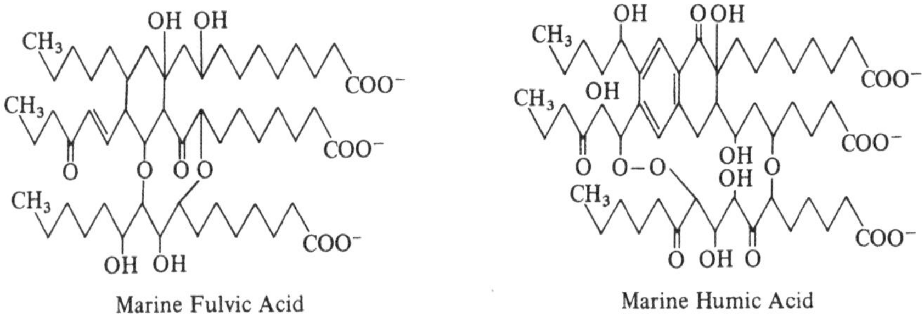

Fig. 2. Structures of marine fulvic acid and marine humic acid proposed by Harvey et al. [25]

Some Indications of Ecological Consequences of Gelbstoff in the Sea

The presence of Gelbstoff in seawater, representing roughly 90% of the dissolved organic matter in terms of organically bound carbon, has stimulated a keen interest in its interactions with and modifications of living and non-living, organic and inorganic components of the marine ecosystem. The literature on such interactions of humic and fulvic material in both fresh and seawater has been reviewed recently [64–68]. In spite of some as yet ill-defined structural differences the similar behaviour in this respect of all components of aqueous humic material justifies a common treatment which includes marine Gelbstoff.

Jackson [69] discussed the ecological significance of humic matter in natural waters and sediments, and many of the questions he raised in this context have not been brought much closer to solution in the meantime. Jackson deals with the following topics: beneficial effects of humic matter cation exchange and metal binding properties which are thought to be instrumental in making essential trace metals available to phytoplankton and detoxifying noxious concentrations, with the immobilization of micronutrients, effects on microbial growth, and with the possible role of humic material in species competition in that it could modify the availability of light energy and growth factors. He comes to the conclusion that "this material offers a promising and richly varied field for further research, both basic and applied".

One aspect presently attracting growing interest is *the photochemistry of Gelbstoff*. Choudhry [70] discusses photosensitizing and quenching effects which apparently have important implications for the environmental degradation of man-made contaminants and could just as well modify or possibly even control the photochemistry of natural constituents of seawater. Momzikoff et al. [48] in studying the photosensitizing properties of seawater found that in the Mediterranean the photosensitizing efficiency (PSE) increased with increasing depths until a maximum is reached near the surface were primary production should be most intense before it decreases again in deeper water layers. This depth dependency would suggest that photosensitizers are either directly excreted by planktonic organisms, or that they are early diagenetic products of excreted substances, i.e. what the fulvic fraction of Gelbstoff is believed to represent.

Considering the many interactions involving marine humic material the former notion that Gelbstoff was a wasteful metabolic byproduct not even fit for consumption by microorganisms appears to undergo a process of revision. It is quite conceivable that the formation as well as the reactions of Gelbstoff play a crucial role in maintaining balanced conditions in seawater with respect to its organic and inorganic trace constituents.

References

1. Kalle, K.: Ann. d. Hydr. *66*, 1 (1938)
2. Kalle, K.: Ann. d. Hydr. *65*, 276 (1937)
3. Kalle, K.: Dt. hydrogr. Z. *9*, 55 (1956)
4. Kalle, K.: Rapports et Proc. Verb. *59*, 98 (1938/39)
5. Jerlov, N.G.: Rep. Sweed. deep sea Exped. *3*, 1 (1951)

6. Kalle, K.: Dt. hydrogr. Z. *2*, 117 (1949)
7. Kalle, K.: Dt. hydrogr. Z. *4*, 92 (1951)
8. Sieburth, J. McN., Jensen, A.: J. exp. mar. Biol. Ecol. *2*, 174 (1968)
9. Sieburth, J. McN., Jensen, A.: in Organic Matter in Natural Waters (D.W. Hood, ed.), Inst. Mar. Sci. Occ. Publ. *1*, 203 (1970)
10. Nissenbaum, A., Kaplan, I.R.: Limnol. Oceanogr. *17*, 570 (1972)
11. Bada, J.L., Lee, C.: Mar. Chem *5*, 523 (1977)
12. Gagosian, R.B., Stuermer, D.H.: Mar. Chem. *5*, 605 (1977)
13. Morris, R.G., Eglinton, G.: Mar. Chem. *5*, 559 (1977)
14. Pocklington, R.: Mar. Chem. *5*, 479 (1977)
15. Skopintsev, B.A.: in Marine Organic Chemistry (E.K. Duursma, R. Dawson, eds.), 125 (1981)
16. Mantoura, R.F.C., Riley, J.P.: Anal. Chim. Acta *76*, 193 (1975)
17. Johnston, R.: J. mar. biol. Ass. U.K. *34*, 185 (1955)
18. Wilson, D.P., Armstrong, F.A.J.: J. mar. biol. Ass. U.K. *31*, 335 (1952)
19. Kerr, R.A., Quinn, J.G.: Deep-Sea Res. *22*, 107 (1975)
20. Kerr, R.A., Quinn, J.G.: Mar. Chem. *8*, 217 (1980)
21. Stuermer, D.H., Harvey G.R.: Nature, Lond. *250*, 480 (1974)
22. Stuermer, D.H., Harvey, G.R.: Deep-Sea Res. *24*, 303 (1977)
23. Stuermer, D.H., Payne, J.R.: Geochim. Cosmoschim. Acta *40*, 1109 (1976)
24. Riley, J.P., Taylor, D.: Anal. Chim. Acta *46*, 307 (1969)
25. Harvey, G.R., Boran, D.A., Chesal, L.A., Tokar, J.M.: Mar. Chem. *12*, 119 (1983)
26. Tokar, J.M., Harvey, G.R., Chesal, L.A.: Deep-Sea Res. *28*, 1395 (1981)
27. Williams, P.J. LeB.: in Chemical Oceanography (J.P. Riley, G. Skirrow, eds.), 301 (1975)
28. Brown, F.S., Baedecker, M.J., Nissenbaum, A., Kaplan, I.R.: Geochim. Cosmochim. Acta *36*, 1185 (1972)
29. Nissenbaum, A.: Mar. Chem. *2*, 59 (1974)
30. Sieburth, J. McN., Jensen, A.: J. exp. mar. biol. Ecol. *3*, 275 (1969)
31. Sieburth, J. McN.: J. exp. mar. biol. Ecol. *3*, 290 (1969)
32. Gillbricht, M.: Dt. hydrog. Z. *33*, 53 (1980)
33. Gayral, P. Algues de la côte Atlantique Marocaine. La Nature au Maroc II. Faculté de Science de Rabat (1958)
34. Pocklington, R., MacGregor, C.D.: Int. J. Environ. Anal. Chem. *3*, 81 (1973)
35. Christan, R.F., Oglesby, R.T.: in Lignins Occurence, Formation, Structure and Reactions (K.V. Sarkanen, C.H. Ludwig, eds.), chapter 18, Wiley Interscience (1971)
36. Gjessing, E.T.: Physical and Chemical Characteristics of Aquatic Humus. Ann Arbor Science Publisher, Ann Arbor, Mich. 120 pp. (1976)
37. Gardner, W.S., Menzel, D.W.: Geochim. Cosmochim. Acta *38*, 813 (1974)
38. Bordovskiy, O.K.: Mar. Geol. *3*, 5 (1965)
39. Hagmeier, E.: Kieler Meeresforsch. *17*, 32 (1961)
40. Duursma, E.K.: in Chemical Oceanography (J.P. Riley, G. Skirrow, eds.), Vo. I., 433, Academic Press, London (1965)
41. Degens, E.T.: in Organic Matter in Natural Waters (D.W. Hood, ed.), Inst. Mar. Sci. Occ. Publ. *1*, 77 (1970)
42. Ehrhardt, M., Bouchertall, F., Hopf, H.-P.: Mar. Chem. *11*, 449 (1982)
43. Gschwend, P.M., Zarifirou, O.C., Mantoura, R.F.C., Schwarzenbach, R.P., Gagosian, R.B.: Environ. Sci. Technol. *16*, 312 (1982)
44. Kalle, K.: Oceanogr. Mar. Biol. Ann. Rev. *4*, 91 (1966)
45. Duce, R.A., Quinn, J.G., Olney, C.E., Piotrowicz, S.R., Ray, B.J., Wade, T.L.: Science *176*, 161 (1972)
46. Garrett, W.D.: Deep-Sea Res. *14*, 661 (1972)
47. Barger, W.R., Daniel, W.H., Garrett, W.D.: Deep-Sea Res. *21*, 83 (1974)
48. Momzikoff, R., Santus, R., Giraud, M.: Mar. Chem. *12*, 1 (1983)
49. Angrick, M., Rewicki, D.: Chemie in unserer Zeit *14*, 149 (1980)
50. Barber, R.T.: Nature (London) *220*, 274 (1968)
51. Williams, P.M., Oeschger, H., Kinney, P.: ibid. *224*, 256 (1969)
52. Sauer, T.C., jr.: Org. Geochem. *3*, 91 (1981)
53. Wheeler, J.: J. Geophys. Res. *77*, 5302 (1972)

54. Stuermer, D.H.: Thesis, Woods Hole Oceanogr. Inst., Massachusetts Inst. Technol. Joint Program, 188 pp (1975)
55. Baker, N., Wilson L.: J. Lipid. Res. *7*, 349 (1966)
56. Zafiriou, O.C.: Mar. Chem. *5*, 497 (1977)
57. Schnitzer, M., Khan, S.U.: Humic Substances in the Environment, Marcel Dekker, 327 pp (1972)
58. Kononowa, M.M.: Soil Organic Matter, Pergamon Press, 450 pp (1966)
59. Kerr, R.A.: Ph. D. thesis, Univ. of Rhode Island, 181 pp (1977)
60. Brown, M.: Estuarine Coastal Mar. Sci. *5*, 309 (1977)
61. Hurst, H.M., Burges, N.: in Soil Biochemistry (A.D. McLaren, G.H. Peterson, eds.), Marcel Dekker, New York, chapter 11 (1967)
62. Ogura, N.: Mar. Biol. *24*, 305 (1974)
63. Stuermer, D.H., Harvey, G.H.: Mar. Chem. *6*, 55 (1971)
64. Reuter, J.H., Perdue, E.M.: Geochim. Cosmoschim. Acta *4*, 325 (1977)
65. Mantoura, R.C.F., Dickson, A., Riley, J.P.: Estuarine Coastal Mar. Sci. *6*, 387 (1978)
66. Mantoura, R.C.F.: in Marine Organic Chemistry (E.K. Duursma, R. Dawson, eds.), Elsevier Oceanography Series, chapter 7 (1981)
67. Daum, K.A., Newland, L.W.: In The Handbook of Environmental Chemistry (O. Hutzinger, ed.), Vol. 2, Part B Springer-Verlag, Berlin, Heidelberg, New York (1982)
68. Saar, R.A., Weber, J.H.: Environ. Sci. Technol. *16*, 510 A (1982)
69. Jackson, T.A.: Soil Science *119*, 56 (1975)
70. Choudhry, G.G.: in The Handbook of Environmental Chemistry (O. Hutzinger, ed., Vol. 2, Part B, Springer-Verlag Berlin, Heidelberg, New York (1982)

The Surface of the Ocean

L. W. Lion

Cornell University
Department of Environmental Engineering
Hollister Hall
Ithaca, NY 14853, USA

Introduction:

Why are we Interested in the Surface of the Ocean?

The surface of the ocean could appropriately be called "the interdisciplinary interface." It has been studied by chemical oceanographers interested in the types and quantities of organic and inorganic species which accumulate and the reasons for their accumulation. it has been studied by biological oceanographers interested in the biological populations which have evolved to habitate a specialized ecological niche. It has been studied by atmospheric scientists interested in tracing the origins of atmospheric aerosols and it has also been studied by fluid mechanicians, hydrodynamicists, and other physical scientists interested in processes such as gas and heat exchange, circulation, and wave generation.

For many years much of the above mentioned research has been confined to disciplinary lines. This approach is certainly valid as long as interdisciplinary interactons may be neglected. However, it appears that in many cases an interdisciplinary perspective is not only useful, but required, in order to adequately understand the ocean's surface. An example of such a case is embodied in our present understanding of the accumulation of pollutant trace metals at the air-sea interface. Interfacial trace metal enrichment most likely results from a combination of transport processes and chemical reactions in which the presence of dissolved organic compounds and particulate matter (including living organisms) plays an important role. These interactions will be discussed in more detail below.

It may first be useful, however, to take a step back and examine the air-sea interface from the perspective of its importance to some of the diverse disciplines indicated above. This discussion is undertaken with the reader's tacit understanding that it is really the author's perspective of a perspective and apologies are given beforehand for any oversimplifications which may ensue.

An Atmospheric Perspective

Among other things, atmospheric scientists are interested in the chemical composition of atmospheric aerosols. These particles are responsible for the atmospheric transport of many elements and play a key role in most geochemical cycles. The size and chemical composition of aerosols can be indicative of their source. For example, comparison of the ratio of an element's concentration to that of a reference element with a predominant marine source (e.g., Na) or a crustal mineral source (e.g., Al) is commonly used to distinguish between marine and terrigenous origins [1–4]. However, some elements in aerosols, such as Cd and Pb, are often found to be highly enriched with respect to both the bulk ocean and the earth's crust. In such cases, anthropogenic pollution and elemental enrichment at the air-sea interface are two (not mutually exclusive) explanations. To evaluate the possible contribution of the ocean's surface to elemental enrichment in aerosols, the atmospheric scientist must come to grips with the enrichment levels that may be expected to occur at the air-sea interface and the extent to which such concentrations can be converted to aerosols by bursting bubbles.

A Chemical Perspective

Many environmental scientists, including chemical oceanographers and geochemists, concern themselves with the fate of pollutant compounds. Studies of the ocean's surface have shown that many such materials may accumulate to high levels at this location. PCB's and DDT serve as an example. These chlorinated organic chemicals are principally introduced into the ocean via the atmosphere [5]. Both PCB's and DDT have low water solubilities but have high lipid solubilities and are strongly partitioned onto a variety of particle surfaces [5]. The atmospheric source and the chemical characteristics of PCB's and DDT are likely to be major factors which help to explain the enrichment which is observed at the ocean's surface. An important and unanswered question is the extent to which ocean-atmosphere exchange of PCB's and DDT is affected by their accumulation at the sea surface and the extent to which the fate of these pollutant compounds in the ocean is controlled by their reaction with the elevated levels of particles, organisms and organic compounds that occur at the air-sea interface.

A Biological Perspective

From the view point of a marine bacterium or planktonic organism the surface of the ocean must be a stressful habitat. Nevertheless, many organisms apparently have evolved physiological or behavioral adaptions designed to keep them at the ocean's surface. There are clearly some potential advantages which may accrue to surface dwellers. Marine surface-dwelling organisms constitute a living example in population ecology. How and why have these organisms adapted to living at surface films? How are they affected by the relatively recent increases in inputs of anthropogenic pollutants? These are among the questions of interest to marine biologists.

Answers to questions asked about the physical, chemical and biological characteristics of the air-sea interface are not easily obtained. Contrary to the bulk ocean and atmosphere, the ocean's surface is extremely difficult to sample and the interfacial concentrations of materials which are determined are consequently extremely uncertain. In the section which follows these difficulties are discussed.

Measurement: The Characterization of the Ocean Surface

The Different Meanings of "Surface"

It is not untrue to say that for many purposes a bucket will suffice to obtain a sample of bulk-ocean surficial water. However, this type of "surface" sample has little to do with what we mean when we discuss the air-sea interface. The region which comprises the air-sea interface has been appropriately termed the "surface microlayer" since it has, at the most, a depth of several hundred micrometers [6]. Thus, a bucket surface sample of the top 20 cm of the ocean would dilute the concentration which occurred in a 200 μm deep microlayer by a factor of 10^5. Clearly, a bucket is an inappropriate device for sampling the surface microlayer. This dilution problem continues on a microscale. If one were to create a synthetic monolayer of some long-chain fatty acid such as palmitic acid $[CH_3(CH_2)_{14}COOH]$ and somehow microscopically observe the monolayer in cross-section, one would see polar-acidic functional groups in the aqueous phase and the hydrocarbon molecular tail predominately in the air. If the polar functional groups of the monolayer were to penetrate into solution to a depth of 20 Å then a special surface sampler which lifted off the top 200 μm of the sea surface would still dilute the true microlayer concentration by a factor of 10^5.

If the surface microlayer of the ocean were occupied by a monolayer of fatty acids, as in the above example, then we could use an appropriate correction factor depending upon the nature of our surface sampler and determine the true surface concentration. Unfortunately, even if the ocean's surface were only populated by fatty-acid molecules there is no reason to assume that they would necessarily be confined to a simple monolayer. Given some degree of turbulence or film compression, multilayers are certainly a possibility and we would probably also want to include any fatty-acid micelles which existed in the top nanometer or so of the ocean. In addition, it is virtually certain that a diverse range of organics exists at the sea surface and that each type may occupy a different effective depth. An example is the proteinaceous materials which may occupy depths more on the order of 1 μm [7]. Of course there are also the surface populations of bacteria to consider which have cellular dimensions of micrometers and larger planktonic organisms which have dimensions of the order of 10 to 100 micrometers. Thus, it is entirely conceivable to obtain a 100 μm deep surface microlayer sample with a dilution correction factor of 1 for diatoms, 100 for bacteria and proteins and 10^5 for fatty acids.

The above discussion provides examples of chemical or biological definitions of the thickness of the air-sea interface but it is also possible to define the microlayer's thickness from the physical perspective of the effective liquid film boundary layer which it poses for gas transfer. Thermal gradients offer yet another physical basis for defining a surface microlayer thickness and there is no reason to suppose that such boundaries should necessarily coincide with those established by chemical gradients or biologial populations.

The point of the above discussion is that the surface microlayer, by virtue of its heterogeneous nature, is difficult to uniquely define. The solution to this proplem for some investigators has been to select a sampling device and to let it operationally define the surface microlayer as the depth sampled. This procedure is analagous to defining particulate matter as anything that will not pass through a filter paper with a pore size of 0.45 μm. However, while there exists a range of smaller sized filters available to permit examination of colloidal-sized particles, a similar range of well-defined surface samplers with differing cut-off depths does not exist.

Sampling and the Nature of the Sample Obtained

The key words in the last sentence of the above discussion are "well-defined" since quite a number of surface microlayer sampling devices do indeed exist. The original microlayer sampler was the screen device developed by Garrett [8]. The screen, when lifted through a surface film, was found to remove a layer of water held by surface tension within the interstices of the screen mesh. Garrett [8] reported varying recovery efficiencies for synthetic fatty acid and alcohol monolayers depending upon the screen mesh size. The sampling depth was 150 μm for a 16 mesh screen made of 140 μm diameter wire [8]. In a later intercomparison of several surface samplers, Hatcher and Parker [9] reported a 369 μm sampling depth for a 16 mesh screen using 260 μm diameter wire. More recently, Van Vleet and Williams [10] compared the sampling efficiencies of screens constructed from a number of different materials (i.e., stainless steel, nylon monofilament and Teflon). The collection efficiencies of these various materials were found to differ significantly for organic surface films and, in addition, individual components in mixed films were found to preferentially adsorb onto some of the screen sampling materials [10]. The coating of sampling screens was observed by Garrett in his original description of the screen device; the wire screen's sampling efficiency was observed to change with repetitious use, finally attaining a constant value of $\simeq 75\%$ [8]. Even Teflon screens, which were reported to have $\simeq 0\%$ efficiency by Van Vleet and Williams [10] (these investigators prevented the coating of their screens by removing all materials obtained in a single sampling by rinsing with methanol), have substantially higher sampling efficiencies once coated with organic film materials [11]. Although, to the author's knowledge an intercomparison of different sampling materials or devices has not been reported after repetitive samplings; it is reasonable to expect that adsorption of film materials would confer a degree of surface uniformity upon different construction materials and reduce or remove any problem of material selectivity. On the other hand, investigators interested in a specific surface microlayer component may wish to exploit the selective adsorption of certain construction materials and to clean their screen after each sample. The variability in reported sampling depths which, for screens, now ranges from 100 to 540 μm [10] remains as a problem which necessitates laboratory or field calibration of each device.

There are a number of alternatives to the screen samplers which are as diverse as the imaginations of their creators. Large volume surface samples may be obtained by using the rolling ceramic drum devised by Harvey [12] in which freshly sampled (≈ 60 μm) surface film is removed from the rotating drum by a scraper as the device is pushed in front of a slowly moving vessel.

Smaller volume samples may be obtained by the simple expedient of dipping a plate through the ocean's surface and scraping off the attached surface film or by rinsing off the film with a solvent. Materials employed as sampling plates have included glass (≈ 60–100 μm sampling depth [13]) and Teflon (≈ 50–100 μm sampling depth [14]). In an interesting variation, germanium prisms have been utilized by Baier [15] who then directly analyzed adsorbed materials by multiple attenuated internal reflection (MAIR) spectroscopy. The thickness of film sampled was determined, using ellipsometric techniques, to range from 100 to 300 Å [16].

The "V-shaped plastic tube" employed by Szekilda et al. [17] made use of surface film compression to collect aggregated film materials at the V's apex. Compressed microlayer materials can literally be spooned out of this device [18]. Unfortunately, calibration of the depth sampled by the V-tube is uncertain although a value of 1.5 μm has been assumed [19].

Floating various materials on the sea surface has also been successfully employed for surface microlayer sampling. Hamilton and Clifton [20] used a sprayed-on PVC film (sampling depth $\simeq 1$ to 16 μm) to sample for solids, and surface-layer organic compounds. Kjellenberg et al. [21] used a Teflon sheet (6.5 μm reported sampling depth) to sample for microorganisms and lipids. Membrane filters have also been used for sampling of surface layer microorganisms by floating them on the surface [21, 22]. Reported sampling depths are 0.9 and 29 μm for hydrophobic and hydrophilic 0.4 μm Nuclepore membranes, respectively [21]. However, the depth sampled is also apparently a function of the time which filters are left on the surface [23]. One of the most interesting innovations in floating samplers is that of Young [24] who sampled for microlayer bacteria by floating electron microscope grids on the surface. Interfacial organisms were then visually observed by transmission electron microscopy after a critical point drying preparatory procedure [24].

For the sake of completeness other surface microlayer sampling tools should be mentioned such as trays (approximately 1 mm sampling depth) [25], funnels [26], and freezing probes (approximately 1 mm sampling depth) [20]. The capture of the jet drops ejected by bursting bubbles is probably one of the more intriguing methods for microlayer sampling. The process of bubble ejection of liquid aerosols is far from simple and has been extensively studied by MacIntyre [27–29]. MacIntyre [27] estimates the top jet drop ejected by a bursting bubble to represent a 10 μm deep surface sample. However, bursting bubbles also scavenge surface active materials from the bulk solution underlying the sea surface and transfer them to the surface as well as to jet drops [30–32]. Thus, the materials found in bubble jet drops represent a mix of those which were already at the ocean's surface when the bubble burst and those delivered to the surface by the bubble.

Despite the uncertainties in sampling depth, with proper calibration the differences in sampling devices may be employed to examine the surface microlayer over a range of depths. A study of this nature has been carried out by Daumas and coworkers [33] who measured particulate (proteins, carbohydrates, organic carbon, organic nitrogen, seston, fatty acids, and alkanes) and dissolved (fatty acids and alkanes) constituents in Mediterranean coastal waters using the rotating drum (50 to 80 μm sampling depth) and screen (440 μm sampling depth). The enrichment for all microlayer constituents relative to underlying bulk water was found to be greater in the shallower drum microlayer samples.

In summary, a wide range of devices currently exists which may be employed to collect a sample from the ocean's surface microlayer. These devices, to a certain extent, may be preselected to give a characteristic sampling depth and perhaps the sampling material used may be chosen for its affinity for certain microlayer components. Unlike the Niskin bottle, the surface microlyer sampler has not achieved any uniformity in its manufacture and the sampling depths and efficiencies which are reported vary widely.

It is interesting to note that the discovery of elevated interfacial concentrations of many pollutant and naturally occurring compounds indicates that (depending upon the compound of interest) a bucket may no longer be considered as an appropriate device for sampling of bulk-ocean surface waters. If a constituent of interest is substantially enriched at the surface microlayer, then care must be taken to exclude the surface microlayer from the bulk-ocean sample in order to avoid the possible measurement of artificially elevated concentrations.

In recognition of the inherent variability of surface-microlayer samples which result, at least in part, from the diversity of sampling techniques described above, I have elected to avoid detailed comparison or listing of numerical values of microlayer partitioning in the discussion which follows. This is in accord with the admonitions of MacIntyre [34] who compared such tabulations to random number tables and because I ignored him and made such comparisons in an earlier review [35]. Several cases of microlayer depletion have been summarized by Hunter [36], however, the evidence relating to particulate materials, microbiota, many classes of organic compounds and most trace metals indicate these substances are indeed enriched at surface microlayers. The actual magnitude of this enrichment is, however, subject to the definition one uses for microlayer thickness.

The Composition of Surface Films

The advent of surface microlayer samplers may have revealed elevated concentrations for many materials at the ocean's surface but these measurements should come as no surprise. The interfacial accumulation of materials which lower the surface tension of water is dictated by the Gibbs adsorption equation:

$$RT\Gamma_i = -(d\gamma/\partial \ln a_i)_T .$$

From this relationship the surface excess of compound i (Γ_i; moles/cm^2) is seen to be positive if surface tension of solution (γ, Joule/cm^2) decreases with increasing activity of species i (a_i; mole/l). A majority of organic solutes decreases the surface tension of water [37] and therefore a positive surface excess may be anticipated.

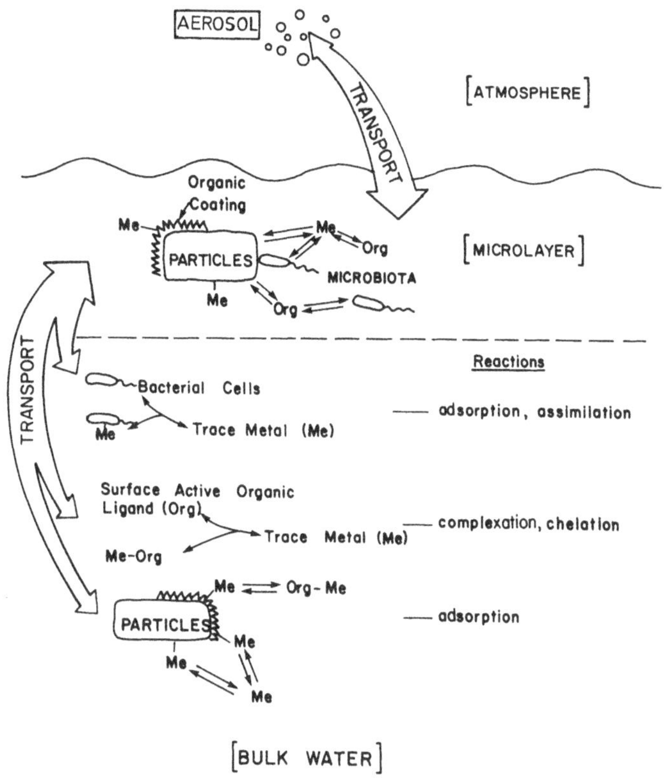

Fig. 1. Transport and reactions of trace metals in sea surface microlayers, and the bulk ocean

Szyszkowski wrote empirical equations to describe the effect of unionized aliphatic organic acids on surface tension in 1908 [38], and early investigators measured surface accumulation of such materials in laboratory systems using microtome devices (i.e., sharp knife blades propelled across the water's surface at a depth of about 0.1 mm). It was therefore clear that, given a good way to sample the ocean's surface, a surface excess of organic solutes could be expected.

Addition of inorganic salts increases the surface tension of water and the Gibbs adsorption isotherm therefore indicates that they should be excluded from the surface microlayer. Nevertheless, some inorganic ions such as trace metal cations are found to accumulate at the sea surface. The explanation of this apparent paradox lies, in part, in a better understanding of the chemical composition of the dissolved and particulate materials which may reside at the surface microlayer. In any discussion of the constituents of the air-sea interface it is useful to keep in mind the dynamic nature of this special layer of the ocean. As Fig. 1 illustrates using trace metals as an example, surface microlayer concentrations result from a continuous exchange between both the atmosphere and bulk solution. The phase distribution (dissolved vs. particulate) and residence time of a substance at the microlayer will reflect both those reactions which occur in situ at the microlayer and those which occur during or prior to transport to the microlayer.

Organics

The chemical structure of organic compounds largely determines their presence at the air-sea interface. A case in point are the petroleum hydrocarbon pollutants most of which have low solubility in water (although various weathering reactions such as photolysis, can result in increased solubility). This lack of solubility and their lower density than water predisposes them to occur at the air-sea interface. In the absence of any polar functional groups (e.g., —COOH, —OH, $—NH_3$, etc.) long chain hydrocarbons will float as a lens on the ocean surface. The presence of a hydrophilic group (hydroxy-, carboxyl, etc.) at the end of the hydrocarbon molecule will increase its water solubility. If the molecule, in addition, has a long hydrocarbon (hydrophobic) tail it will then be "surface-active" in that the best balance between the forces which exclude the non-polar tail from solution and the forces which keep the polar functional head group in the aqueous phase is achieved at the air-sea interface. At high concentrations such surface-active molecules form coherent films or "slicks" replacing water molecules in the boundary layer between the atmosphere and the ocean.

Both pollutant and naturally occurring hydrocarbons have been detected on numerous occasions at the ocean's surface [39]. The former category consists primarily of petroleum products released as a part of the routine operation of maritime vessels and, on occasion, as catastrophic spills. Natural hydrocarbons are of biogenic origin and include lipids, fatty acids, alcohols and their esters.

A second general class of organic materials found to accumulate at the air-sea interface are the polymeric materials: proteins and polysaccharides [16, 40]. These molecules are appropriately termed "wet surfactants" by MacIntyre [7], and possess a large number and more uniform distribution of functional groups relative to long chain fatty acids or alcohols. Wet surfactants therefore reside mostly in the aqueous phase.

There is some dispute over which of the above (lipids vs. polymeric compounds) are the dominate constituents of surface films. In the special case of a condensed film, where the available interfacial surface area is limited, one might expect to find more soluble materials to be excluded from the interface; this is illustrated by the ability of artificial films to displace natural film constituents [41]. However, available evidence indicates that in the open ocean condensed surface films or slicks are unlikely to occur because of the relatively heterogeneous nature of organics at the air-sea interface and because the surface microlayer concentrations of organics are too low [42].

Continuous disturbance by physical processes such as wind or local variations in biological productivity (e.g., within vs. outside kelp beds) ensure that the concentrations of organics in surface films are likely to vary both temporally and spatially. Nevertheless, the results of Baier [15, 16] suggest a relatively consistent composition of proteinaceous materials (proteoglycans and glycoproteins) and carbohydrates. Unfortunately, the sensitivity of the MAIR analytical procedure to the detection of lipid materials is uncertain [42, 43]. In their review of available data, Hunter and Liss [42] indicate that simple lipids are likely to comprise only a small fraction of surface microlayer organics and suggest (based on the data of Baier et al. [16] and Sieburth et al. [40]) that surface microlayer organics must have a substantial number of carboxyl and hydroxyl functional groups. This composition is consistent with proteinaceous and/or carbohydrate material.

A third class of compounds which have been identified at the air-sea interface are the pollutant chlorinated compounds such as the polychlorinated biphenyls (PCB's) and DDT. There is some concern over the presence of elevated concentrations of these pollutants at the sea surface since the microlayer also includes high populations of microbiota (see discussion below). The first opportunity for biological uptake and transfer of chlorinated hydrocarbon pollutants of atmospheric origin into the marine food web is at the surface microlayer. Reported enrichments for compounds such as DDT and PCB's at the sea surface are one order of magnitude or higher in surface samples obtained by a spectrum of techniques including the screen and rotating drum. Polynuclear aromatic hydrocarbons (PAH) represent yet another class of pollutants of an anthropogenic origin for which elevated concentrations are found at fresh water microlayers [44]. These materials, like the PCB's, have primarily an atmospheric source.

Surface microlayer organics obviously include both dissolved and particulate materials. What is less obvious is the apparent dynamic interrelationship between these two phases which may be unique to the interfacial environment. For example, interfacial dissolved organic carbon (DOC) may apparently be converted into particulate organic carbon (POC) when condensed surface films are compressed and collapse [45]. Film compression may be brought about by wind or wave forces or Langmuir circulation in the open ocean and the author has observed similar effects created by tidal action in an estuarine system. DOC at the surface microlayer may also be polymerized into insoluble material by UV radiation [46,47]. In fact the high intensity UV radiation at the sea surface and the ensuing reactions of organic materials may ultimately be responsible for the evolution of life on earth [48,49].

DOC-POC interconversion is apparently also mediated by rising bubbles. Although the validity of this statement has been subject of considerable controversy, the photographic evidence provided by Johnson [50] is persuasive. Johnson constructed a device which allowed him to maintain a bubble in constant position while the experimental system was pressurized resulting in bubble dissolution. DOC adsorbed on the bubble surface and/or colloidal POC intercepted by the bubble was shown to be converted into filterable POC in the course of this process. Presumably, similar reactions constantly occur in the sea as waves break and introduce small sized bubbles. Bubbles which happen to be too large to dissolve are excellent scavengers for any POC which may be produced and can deliver this material to the sea surface [51].

Observed levels of enrichment of DOC and POC at the surface microlayer have been summarized in a review by Lion and Leckie [35]. Although the array of sampling methods used by the various investigators and the locations sampled prohibit meaningful detailed intercomparison it is safe to say that both forms of carbon are commonly found to have elevated concentrations at the sea surface. As a consequence of this surface enrichment we may speculate that the ocean's surface layer has an elevated concentration of organic binding sites for trace metal cations and that trace metal concentrations may therefore be elevated at the surface microlayer in spite of the previous predictions for their exclusion based on the Gibbs adsorption equation. In addition, DOC and POC may serve as a substrate for surface-dwelling microbiota so that it is conceivable that organism concentrations may also be elevated at the sea surface. The enrichment of trace metals and microbiota at the sea surface are discussed in the sections which follow.

Inorganics

Although the Gibbs adsorption isotherm provides a useful starting point, the excellent review by MacIntyre [34] summarizes calculations which demonstrate that surface exclusion of charged ions by negative adsorption may only be expected to extend to a depth of approximately 5 Å. This is of the order of the diameter of a single water molecule and substantially less than the sensitivity of even the shallowest surface sampling device. It follows that ion concentration gradients arising from negative adsorption are likely to be undetectable and of dubious geochemical importance since the jet drops ejected by bursting bubbles remove surface layers $\simeq 10$ µm in depth. It is therefore necessary to invoke a specific causal mechanism to explain observed enrichments of some inorganic compounds at the air-sea interface. The discussion which follows will focus on chemical reactions which may be related to the partitioning of toxic trace metals at the surface microlayers.

As previously suggested, metal binding to surface-active dissolved organic ligands and adsorption onto particle surfaces at the air-sea interface are two mechanisms which may result in trace metal enrichment and which would therefore appear to warrant further consideration. One approach to evaluate the role of adsorptive surfaces and surface-active ligands is from the perspective of chemical equilibrium. Given an estimate of the surface concentration of trace metals and the other constituents of the air-sea interfacial solution and a set of reasonable equilibrium constants for both complex formation and metal adsorption, one may then calculate the equilibrium chemical speciation of trace metals of interest. Such calculations have been performed for simulated surface-microlayer solutions with and without the presence of adsorbing solid surfaces (i.e., SiO_2 and amorphous $Fe_2O_3 \cdot H_2O$) and the results are illustrated for the metal Pb in Fig. 2. The calculations suggest that it is reasonable to anticipate that both adsorption of Pb by solid surfaces and complexation of Pb with surface-active ligands (at high ligand concentrations) may occur at the air-sea interface [52]. In addition, a substantial fraction of total Pb is calculated to be adsorbed in the bulk ocean. This may favorably influence the transport of Pb to the surface microlayer since rising bubbles are effective scavengers for particulate materials. Similar calculations suggest that solid adsorption and organic complexation reactions are also plausible explanations for enrichment of Cu in surface microlayers. The available field data for Cu and Pb agree in substance with these conclusions, with partitioning of organically bound metals (typically measured by chloroform extraction) or of the particulate metals being frequently reported [11, 36, 53–55].

Cd and Hg are two metals which, based on equilibrium calculations, are expected to be strongly complexed by chloride in both the bulk ocean and in surface microlayers. The chloride ligand effectively competes for these metals even in the microchemical environment of the air-sea interface in which concentrations of adsorbing solids and complexing ligands may be orders of magnitude above the levels in the bulk ocean. It is noteworthy and perhaps less than coincidental in light of

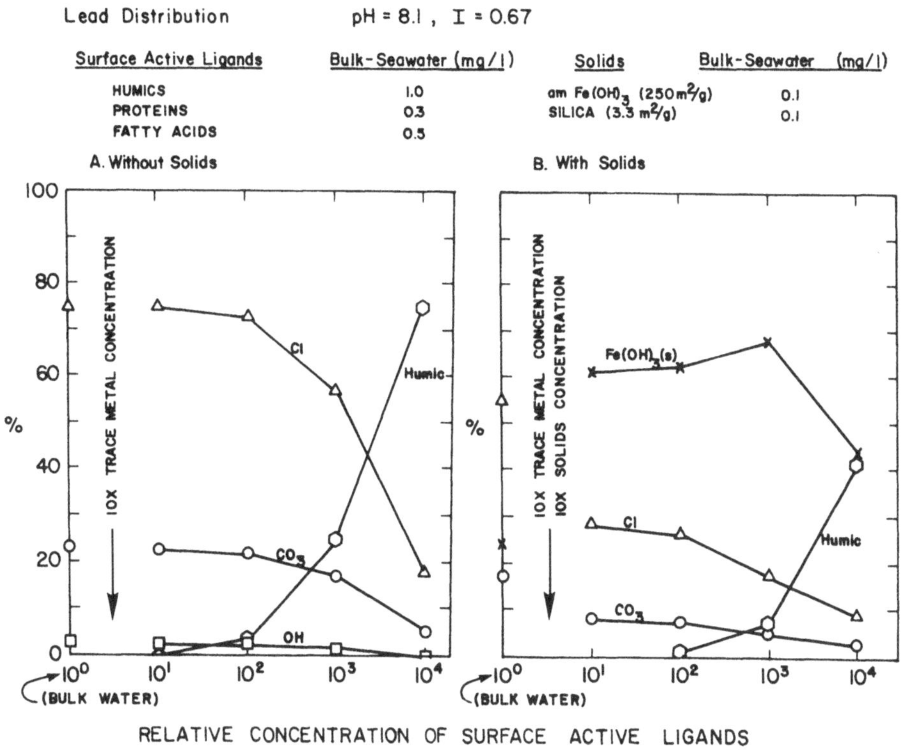

Fig. 2. Equilibrium speciation of Pb in the bulk ocean and simulated surface microlayers with varying concentrations of surface active ligands

these calculations, that Fitzgerald and Hunt [56] report no measurable enrichment of Hg in microlayer samples from open ocean and continental shelf waters. Similarly in our measurements conducted within the San Francisco Bay estuary we found more than 91% of total Cd to be dissolved. Although particulate Cd was found to be partitioned at the air-sea interface, the observed surface enrichment of the metal could be accounted for by the natural abundance of Cd in crustal mineral particles which accumulated at the estuarine surface microlayer [11].

In all fairness, it should be noted that the results of equilibrium calculations and field observations do not in all cases agree as nicely as the above discussion suggests. For example, Fitzgerald and Hunt [56] report significant concentrations of organically associated Hg in coastal waters, while chloro complexes predominate in all of our calculations (which considered a wide range of organic ligand concentrations and complex formation constants). There is no ready explanation for this disparity except to note that the calculations performed were for abiotic systems so that Hg reactions such as assimilation and methylation are not accounted for, nor would the calculations account for a local source of organo-mercuricals which was offered by Fitzgerald and Hunt as a possible reason for the differences in concentration observed between open ocean and coastal waters [56].

The substantial agreement between the results of equilibrium calculations and field observations for partitioning of particulate and organically associated trace metals [11, 52, 57] suggest that metal complexation with surface-active organic ligands and adsorption to particle surfaces are reasonable explanations for the observed enrichment of these inorganic elements at the air-sea interface. Qualitatively we may predict that the binding characteristics of metals may be reflected in the extent of their enrichment at surface microlayers. For example, we have observed that the partitioning of total Cd, Cu, and Pb in an estuarine salt marsh surface microlayer with high particle concentrations followed the order Pb > Cu > Cd, which is consistent with our calculated estimates for adsorption of these metals onto particle surfaces in simulated marine microlayers.

The role of surface microbiota in the microlayer enrichment of trace metal cations is uncertain. Certainly all surfaces may constitute binding sites for metals, but our observations in estuarine systems indicated that the magnitude of bacterial enrichment at the surface was insufficient to account for observed levels of particulate metal enrichment [11, 57]. This conclusions was based on the assumption that the surface microlayer bacterial surface characteristics for metal adsorption and levels of metal assimilation are comparable to those of bacteria in the bulk ocean. However, there is evidence that surface-dwelling bacterial populations may be different from subsurface populations in both species composition and activity (see discussion below) and the effect of these differences on metal association with cells is uncertain. In any event, the surfaces of bacterial isolates from the surface microlayer have been shown to strongly adsorb metals such as Pb [58]. Metals bound to surface microlayer bacteria may account for a substantial fraction of particulate metal at the ocean's surface, particularly in instances where the concentrations of inorganic particles (which would complete with bacterial surfaces for trace metal adsorption) are relatively low [58].

An additional contribution of bacteria to metal partitioning is likely to occur through bacterial modification of particle surfaces with exocellular polymers. These exudate materials have been demonstrated to strongly bind trace metals [59] and to enhance metal adsorption onto particle surfaces [60]. Exocellular polymer production is characteristic of marine film-forming bacteria [61], and polymer binding is believed to act as an attachment mechanism for suspended cells [62]. In estuarine systems a high percentage of microlayer bacteria has been observed to be attached to particle surfaces [63], and it is therefore likely that these surfaces have been modified by bacterial exopolymers. Therefore, as a consequence of cell attachment, the trace metal sorptive properties of particles at the estuarine air-sea interface may be higher than those in bulk solution. If such phenomena do occur, and if they are not restricted to estuarine systems, it is interesting to imagine what the consequences might be on exchange of trace metals between the earth's oceans and atmosphere.

Organisms

Considering the elevated concentrations of many inorganic and organic pollutant compounds found at the surface microlayer, one can not help but be surprised at the diverse array of organisms which are adapted for living in what must be a stressful habitat. Selected organisms throughout the marine food chain display physiological and behavioral modifications designed to maintain them within the biological community at the surface microlayer (collectively termed "neuston"). Many bacterioneuston are encapsulated with exocellular polymers [24] which may improve their surface activity or which they may use to bind themselves onto particles at the surface microlayer. Also, many bacterioneuston appear to be pigmented which may confer resistance to the relatively high intensity UV radiation at the sea surface [49].

Algae may use gas vacuoles as bouyancy controls to enhance their concentration at the surface microlayer [65]. Phototaxis of phytoplankton has been shown to result in their mid day accumulation in surface films [66].

Zooneuston can attach themselves to surface films by appendages or protoplasmic proturbances [67]. Others use bouyant floats or construct rafts of bubbles [68] and still others may migrate in and out of the surface layer for the purpose of feeding [69]. Many periphitic benthic organisms are apparently capable of dwelling at the surface microlayer [70]. We have observed extremely high concentrations of benthic diatoms at the surface microlayers in an intertidal salt marsh. Microlayer concentrations of diatoms (sampled by screen) exceeded those in bulk solution by up to four orders of magnitude on the incoming tide, and apparently originated from tidal skimming of exposed mud flats [71].

The larval stages and small post-larval stages of many commercially important fish such as the tuna, and marlin are commonly collected by neuston nets which sample approximately the upper 10 cm of the ocean's surface [68, 72]. The presence of such organisms in surface microlayers suggests that enriched interfacial pollutant concentrations may be directly introduced at the microlayer into species which constitute a portion of man's food supply.

The reason for the selection of the surface microlayer as a habitat by so many different organisms is likely to be one held in common by all of the neuston; i.e., the greater availability of food and/or nutrients at the surface microlayer. In the case of bacteria, the reported surface partitioning of dissolved and particulate carbon, nitrogen, and phosphorous suggests that the surface microlayer may provide an excellent growth medium [73]. Phytoneuston can presumably make use of the high intensity unfiltered sunlight at the sea surface and, in addition, nutrients such as NH_3 and PO_4^{-3} are reported to be partitioned [74–77]. Finally, zooneuston are likely to find abundant populations of bacterioneuston and phytoneuston which may serve as a food source. The reader should note that the presence of significantly elevated concentrations of the nutrients NH_4^+, NO_2^-, NO_3^-, and PO_4^{-3} in sea surface microlayers have recently been questioned by Chapman and Liss [78]. These investigators were unable to measure any PO_4^{-3} partitioning in screen (300 cm) samples taken from productive coastal waters and report lower levels of NH_4^+ and NO_3^- enrichment than those in previous investigations [75, 76].

The toxicity of the pollutant compounds which accumulate at the air-sea interface should perhaps be questioned, particularly in light of what appears to be a robust surface community. The fact that tar balls and plastic floatsam are readily colonized by a wide array of marine organisms [79, 80] suggests that these particulate hydrocarbon materials are not exerting an accute toxic effect. Trace metals at the surface microlayer may be toxic to the neuston, however available data suggest that this toxicity should be dependent upon chemical speciation. Typically, metal toxicity has been shown to vary with the concentration of free aquo metal ion and may occur at extremely low concentration levels [81–83]. Recent experiments have shown that the metals solubilized from atmospheric particles can inhibit ^{14}C assimilation by marine plankton at high particle concentrations [84]. Estimates of particle loading to surface microlayers indicate that inhibitory levels may in fact be achieved under some circumstances [84].

The effects of chlorinated hydrocarbons such as PCB's and DDT on the neuston have not been evaluated. The microlayer environment certainly offers the potential for exposure to elevated concentrations of these compounds and phenomena such as biomagnification through the various trophic levels of the neustonic community may ensue. The long term effects of this uptake represent a hazard of uncertain magnitude.

One method for assessing the effects of pollutant and environmental stress at the surface microlayer is through comparison of species diversity in this habitat to that of the bulk ocean. The presumption in this evaluation is that decreased diversity at the surface microlayer may be indicative of difficult living conditions. Available data which examine this question are in conflict. Decreased diversity of surface microlayer planktonic communities has been reported by some investigators [85, 86] and increased diversity by others [87].

Another test for the possible effect of elevated pollutant concentrations at the surface microlayer is through comparison of the metabolic activity of neustonic organisms to that of bulk solution populations. It is assumed that toxic effects may be reflected in lower activity levels in the surface microlayer. Again available data are contradictory. Microbial activity as measured by plate counts, ATP concentration and uptake of ^{14}C labeled substrate is reported to be depressed in the surface microlayer by some investigators [88, 89] and to be elevated (ATP concentration) by others [40].

On the basis of the available data it does not appear that one may state with overwhelming certainty that the effects of pollutant compounds are evident in either decreased diversity or diminished activity of the surface microlayer biological community. Alternately, the available evidence clearly does not warrant dismissing this concern as unimportant. The inconsistencies noted above may reflect the level of noise in the system, or differences in important variables which this superficial comparison has neglected (such as location, time of day, DOC and pollutant levels in the samples, etc.). Additional research may provide a clearer picture of the true benefits and detriments of the microlayer as a biological habitat.

Questions of diversity and activity aside, there is little disagreement that the sheer population numbers of bacterioneuston are substantially higher than those which occur in bulk solution [12, 22, 63, 64, 88–90]. These elevated numbers, however, may reflect high concentrations of certain organisms specifically adapted for living at the surface microlayer such as those with pigmentation [30, 64]. Pigmentation of bacterial cells is apparently reflected in their surface characteristics which in turn influences their affinity for the air-water interface. Cells with hydrophobic surfaces are readily separated from suspension by rising bubbles and pigmented cells of the marine bacterium *S. Marcescens* have been shown to appear in much higher concentrations (relative to unpigmented cells) in the jet drops produced by bursting bubbles [91]. Bubble transport (see discussion below) is thought to be one of the most effective mechanisms for fractionation of surface-active materials from the bulk ocean to the surface microlayer.

The Origin of Surface Films

In the vast majority of circumstances there are only two possible sources for the materials which accumulate at the air-sea interface: The bulk ocean and the atmosphere. Transport from each of these sources is discussed below. It should be noted, however, that in intertidal regions a third source for materials at the air-sea interface is direct transfer from the sediments. Our observations in the south San Francisco Bay estuary, which has extensive intertidal mud-flats, suggest that the skimming action of the incoming tide may result in tremendous interfacial concentrations of particulate materials including associated bacteria, diatoms, and trace metals.

We suspect (but have not established) that both surface tension effects and buoyancy of non-wettable or slow to wet surfaces may be responsible for this tidal skimming process. The net result of the observed skimming was transport of particles and particulate metals accumulated at the surface microlayer into the salt marsh systems which border the Bay [92, 93]. [To avoid misleading the reader, it should be noted that we also observed that export of particles and particulate metals from the salt marsh systems in bulk suspension outweighed (by far) the microlayer import, so that the salt marsh system appears to act as a source of particles to the Bay [92, 93]].

To learn more about the factors which influenced origin of particles in the intertidal estuarine system the particle-size distributions in paired surface microlayer and bulk solution samples was [94] compared. Analysis of marine particle-size distributions has been used as a means to evaluate the physical and/or chemical changes which occur with particle transport [95] and as a means for evaluating particle aggregation mechanisms [96]. Results for the estuarine particles in the 0.06 to 50.8 μm radius size range indicated that particle size distributions in both the microlayer and bulk solution were comparable. These data suggest that the processes which resulted in particle enrichment at the microlayer were not selective for a specific particle size. Any alteration in particle size, such as fragmentation or dissolution, which may have occurred was insufficient to measurably alter the observed size distributions [94].

Transport from the Ocean

There are many transport mechanisms which may operate in carrying materials between the air-sea interface and the bulk ocean. These include difffusion and eddy dispersion which operate in the direction of chemical concentration gradients, thermal diffusion which operates in the direction of thermal gradients [34], wind driven circulation cells which result in zones of material accumulation parallel to the wind direction (Langmuir Circulation) [97] and transport by bubble flotation. Of these processes, transport by rising bubbles is thought to be the major vehicle for scavenging of surface-active materials from bulk solution. Rising bubbles have been shown to effectively transfer marine phytoplankton, POC and particulate trace metals [98, 99], marine bacteria [30, 100], and surface active organic compounds such as proteins or exocellular exudates [101, 102]. The ring of foam which may typically be observed to encircle wave battered rocky pinnacles or shore lines provides visual evidence to the casual observer of the bubble fractionation process at work.

Bubble densities in the ocean have been measured acoustically by Medwin [103, 104]. The reported results indicate season, time of day, wind velocity, the presence or absence of sea slicks, and depth to be parameters which influence the number density of bubbles of a given size. Large populations of small (< 100 μm) bubbles have been observed even under quiescent conditions. Bubble sources in the oceans' surface waters include bubbles which originate with breaking waves, bubbles introduced by atmospheric precipitation and entrained with atmospheric aerosols as well as bubbles which are formed at depth from biological activity or from pressure changes caused by surface waves [105].

Bubbles originating with biological activity in the sediments constitute another source. Small rising bubbles may decrease in size through gas diffusion and surface tension effects although the presence of organic surface coatings may inhibit this process. Larger bubbles have more rapid rise velocities, and may grow in size as they rise in response to the decrease in hydrostatic pressure.

As prevously discussed, the complete dissolution of small bubbles has been shown to consistently result in the formation of POC in filtered seawater. In addition, recent experiments have shown that dissolution of small (< 100 μm) bubbles can also produce stable microbubbles (1 to 13.5 μm diameter) which are resistant to further dissolution [106]. Some of the microbubbles formed were stable for time periods as long as 30 hours and most were stable over a 22 hour period [106]. Although the mechanism of such bubble stability is unproven, it is likely to reflect the organic materials concentrated at the bubble surface. The fact that bubbles are such effective scavengers of surface-active materials is indicative of their important role in the delivery of such materials to the air-sea interface.

The effectiveness of the bubble scavenging process is clearly related to the surface-active nature of the material to be scavenged. Thus, while bubbles may efficiently strip surface-active organic molecules such as proteines and polysaccharides from the ocean, they will be ineffective at removing free trace metal cations. If, however, the metal ions are bound to a surface active ligand they may be effectively removed. This process has been demonstrated in laboratory studies in which addition of an artificial organic "collector" such as sodium lauryl sulfate is used to scavenge a metal such as lead [107].

An alternate process for scavenging of a dissolved metal cation is through adsorption of the metal onto a colloidal-solid collector and subsequent adsorption of a surface-active organic. The resulting metal-colloid-surfactant assemblage may be readily scavenged by rising bubbles. Metals such as Cu and Zn have been scavenged from seawater solutions in the laboratory using amorphous ferric hydroxide precipitates as the adsorbing colloid and dodecylamine as the organic surfactant [108]. Laboratory bubble separation experiments performed by Wallace and Duce [51], using surface water samples from the Narragansett Bay estuary showed that approximately 50% of the particulate form of several metals including Pb, Cu, and Zn could be separated from solution by rising bubbles along with POC. The metal to carbon ratio was reported to be enriched in the foams obtained relative to that in the residual (stripped) solution, suggesting selective removal of particles with a higher ratio [51].

Bubble scavenging experiments have also been conducted in-situ by Piotrowicz et al. [32] in the Narragansett Bay estuary. The aerosal produced by bursting bubbles was collected and analyzed along with the underlying bulk solution. Results showed the Zn and Cu concentrations in aerosols [reported as enrichment factors $EF_{sea} = (X/Na)air/(X/Na)sea$] to increase in proportion to the height of bubble rise, suggesting that both elements were scavenged from bulk solution. Zn in bulk solution was predominately dissolved and it is therefore possible that this metal was scavenged through an association with a dissolved surface-active ligand. Cu enrichment in the aerosol was positively correlated with chlorophyll and ATP concentrations suggesting that its scavenging may be related to biological processes (which could presumably include assimilation, association and/or binding of Cu with dissolved surface-active exocellular exudates).

The relative importance of dissolved vs. particulate collectors in the bubble scavenging of trace metals in marine systems is difficult to establish based on the available evidence. If inorganic particulate surfaces are functioning as metal-adsorbing collectors, then their surface properties must be further altered by the adsorption of organic surfactants in order to promote efficient bubble scavenging. There is ample evidence that the latter process occurs extensively in marine systems. Observations reported by both Neihoff and Loeb [109] and Hunter and Liss [110] indicate adsorption of organic solutes from seawater is capable of reversing the charge of uncoated surfaces introduced into solution.

A possible effect of adsorbed organic coatings on inorganic particle surfaces may be to decrease the wetability of the original surface, particularly if it is a clay mineral or hydrous oxide. A decreased wetability of the solid surface is likely to be reflected in an increased attachment efficiency should the surface be contacted by a rising bubble. In addition, adsorbed organics may also alter the metal binding characteristics of the surface; metal binding constants for adsorbed marine organic coating materials have been deduced from electrophoretic measurements by Hunter [111].

The mechanisms by which a rising bubble may collect suspended particles have been summarized by Weber et al. [112] and include: gravitational deposition, inertial impaction, convective diffusion and interception. For small, colloidal-sized particles the first two of these processes may be neglected. The relative importance of diffusion vs. interception are dependent on the dimensions of the colloids. The collection efficiency resulting from particle diffusion can be shown to increase with decreasing particle size while the collision efficiency resulting from particle interception increases with increasing particle size [112, 113]. The net result of these processes is that diffusion is likely to control the collection of very small-sized particles, such as virus, and interception of the collection of bacterial sized particles. Interception alone is apparently sufficient to account for the observed scavenging of bacteria by rising bubbles [112].

The surfaces of clean uncoated bubbles have been described as being "mobile" as opposed to the bubble surfaces coated with surface-active materials which are "rigid" or immobile [114]. Mobile bubble surfaces are apparently the most effective scavengers for particles. Collection efficiency, as expressed by the enrichment of bacteria in the jet drops produced by a bursting bubble, has been shown to increase linearly during the first few centimeters of bubble travel and then to plateau, or even decrease, as the bubble surface becomes coated [114].

The magnitude of the bubble-induced flux of particulate materials from bulk solution to the ocean's surface has been estimated by Wallace and Duce [51, 99]. Estimated values are a function of the particle concentration in solution and for POC range from 0.2 $\mu g\ C \cdot m^{-2} \cdot s^{-1}$ for the Sargano Sea to 4.0 $\mu g\ C \cdot m^{-2} \cdot s^{-1}$ for the estuarine waters of Narragansett Bay. These fluxes are estimated to be sufficient to establish a ten-fold surface microlayer partitioning of POC in a time frame of ten minutes or less [99]. Trace metals are also associated with the particles delivered to the sea surface and estimates for the bubble induced flux of particulate trace metals to the sea surface indicate that it may be of the same magnitude as the atmospheric depositional flux of trace metals to the sea surface [36, 51].

Transfer from the Atmosphere

Both wet and dry deposition of materials in particulate, aqueous and gaseous phases constitute inputs from the atmosphere to the surface of the ocean. Modeling considerations for wet and dry deposition have been discussed by Scott [115] and Slinn and Slinn [116], respectively. In the case of pollutant trace metals, those associated with atmospheric aerosols are the primary form of concern although input from the vapor phase may be important for metals which have species of substantial volatility such as methyl mercury. We may classify the origin of elements in atmospheric particles as anthropogenic (for those which are mobilized though societal activities such as the combusion of fossil fuels), crustal (for those which originate from weathering) and marine (for those aerosols which originate with sea spray). These distinctions are typically made by evaluating elemental enrichment factors which, relative to crustal and marine sources, are respectively defined as:

$$EF_{crust} = (X/Al)air/(X/Al)crust \ ,$$

$$EF_{sea} = (X/Na)air/(X/Na)sea \ ,$$

where Al, Na, and X are the concentrations measured for reference elements and the element of interest (X) in the earth's crust, ocean, or atmosphere. Enrichment factors for trace elements substantially greater than a value of one for EF_{crust} or EF_{sea} are indicative of a pollutant source, although natural processes such as fractionation at the surface microlayer may also explain such observations. Regardless of their source, all atmospheric aerosols act as a source to the sea surface microlayer. However, it is often instructive to evaluate the magnitude of the flux which may be attributed to natural sources relative to those which arise from man's activities. A case in point is the current $\simeq$100-fold increase in atmospheric Pb/Al ratios relative to those of crustal materials [1]. This increase has been attributed to combustion of leaded gasoline, metal smelting, industrial dusts introduced into the atmosphere by resuspension, and combustion of dusts deposited on plant materials during forest fires [117]. Preliminary estimates indicate that the input of such lead-rich aerosols have resulted in approximately a sixteen-fold increase in the Pb concentration of the mixed layer in the world's oceans [117].

Estimates of the flux of particulate trace metals from the atmosphere may be made by various means inluding direct observation of dry and wet deposition fluxes through exposing plate samplers and analysis of rainwater, calculations based on analysis of atmospheric aerosol concentrations and an aerosol deposition velocity, or inferred from sediment deposition records (particularly in lake waters).

Estimates for atmospheric input of trace metals to the ocean using the first and second of these methods have been given by Duce et al. [1], and Wallace et al. [118]. The various estimates agree within an order of magnitude. The elements Pb, Zn, Cu, Hg, Cd, Se, As, Sb, and Ag are identified as being anonymously enriched with respect to both the sea and crustal sources and therefore possibly of anthropogenic origin [1].

Similar estimates indicate that the atmospheric flux of some particulate trace metals, e.g., Ni, Cu, Zn, and Pb, is sufficient to achieve a two-fold surface microlayer enrichment given microlayer residence times of an hour or less in North Sea Coastal waters [36]. Calculations for ocean waters near the African continent and close to the Saharan dust veil suggest a microlayer residence time on the order of seconds is sufficient to explain observed enrichments of Fe (considering the atmosphere as the only source) [119]. In contrast, Fe was found by Hunter to be depleted in the North Sea microlayer samples [36]. The explanation for this difference was suggested to be a bulk water (vs. atmospheric) source for large Fe bearing particles in the North Sea with a net microlayer depletion resulting from a greater removal rate through sedimentation relative to the rate of input from turbulent mixing.

Just exactly what an appropriate residence time should be for atmospheric particles at the sea surface microlayer is not clear. One would expect that physical conditions such as the extent of turbulence in the system to play an important role. In addition, the diameter and density of deposited aerosols will certainly influence removal rates by processes such as sedimentation. The wetability of the surface should also play a role, even easily-wet inorganic mineral surfaces introduced into the organic rich sea surface may be coated and thereby rendered more hydrophobic. The kinetics of such reactions and the extent to which they must occur in order to have an appreciable affect on particle residence time at the sea surface are not well-known.

The exchange of vapor-phase pollutants between the oceans and atmospheres is typically modeled using two-film theory as described by Liss and Slater [120]. In this model, gas and liquid films at the sea-surface are considered as resistances to the flux of a given compound, with the relative importance of gas vs. liquid film resistances being determined by the equilibrium partitioning of the constituent as described by its Henry's Law constant (H), and its mass transfer coefficient in each phase (k_l and k_g) where:

$$H = \frac{\text{concentration or partial pressure in the vapor phase}}{\text{concentration in the aqueous phase}}$$

and

$$\text{Total Resistance to Transfer on a liquid phase basis} = \frac{1}{K_l} = \frac{1}{k_l} \frac{1}{Hk_g}.$$

For compounds with large Henry's Law constants (i.e., $\geqq 10^{-3}$ atm.-m^3/mole) liquid-film resistance predominates and the compounds preferentially partition into vapor phase [121]. The liquid film controlled flux (F) of a compound is described by:

$$F = K_l[C_g/H - C_l],$$

where C_g and C_l are the gas and liquid phase concentrations, respectively.

The application of such equations to environmental pollutants of interest is frequently impaired by uncertainty in gas and liquid phase concentrations as well as Henry's Law constants. A case in point is the effort of Doskey and Andren to model PCB flux across the air/water interface in the Great Lakes [122]. These investigators concluded that neither gas nor liquid phase control, nor the direction of PCB flux, could be established given the uncertainty in the Henry's Law constants for PCB isomers. In addition, the concentrations of PCB's in atmospheric vapor and particulate phases were found to suffer from the necessarily operational definition of what constitutes a particle and non-uniform sample collection efficiencies for different PCB isomers [122]. Preliminary estimates of the atmospheric flux to marine and fresh water systems have been given for PCB's [5, 123–126], PAH [44], organochlorine pesticides [126], and some halocarbons [120].

An interesting and unanswered question is the effect of organics in surface microlayers on the fluxes of volatile pollutant organics across the air-sea interface. Coherent organic films are known to impede the exchange of CO_2, O_2, N_2, and H_2O [127, 128]. With the possible exception of oil slicks and waters in the vicinity of kelp beds which may be coated with surface-active alginates, available evidence indicates that the heterogeneous nature and concentration of organics at the surface of the world's oceans are unlikely to directly effect gas exchange [129]. However, an indirect effect may result from the damping of capillary waves by surface films [130] since such waves may promote gas transfer [131].

Characteristics of Surface Films Which May Have Geochemical Impact

The above discussion has touched on the interesting and occasionally unique chemical and biological characteristics of surface films. In this section potential impact of these characteristics on the composition of the oceans and atmosphere will be briefly discussed.

Transfer to the Atmosphere

As previously indicated, bursting bubbles eject jet drops which effectively transfer materials from the sea to the atmosphere. Recent results which examine the size spectrum of marine bubbles and sea spray, show a striking similarity consistent with a causal relationship [132]. This sea spray provides a ready explanation for many of the constituents of atmospheric aerosols. The presence of elements such as Na^+, K^+, and Cl^- in concentration ratios comparable to that of sea water is not unexpected, but of greater interest (for the purposes of this discussion) are the elevated atmospheric concentrations of some trace metal species and organic compounds. It is not clear, for example, that the origins of such enrichments results from an anthropogentic pollutant source, from bubble ejection of materials enriched at the sea surface, or from a combination of the two processes. The materials bubbles eject may originate both from bulk solution and the surface microlayer since rising bubbles effectively scavenge materials from bulk solution. In the case of bacteria ejected by rising bubbles, it has been argued that observed concentrations in jet drops can be explained solely by bubble scavenging from bulk solution vs. removal of bacterioneuston [114]. In the case of trace metals, the results obtained using the Bubble Interfacial Microlayer Sampler (BIMS) developed by Fasching et al. [133] indicate both scavenging from bulk solution and ejection of surface microlayer waters may explain the observed enrichments of Cu while bubble scavenging alone may explain the enrichment of Zn [32]. Evaluation of the relative importance of the possible sources for enriched elements is among the objectives of a large Sea-Air Exchange (SEAREX) study currently in progress.

In the case of marine organics, proteinaceous materials characteristic of surface microlayers are found in sea fog [134] and fatty acids are found in marine aerosols [135]. The origin (microlayer vs. bulk solution scavenging) of such materials is not established. Nevertheless, several investigators have observed bubble induced removal of organic surface films [136, 137].

Transfer to the Ocean

A major unresolved question is the extent to which particle aggregation at the surface microlayer may influence the flux of particles to the bulk ocean. Wallace et al. have provided an estimate of the POC mediated removal of particulate trace metals in the Northwest Atlantic [118]. It is perhaps noteworthy that this estimate was very similar to the estimated atmospheric flux of particulate trace metals to the sea surface which was calculated independently. One implication is that particle aggregation at the microlayer may play a role in the fate of atmospheric pollutants introduced into the sea surface. Particulate metals introduced via atmospheric aerosols may be solubilized [84], but soluble metals may in turn be adsorbed onto the surface of organic aggregates or complexed with surface-active ligands. The relative roles of the particle scavenging of trace metals in the ocean surface layer vs. those in the ocean's mixed layer are not established.

Recent data for particle flux in the oceans stress the short residence times for elements such as Al which must be accounted for by packaging of small clay particles into large rapid settling agglomerates [138]. Although the fecal pellets produced by marine organisms constitute a major source for such large particles, the potential role of particle agglomeration at the surface microlayer also merits consideration.

Special Opportunities for Biological-Chemical Interactions

The affinity of marine bacteria for surfaces is well established and it is therefore not surprising (given the high surface area to volume ratios encountered at the air-sea interface) that attached bacteria are predominately observed in surface microlayers [63]. Attached bacteria at the surface microlayer have been shown to be metabolically active [139] and may utilize surface adsorbed organic compounds as substrate [140]. Of concern are the elevated concentrations of the particulate form of toxic trace metals also identified at surface microlayers [36, 54] and the opportunity for incorporation of pollutant metals into the marine food web.

In addition to trace metals, many hydrophobic organic pollutants are introduced into the surface microlayer from the atmosphere. The sorption of such compounds is typically found to depend, in part, on the organic content of the sorbent [141, 142] and one would anticipate that organic aggregates or organically coated particles at the sea surface would therefore function as effective binding sites. Dissolved organic materials such as humic compounds are also effective binding agents for hydrophobic pollutants [143] and it is not unreasonable to speculate that the dissolved surface active organics at the air-sea interface may play a similar role. The extent to which pollutant organics at the surface microlayer (either free, sorbed, or bound to dissolved organics) are introduced into the marine food chain is unknown.

Conclusions

The above discussion has been undertaken to give the reader a "flavor" for the surface of the ocean as differentiated from the bulk underlying solution. For a greater overview, several review articles have been published which variously focus on transport processes [34], air-sea exchange [144], and on the chemical and biological components of the microlayer and their interactions [6, 35, 145]. Although the surface microlayer and bulk ocean share in common the major chemical constituents of the marine salt matrix, the surface microlayer is seen to have enriched concentrations of many toxic trace metals, surface-active and pollutant organic compounds, particles and microbiota. The concentration of these materials at the sea surface provides the opportunity for interactions which may influence the form and fate of many compounds of concern in both the earth's atmosphere and the oceans.

List of Symbols and Abbreviations

a_i chemical activity of a solute, i, in solution
ATP adenosine triphosphate
Cg concentration (or partitial pressure) of a substance in the gaseous phase
C_l concentration of a substance in the liquid phase
DDT dichlorodiphenyltrichloroethane, a pesticide
DOC dissolved organic carbon
EF enrichment factor, usually given relative to the earth's crust, EF_{crust}, or the ocean, EF_{sea}

F flux of a substance through the air-water interface
H Henry's Law constant
k_g mass transfer coefficient for the gas film at an air-water interface
k_l mass transfer coefficient for the liquid film at an air-water interface
K_l overall mass transfer coefficient (liquid film basis) at an air-water interface
PAH polyaromatic hydrocarbons, a by-product of combustion
PCB polychlorinated biphenyls, previously used as wood preservatives, paint additives, plasticizers, and in carbonless-carbon paper, and now primarily employed in the manufacture of electrical capacitors and transformers
POC particulate organic carbon
PVC polyvinyl chloride, a plastic
R the universal gas constant
T absolute temperature, °K
γ surface tension of an aqueous solution
Γ_i surface excess of compound, i.

References

1. Duce, R.A., Hoffman, G.L., Ray, B.J., Fletcher, I.S., Wallace, G.T., Fasching, J.L., Piotrowicz, S.R., Walsh, P.R., Hoffman, E.J., Miller, J.M., Heffter, J.L.: Marine Pollutant Transfer, p.77, H.L. Windon and R.A. Duce (eds.), Lexington, Mass., D.C. Heath & Co. (1976)
2. Peirson, D.H., Cawse, P.A., Cambray, R.S.: Nature *251*, 675 (1974)
3. Rancitelli, L.A., Perkins, R.W.: J. Geophys. Res. *75*, 3055 (1970)
4. Yano, N., Katsuragawa, H., Maebashi, K.: J. Rech. Atmos. *13*, 807 (1974)
5. Bidleman, T.F., Rice, C.P., Olney, C.E.: Marine Pollutant Transfer, p.323, H.L. Windom and R.A. Duce (eds.), Lexington, Mass., D.C. Heath & Co. (1976)
6. Liss, P.S.: Chemical Oceanography, p. 193, J.P. Riley and G. Skirrow (eds.), London, N.Y. and San Francisco, Academic Press (1975)
7. MacIntyre, F.: Sci. Amer., p. 62 (May, 1974)
8. Garrett, W.D.: Limnol. & Oceanogr. *10*, 602 (1965)
9. Hatcher, R.F., Parker, B.C.: Limnol. & Oceanogr. *19*, 162 (1974)
10. Van Vleet, E.S., Williams, P.M.: Limnol. & Oceanogr. *25*, 764 (1980)
11. Lion, L.W., Harvey, R.W., Leckie, J.O.: Marine Chemistry *11*, 235 (1982)
12. Harvey, G.W.: Limnol. & Oceanogr. *11*, 608 (1966)
13. Harvey, G.W., Burzell, L.A.: Limnol. & Oceanogr. *17*, 156 (1972)
14. Larsson, K., Odham, G., Sodergren, A.: Mar. Chem. *2*, 49 (1974)
15. Baier, R.E.: J. Geophysical Res. *77*, 5062 (1972)
16. Baier, R.E., Goupil, P.W., Perlmutter, S., King, R.: J. Rech. Atmos. *8*, 571 (1974)
17. Szekielda, K.H., Kupferman, S.L., Klemas, V., Polis, D.F.: J. Geophys. Res. *77*, 5278 (1972)
18. Pellenbarg, R.E.: Ph.D. Thesis, Univ. of Delaware, 166pp. (1976)
19. Pellenbarg, R.E., Church, T.M.: Science *203*, 1010 (1979)
20. Hamilton, E.I., Clifton, R.J.: Limnol. & Oceanogr. *24*, 188 (1979)
21. Kjellenberg, T., Stenstrom, A., Odham, G.: Mar. Biol. *53*, 21 (1979)
22. Crow, S.A., Ahearn, D.G., Cook, W.L., Bourquin, A.W.: Limnol. & Oceanogr. *20*, 644 (1975)
23. Syzdek, L.D.: Limnol. & Oceanogr. *27*, 172 (1982)
24. Young, L.Y.: Microbial Ecology *4*, 267 (1978)
25. Parker, B.C., Woodehouse, E.B.: Water For Texas, Water Resour. Inst., 15th Ann. Conf. Texas A&M Univ. (1971)
26. Morris, R.J.: Mar. Poll. Bull. *5*, 105 (1974)
27. MacIntyre, F.: J. Phys. Chem. *72*, 589 (1968)
28. MacIntyre, F.: Tellus *22*, 451 (1970)
29. MacIntyre, F.: J. Rech. Atmos. *8*, 515 (1974)
30. Carlucci, A.F., Williams, P.M.: J. Cons. Perma. Int. Explor. Mer. *30*, 28 (1965)
31. Blanchard, D.C., Syzdek, L.D.: J. of Geophysical Res. *77*, 5087 (1972)
32. Piotrowicz, S.R., Duce, R.A., Fasching, J.L., Weisel, C.P.: Mar. Chem. *7*, 307 (1979)

33. Daumas, R.A., Laborde, P.L., Marty, J.C., Saliot, A.: Limnol. & Oceanogr. *21*, 319 (1976)
34. MacIntyre, F.: The Sea, Vol. 5, Marine Chemistry, p. 245, E.D. Goldberg (ed.) N.Y., J. Wiley & Sons (1974)
35. Lion, L.W., Leckie, J.O.: Ann. Rev. Earth Planet. Sci. *9*, 449 (1981)
36. Hunter, K.A.: Mar. Chem. *9*, 49 (1980)
37. National Research Council: International Critical Tables of Numerical Data, Physics, Chemistry and Technology; Vol. IV., p. 466, N.Y., McGraw-Hill, Inc. (1928)
38. Szyszkowski, B.: Zeitschrift für Physikalische Chemie *64*, 385 (1908)
39. Hardy, R., Mackie, P.R., Whittle, K.J.: Rapp. P.-V. Reun. Cons. Int. Explor. Mer. *171*, 17 (1977)
40. Sieburth, J. McN., Willis, P.-J., Johnson, K.M., Burney, C.M., Lavoie, D.M., Hinga, K.R., Caron, D.A., French III, F.W., Johnson, P.W., Davis, P.G.: Science *194*, 1415 (1976)
41. Brockmann, U.H., Huhnerfuss, H., Kattner, G., Broecker, H.-C., Hentzchel, G.: Limnol. & Oceanogr. *27*, 1050 (1982)
42. Hunter, K.A., Liss, P.S.: Mar. Chem. *5*, 361 (1977)
43. Gucinski, H., Goupil, D.W., Baier, R.E.: Atmospheric Pollutants in Natural Waters, S.J. Eisenreich (ed.), p. 165, Ann Arbor, MI, Ann Arbor Sci. Publ. (1981)
44. Andren, A.W., Strand, J.W.: Atmospheric Pollutants in Natural Waters, S.J. Eisenreich (ed.), p. 459, Ann Arbor, MI, Ann Arbor Sci. (1981)
45. Wheeler, J.R.: Limnol. & Oceanogr. *20*, 338 (1975)
46. Wangersky, P.J.: Chimia *26*, 559 (1972)
47. Wangersky, P.J.: Ann. Rev. Ecol. Systematics *7*, 161 (1976)
48. Goldacrce, R.J.: Surface Phenomena in Chemistry and Biology, J.F. Danielle, H.G.A. Parkhurst, A.C. Riddiford (eds.), p. 278, N.Y., Pergamon Press, N.Y. (1958)
49. Wangersky, P.J.: Amer. Scientist *53*, 358 (1965)
50. Johnson, B.D.: Limnol. & Oceanogr. *21*, 444 (1976)
51. Wallace, G.T., Duce, R.A.: Mar. Chem. *3*, 157 (1975)
52. Lion, L.W., Leckie, J.O.: Env. Geol. *3*, 293 (1981)
53. Duce, R.A., Quinn, J.G., Olney, C.E., Piotrowicz, S.R., Ray, B.J., Wade, T.L.: Science *176*, 161 (1972)
54. Piotrowicz, S.R., Ray, B.J., Hoffman, G.L., Duce, R.A.: J. Geophys. Res. *77*, 5243 (1972)
55. Barker, D.R., Zeitlin, H.: J. Geophys. Res. *77*, 5076 (1972)
56. Fitzgerald, W.F., Hunt, C.D.: J. Rech. Atmos. *13*, 629 (1974)
57. Lion, L.W., Leckie, J.O.: Atmospheric Pollutants in Natural Waters, S. Eisenreich (ed.), p. 143, Ann Arbor, MI, Ann Arbor Sci. Publ. (1981)
58. Harvey, R.W., Lion, L.W., Young, L.Y., Leckie, J.O.: J. Mar. Res. *40*, 1201 (1982)
59. Corpe, W.A.: Devel. Indust. Microbiol. *16*, 249 (1975)
60. Harvey, R.W.: Ph.D. Thesis, Dept. of Civil Engr., Stanford Univ. (1981)
61. Corpe, W.A.: Dev. Ind. Microbiol. *11*, 402 (1970)
62. Roper, M.M., Marshall, K.C.: Microb. Ecol. *4*, 279 (1978)
63. Harvey, R.W., Young, L.Y.: Appl. Environ. Microbiol. *39*, 894 (1980)
64. Tsyban, A.V.: J. Oceanogr. Soc. Japan *27*, 56 (1971)
65. Parker, B., Barsom, G.: Bioscience *20*, 87 (1970)
66. Wandschneider, K.: Mar. Biol. *52*, 105 (1979)
67. Norris, R.E.: J. Protozoology *12*, 589 (1965)
68. David, P.M.: Endeavor *24*, 95 (1965)
69. Hemple, G., Weikert, H.: Mar. Biol. *13*, 70 (1972)
70. Maynard, N.G.: Z. Allg. Mikrobiol. *8*, 119 (1968)
71. Harvey, R.W., Lion, L.W., Young, L.Y.: Est. Coastal Shelf Sci. in press (1983)
72. Barrtlett, M.R., Haedrich, R.L.: Copeia *938*, 469 (1968)
73. Drachev, S.M., Bylinkina, V.A., Sosunova, I.N.: Biol. Abstr. *46*, 101500 (1965)
74. Goering, J.J., Menzel, D.W.: Deep Sea Res. *12*, 839 (1965)
75. Williams, P.M.: Deep Sea Res. *14*, 791 (1967)
76. Nishizawa, S.: Bull. Plankton Soc. Japan *18*, 42 (1971)
77. Lyons, W.B., Pybus, M.J., Coyne, J.: Oceanol. Acta *3*, 151 (1980)
78. Chapman, P., Liss, P.S.: Limnol. & Oceanogr. *26*, 387 (1981)
79. Horn, J.M., Teal, J.M., Backus, R.H.: Science *168*, 245 (1970)
80. Carpenter, E.J., Smith, K.L.: Science *175*, 1240 (1972)

81. Anderson, D.M., Morel, F.M.M.: Limnol. & Oceanogr. *23*, 283 (1978)
82. Sunda, W., Guillard, R.R.L.: J. Mar. Res. *34*, 511 (1976)
83. Sunda, W.G., Engel, D.W., Thuotte, R.M.: Env. Sci. & Tech. *12*, 409 (1978)
84. Hardy, J.T., Crecelins, E.A.: Env. Sci. & Tech. *15*, 1103 (1981)
85. Hardy, J.T.: Limnol. & Oceanogr. *18*, 525 (1973)
86. MacIntyre, W.G., Smith, C.L., Munday, J.C., Gibson, V.M., Lake, J.L., Windsor, J.G., Dupuy, J.L., Harrison, W., Oberholtzer, J.D.: U.S.E.P.A., Env. Prot. Tech. Series EPA-670/2-73-099 (1974)
87. Taguchi, S., Nakajima, K.: Bull. Plankton Soc. Japan *18*, 20 (1971)
88. Marumo, R., Taga, N., Nakai, T.: Bull. Plankton Soc. Japan *18*, 36 (1971)
89. Dietz, A.S., Albright, L.J., Tuominen, T.: Can. J. Microbiol. *22*, 1699 (1976)
90. Sieburth, J.M.: Deep Sea Res. *18*, 1111 (1971)
91. Blanchard, D.C., Syzdek, L.D.: Limnol. & Oceanogr. *23*, 389 (1978)
92. Lion, L.W., Leckie, J.O.: Limnol. & Oceanogr. *27*, 111 (1982)
93. Lion, L.W.: Ph.D. thesis, Stanford University, Dept. of Civil Engr., Stanford, CA (1980)
94. Lion, L.W., Leckie, J.O.: Est. Coastal and Shelf Sci., in press (1983)
95. Lal, D., Lerman, A.: J. Geophys. Res. *80*, 423 (1975)
96. Hunt, J.R.: Env. Sci. & Tech. *16*, 303 (1982)
97. Langmuir, I.: Science *87*, 119 (1938)
98. Wallace, G.T., Jr., Loeb, G.I., Wilson, D.F.: J. Geophys. Res. *77*, 5293 (1972)
99. Wallace, G.E., Jr., Duce, R.A.: Limnol. & Oceanogr. *23*, 1155 (1979)
100. Blanchard, D.C., Syzdek, L.D.: J. Rech. Atmos. 529 (1974)
101. Wilson, W.B., Collier, A.: J. Mar. Res. *30*, 15 (1972)
102. Charm, S.E.: Adsorptive Bubble Separation Techniques, R. Lemlich (ed.), p. 157, N.Y. & London, Academic Press (1972)
103. Medwin, H.: J. Geophys. Res. *75*, 599 (1970)
104. Medwin, H.: J. Geophys. Res. *82*, 971 (1977)
105. Garrettson, G.A.: J. Fluid Mech. *159*, 187 (1973)
106. Johnson, B.D., Cooke, R.C.: Sci. *213*, 209 (1981)
107. Rubin, A.J.: Adsorptive Bubble Separation Techniques, R. Lemlich (ed.), p. 199, N.Y., Academic Press (1972)
108. Kim, Y.S., Zeitlin, H.: Separation Science *7*, 1 (1972)
109. Neihof, R.A., Loeb, G.I.: Limnol. & Oceanogr. *17*, 7 (1972)
110. Hunter, K.A., Liss, P.S.: Nature *282*, 823 (1979)
111. Hunter, K.A.: Limnol. & Oceanogr. *25*, 807 (1980)
112. Weber, M.E., Blanchard, D.C., Syzdek, L.D.: Limnol. & Oceanogr. *28*, 101 (1983)
113. Yao, K.-M., Habibian, M.T., O'Melia, C.R.: Env. Sci. & Tech. *5*, 1105 (1971)
114. Blanchard, D.C., Syzdek, L.D., Weber, M.E.: Limnol. & Oceanogr. *26*, 961 (1981)
115. Scott, B.C.: Atmospheric Pollutants in Natural Waters, S.J. Eisenreich (ed.), p. 3, Ann Arbor, MI, Ann Arbor Sci. Publ. (1981)
116. Slinn, S.A., Slinn, W.G.N.: Atmospheric Pollutants in Natural Waters, S.J. Eisenreich (ed.), p. 23, Ann Arbor, MI, Ann Arbor Sci. Publ. (1981)
117. Schale, B., Patterson, C.C.: Lead in the Marine Environment, M. Branica and Z. Konrad (eds.), p. 31, N.Y., Pergamon Press (1980)
118. Wallace, G.T., Jr., Hoffman, G.L., Duce, R.A.: Mar. Chem. *5*, 143 (1977)
119. Hoffman, E.J., Hoffman, G.L., Duce, R.A.: J. Rech. Atmos. *13*, 675 (1974)
120. Liss, P.S., Slater, P.G.: Nature *247*, 181 (1974)
121. Mackay, D., Yuen, A.T.K.: Atmospheric Pollutants in Natural Waters, S.J. Eisenreich (ed.), p. 55, Ann Arbor, MI, Ann Arbor Sci. Publ. (1981)
122. Doskey, P.V., Andren, A.W.: Env. Sci. & Tech. *15*, 705 (1981)
123. Eisenreich, S.J., Hollod, G.J., Johnson, T.C.: Atmospheric Pollutants in Natural Waters, S.J. Eisenreich (ed.), p. 425, Ann Arbor, MI, Ann Arbor Sci. Publ. (1981)
124. Eisenreich, S.J., Looney, B.B., Thornton, J.D.: Env. Sci. & Tech. *15*, 30 (1981)
125. Murphy, T.J., Schinsky, A., Paolncci, G., Rzeszutko, C.P.: Atmospheric Pollutants in Natural Waters, S.J. Eisenreich (ed.), p. 445, Ann Arbor, MI, Ann Arbor Sci. Publ. (1981)
126. Bidleman, T.F., Christensen, E.J., Harder, H.W.: Atmospheric Pollutants in Natural Waters, S.J. Eisenreich (ed.), p. 481, Ann Arbor, MI, Ann Arbor Sci. Publ. (1981)

127. Garrett, W.D.: The Changing Chemistry of the Oceans, D. Dyrssen and D. Jagner (eds.), p. 75, Wiley Interscience (1971)
128. Lou, Yu.S., Rasmussen, G.P.: Water Res. Research *9*, 1258 (1973)
129. Liss, P.S.: Rapp. P.-V. Reu. Const. Int. Explor. Mer. *17*, 120 (1977)
130. Garrett, W.D., Bultman, J.D.: J. Colloid. Sci. *18*, 798 (1963)
131. MacIntyre, F.: Physics Fluids *14*, 1596 (1971)
132. Wu, J.: Sci. *212*, 324 (1981)
133. Fasching, J.L., Courant, R.A., Duce, R.A., Piotrowicz, S.R.: J. Rech. Atmos. *8*, 649 (1974)
134. Baier, R.E.: U.S. Naval Res. Lab., Summary Rept., Washington, D.C., Project No. VA-5788-M (1975)
135. Barger, W.R., Garrett, W.D.: J. Geophys. Res. *81*, 3151 (1976)
136. Baier, R.E.; Proc. 13th Conf. Great Lakes Res., p. 114 (1970)
137. Bezdek, H.F., Carlucci, A.F.: Limnol. & Oceanogr. *19*, 128 (1974)
138. Denser, W.G., Brewer, P.B., Jickells, T.D., Commeau, R.F.: Sci. *219*, 388 (1983)
139. Harvey, R.W., Young, L.Y.: Appl. Env. Micro. *40*, 156 (1980)
140. Li, A.Y.L., Giano, F.A.: J. Water Poll. Cont. Fed. *55*, 392 (1983)
141. Sullivan, K.F., Atlas, E.L., Giam, C.S.: Env. Sci. & Tech. *16*, 428 (1982)
142. Means, J.C., Hassett, J.J., Wood, S.G., Banwart, W.L.: Polynuclear Aromatic Hydrocarbons: Third International Symposium on Chemistry and Biology – Carcinogenesis and Mutagenesis, P.W. Jones and P. Leber (eds.), p. 327, Ann Arbor, MI, Ann Arbor Sci. Publ. (1979)
143. Carter, C.W., Suffet, I.H.: Env. Sci. & Tech. *16*, 735 (1982)
144. Duce, R.A., Hoffman, E.J.: Ann. Rev. Earth & Planetary Sci. *4*, 187 (1976)
145. Hardy, J.T.: Progress in Oceanography *11*, 307 (1982)

Atmospheric Nitrogen Chemistry, Nitrification, Denitrification, and their Interrelationships

R. D. Hauck

Division of Agricultural Development
Tennessee Valley Authority
Muscle Shoals, AL 35660, USA

Introduction

Nitrogen compounds are produced from atmospheric dinitrogen (N_2) by the following processes:

(1) naturally occurring biochemical reactions (biological dinitrogen fixation);

(2) naturally occurring abiological reactions (ozonation and combustion, e. g., via fire, lightning, and coronal discharges); and

(3) anthropogenic abiological reactions (combustion and industrial synthesis).

These processes result in the addition of fixed N to soils and waters. This added N, after residence times as short as minutes or as long as eons, cycles back to the atmosphere. The global transfers of N between the atmosphere and soils and waters involve two major subcycles [19], designated the N_2 fixation-denitrification cycle and the precipitation-volatilization cycle. The former cycle is mainly biological in character; the latter is mainly abiological. This discussion is concerned with portions of both cycles, with special emphasis on the role of nitrification and denitrification. Biological dinitrogen fixation will not be discussed, but the main abiological dinitrogen fixation processes will be summarized. – General aspects of the Nitrogen Cycles are discussed by R. Söderlund and T. Rosswall in This Handbook, Vol I. B, p. 61.

Abiological Reactions of Atmospheric Nitrogen

The Atmosphere

The various regions of the atmosphere are designated as follows:

(1) The troposphere extends from the earth's surface to an altitude of about 8 km at the poles to 16 km at the equator. It is characterized by a decrease in temperature with increase in altitude and can be as cold as −80 °C at tropical latitudes at the upper boundary. Therefore, almost all water vapor and other condensable gases condense or freeze in the troposphere. Most clouds formed in the troposphere evaporate before forming precipitation, and some condensable gases may pass into the stratosphere, although experimental evidence for this is lacking. In time, all of the water vapor and condensed gases are removed as precipitation.

(2) The tropopause is the upper boundary of the troposphere and the lower boundary of the stratosphere.

(3) The stratosphere extends from the tropopause to an altitude of about 50 km. It contains about 20% of the mass in the total atmosphere. Absorption of ultraviolet radiation causes a temperature rise between the lower and upper boundaries (from 220 K to 270 K). This temperature inversion retards vertical mixing. The stratosphere is extremely dry, giving evidence for the capacity of the troposphere to remove condensable gases.

(4) Between the upper boundary of the stratosphere and the ionosphere is the mesosphere, a region of falling temperatures. It extends from about 50 km to 85 km, above which ionic chemical processes predominate.

(5) The ionosphere is a region of high ionic activity which can extend from over 50 km to more than 400 km altitude. It is not known whether ionosphere reaction products diffuse downward into the lower stratosphere.

Nitrous Oxide

The atmosphere contains about 1.98% of the earth's nitrogen [80], N_2 being by far the dominant species. Calculations based on average atmospheric mass and composition indicate an inventory of 3865×10^6 Tg[1] [19]. Of the nitrogen gases, N_2O is present in the second largest amount (about 2000 ± 200 Tg [39]). The fractional volume abundance of N_2O in the troposphere is 3.3×10^{-7} [72]; the gas is uniformly mixed to about 10 to 12 km altitude [8, 37], but decreases rapidly in abundance to 1×10^{-7} at 30 km [30]. The decrease in concentration extends into the stratosphere, which provides a sink for N_2O removal. Photodissociation is the main destruction process:

$$N_2O + h\nu \rightarrow N_2 + O \quad (1)$$

$$N_2O + h\nu \rightarrow NO + N. \quad (2)$$

Destruction in the stratosphere occurs also by reaction with atomic oxygen, as described later. Crutzen [29] estimated that 3.4×10^9 N_2O molecules cm^{-2} s^{-1} cross the tropopause at 15 km altitude and are either photolyzed or react with nascent oxygen, resulting in a net consumption in the stratosphere of 3.1×10^9 molecules cm^{-2} s^{-1}. This amount corresponds to an annual removal of 33 Tg of N_2O [25], which is somewhat higher than an earlier estimate of about 10 Tg $year^{-1}$ [8]. Other estimates range from 14 to 28 Tg $year^{-1}$ [39].

1 One Tg $= 10^{12}$ g = one million metric tons

Nitrous oxide is formed in the atmosphere by the reaction:

$$N_2+O_3 \rightarrow N_2O+O_2 \quad (3)$$

This reaction is considered too slow by a factor of 40 to account for the observed concentration of atmospheric N_2O [37]. To balance the loss by photolysis, there must be sources of N_2O that provide about 3×10^9 molecules cm^{-2} s^{-1} [8]. Based on the earth's land and water surfaces, this rate of emission would contribute annually about 24 Tg of N_2O to the global atmosphere. This value is comparable to those of Söderlund and Svensson [79], who reported estimates of annual net emissions of N_2O in the range of 16 to 69 Tg. Biological denitrification, nitrification, and perhaps chemodenitrification are the main processes giving rise to these emissions. These processes will be discussed later.

Recent evidence from laboratory experiments points to the possibility that significant amounts of nitrogen oxides, including N_2O, can be formed in the atmosphere during lightning and coronal substorms [91]. Mestastable N_2 molecules are produced by the impact of low energy electrons on N_2 and by the absorption of cosmic rays and solar x-rays. A possible mechanism would involve the interaction of the energy-rich molecules $N_2(A^3\Sigma_u^+)$ and $O_2(A^3\Sigma_g^-)$ (symbols within parentheses are spectroscopic notations designating charge, symmetry, and angle of momentum). These molecules are metastable (e.g., the N_2 species is dissipated within two seconds after formation) and highly reactive. If processes such as those observed in the laboratory with excited N_2 and O_2 molecules occur in natural lightning, then a typical discharge could produce 3×10^{26} NO_x ($NO+NO_2$) molecules and comparable amounts of N_2O [91]. Atmospheric electrical discharges would then be an important source of N_2O and other nitrogen oxides in the troposphere.

Unlike the stratosphere sink, tropospheric sinks for N_2O are poorly understood. Soils and oceans are obvious sinks. Although direct evidence of N_2O absorption by soils is available [9], the scarcity of quantitative data does not permit justifiable extrapolation of point measurements to global dimensions. However, in the absence of significant destruction of N_2O in the troposphere, and assuming a steady state equilibrium, sinks large enough to remove at least 180 Tg of N_2O annually need to be identified [39]. Estimates of annual fluxes of N_2O between the atmosphere and terrestrial surfaces range from 180 Tg [39] to 592 Tg [76]. There is little basis for calculating N_2O fluxes between the atmosphere and oceans. However, calculations based on measurements made in waters of the North Atlantic and Eastern tropical Pacific oceans supersaturated with N_2O indicate that annual emissions of marine origin should be at least 25 Tg of N_2O [39].

Various flux estimates indicate a residence time for tropospheric N_2O from as short as 5 years to as long as 160 years. This range in values can be compared with residence times for NO, NO_2, and N_2, which are estimated to be 35 days, 36 days, and 17×10^6 years, respectively [19].

Tropospheric Nitric Oxide and Nitrogen Dioxide

Tropospheric and stratospheric reactions of NO_x (NO and NO_2) differ in their chemistry and environmental effects. In the troposphere, NO_x is formed by (1) electrical discharges, which, as already indicated, also may form N_2O; (2) high temperature (2300 K or about 2000 °C) combustion of N_2; and (3) a probable series of reactions that produces ozone from hydrocarbons and carbon monoxide (CO) oxidation at lower temperatures (e.g., 300 K):

$$2NO + O_2 \rightleftarrows 2NO_2 \tag{4}$$

Photolysis of NO_2 by photons with wavelengths <400 nm occurs throughout the troposphere, from tropopause to the earth's surface.

$$NO_2 + h\nu \rightarrow NO + O. \tag{5}$$

The reactions of NO_x in the atmosphere can be placed in two groups: (1) those that participate in the mechanisms that regulate ozone (O_3) concentrations in the stratosphere; and (2) those that are involved in smog formation.

The net effect of NO_x reactions in the stratosphere is ozone destruction. Reactions in the troposphere can result in either ozone formation or destruction. The basis of proposed mechanisms for ozone regulation and aerosol formation are reactions (adapted from [25, 27, 28]) involving the free radicals hydroperoxyl (HO_2), alkylperoxyl (RO_2), and acylperoxyl ($RCOO_2$), where R equals methyl, ethyl, and longer-chain alkyl groups. For example:

$$HO_2 + NO \rightarrow HO + NO_2 \tag{6}$$

$$NO_2 + h\nu \rightarrow NO + O \tag{7}$$

$$O + O_2 \rightarrow O_3 \tag{8}$$

$$\text{net: } HO_2 + O_2 \rightarrow HO + O_3. \tag{9}$$

The free radical HO is involved in the production of additional HO_2 through interaction with CO:

$$CO + OH \rightarrow H + CO_2 \tag{10}$$

$$H + O_2 \rightarrow HO_2. \tag{11}$$

The overall effect of the reactions given in Eqs. (6) through (11) is:

$$CO + 2\,O_2 + h\nu \rightarrow CO_2 + O_3. \tag{12}$$

This process causes the indirect dissociation of O_2, which is characteristic of O_3 formation in smog. Photons with wavelengths <240 nm needed for direct photolysis cannot penetrate below about 20 km altitude.

Typical reactions contributing to smog are those that form nitrous acid (HONO), nitric acid ($HONO_2$), dinitrogen pentoxide (N_2O_5), and peroxyacyl nitrate ($CH_3COO_2NO_2$).

$$2\,NO_2 + H_2O \rightleftarrows HONO + HONO_2 \tag{13}$$

$$NO_2 + NO + H_2O \rightleftarrows 2\,HONO \tag{14}$$

$$2\,NO_2 + O_3 \rightarrow N_2O_5 + O_2 \tag{15}$$

$$NO_2 + CH_3COO_2 + h\nu \rightarrow CH_3COO_2NO_2. \tag{16}$$

Reactions of this type are well documented (e.g., see [21]). The aerosol constituents vary in concentration throughout any given day, depending on the intensity of sunlight and the concentration of the reactants. Eventually, the intermediate compounds form nitrate salt aerosols, which may be deposited in rainfall. Crutzen [28] listed 12 reactions which define the relative concentrations of N, NO, NO_2, NO_3 (radical), and N_2O_5. The list does not include reactions with acyl and alkyl free radicals of importance in aerosol formation. Reactions needing further study are: NO or NO_2+RO_2, $NO+NO_2$, NO_2+NO_3, and $NO_3+h\nu$ [25].

Stratosphere: Nitrogen Oxides and Ozone Balance

The natural abundance of ozone in the stratosphere is determined largely by the balance between its formation from solar radiation and its destruction by catalytic reactions, the principal agents being NO and NO_2. Maintaining this balance appears to be of vital importance because stratospheric ozone is the only effective shield for life on the Earth's surface against ultraviolet radiation at wavelengths between 250 and 300 nm. Ozone is formed in the stratosphere by the action of photons with wavelengths <240 nm on O_2:

$$O_2 + h\nu \rightarrow 2\,O \tag{17}$$

$$O + O_2 + M \rightarrow O_3 + M. \tag{18}$$

(where M refers to a third molecule, e.g., N_2)

$$\text{net: } 3\,O_2 \longrightarrow 2\,O_3. \tag{19}$$

The reverse sequence ($O_3+h\nu\rightarrow O_2+O$; $O+O_3\rightarrow 2O_2$) is too slow to account for the observed abundance of ozone in the stratosphere. However, several substances have been identified which catalyze the conversion of ozone to molecular oxygen. The general catalytic cycle is:

$$X + O_3 \rightarrow XO + O_2 \tag{20}$$

$$XO + O \rightarrow X + O_2 \tag{21}$$

$$\text{net: } O + O_3 \longrightarrow 2\,O_2, \tag{22}$$

where X = NO, H, OH, or Cl [28].

The main catalysts in this cycle, NO and NO_2, are formed directly or indirectly from N_2O that has diffused from the troposphere to the stratosphere. Nitric oxide is formed as described in equation *2* and by the reaction:

$$N_2 + 2\,O('D) \rightarrow 2NO. \tag{23}$$

The term O('D) in Eq. [23] is a spectroscopic designation for atomic oxygen with a single charge and D spin state. This species has a maximum life of 110 s. Ozone destruction is by the reaction sequence:

$$NO + O_3 \rightarrow NO_2 + O_2 \tag{24}$$

$$NO_2 + O \rightarrow NO + O_2 \tag{25}$$

$$\text{net: } O + O_3 \longrightarrow 2\,O_2. \tag{26}$$

About 5% of the N_2O that diffuses into the stratosphere forms NO. The remainder is photolyzed to N_2 or undergoes the reaction:

$$N_2O + O(^1D) \rightarrow N_2 + O_2 . \quad (27)$$

Small amounts of NO are injected directly into the stratosphere as emissions from supersonic aircraft or volcanic eruptions [54]. Minor but possibly significant amounts of NO probably were injected into the stratosphere during the testing of nuclear bombs before 1965. Aurora discharges and photoionization reactions in the ionosphere (about 80 km altitude) also produce NO, but photodissociation probably destroys it before it can enter the stratosphere.

The nitrogen oxide-ozone subcycle involves a relatively small amount of NO, estimated to be between 0.26 and 1.2 Tg in the global stratosphere [84]. The main source of this NO appears to be N_2O originating from the earth's land and water surfaces. Only one other catalyst is known to significantly deplete ozone concentrations in the lower stratosphere, i.e. chlorine. Recent evidence implicates the chlorofluoromethanes, trade-named the freons and used as refrigerants and aerosol propellants. These usually inert compounds diffuse into the stratosphere where photolysis liberates chlorine, which then catalyzes ozone decomposition according to Eqs. *20* and *21*.

Atmospheric Ammonia and its Salts

As far as is known, all of the ammonia (NH_3) and most of the nitrate (NO_3^-) in the atmosphere are of biological and/or industrial origin. The atmosphere also contains relatively small amounts of nitrite (usually included with NO_3^-) and organic nitrogen compounds associated with dust of terrestrial origin. Chemical reactions in the troposphere include the hydration and acid aerosol neutralization of NH_3:

$$NH_3 + H_2O \rightleftarrows NH_4^+ + OH^- \quad (28)$$

$$NH_3 + HNO_3 \rightarrow NH_4NO_3 \quad (29)$$

$$2\,NH_3 + SO_3^{2-} + H_2O \rightarrow (NH_4)_2SO_4 . \quad (30)$$

Most of the atmospheric NH_3 is converted into aerosol by reaction *28* followed by neutralization (e.g., reactions *29* and *30*) or by NH_3-acid neutralization in the vapor state to produce NH_4^+ salts. These scavenging processes clean the atmosphere of NH_3, NO, and NO_2 by causing them to fall to the earth's surface in the form of dry and wet depositions.

Soils and plants readily absorb NH_3 from the atmosphere. The amount absorbed may have little significance in intensive agricultural systems, but may be of considerable importance to the nitrogen economy of uncultivated nonagricultural ecosystems. Estimates of annual global depositions of NH_3/NH_4^+ recently compiled [52] range between 65 and 186 Tg. Emissions from all sources range between 18 and 244 Tg. Based on numerous measurements made over the United States and Europe, Eriksson estimated NH_4^+ plus NO_3^- fallout ranging from 0.8 to 22 kg of N ha^{-1} yr^{-1} [32]. Using these data and assuming a mean atmospheric NH_3 concentration of 6 ppb, a gaseous deposition velocity of 1 cm s^{-1}, and allowing for latitudinal variations (tropical air contains 10 to 30% more mineral nitrogen), Robinson and Robbins [76] calculated a total NH_3/NH_4 flux of 1179 Tg, but did not specify sources. It is evident that mass balances for global NH_3 emissions and depositions need considerable refinement.

Total Nitrogen Fluxes

The preceding sections have emphasized the chemistry of atmospheric reactions more than fluxes of atmospheric constituents. A quantitative summary of known inputs into the atmosphere and further assessment of their significance will be attempted in the final portion of this chapter. Next to be addressed are two major nitrogen cycle processes that directly or indirectly give rise to or affect most of these inputs.

Nitrification

Nitrification is usually defined as the biological oxidation of ammonium to nitrate with nitrite as an intermediate in the reaction sequence. This definition has some limitations where heterotrophic microorganisms are involved [4], but is adequate for the autotrophic and dominant process. Nitrification is an energy-producing process involving a net transfer of eight electrons, causing a valence change for N from -3 to $+5$. The process is carried out in soils by chemosynthetic autotrophs (*Nitrobacteriaceae*), which derive from it the energy needed for growth and metabolism. The energy, stored in the energy-rich bond of adenosine triphosphate (ATP) is used in part to produce the nicotinamide adenine dinucleotide (NAD) required by the nitrifying bacteria for CO_2 fixation.

Ammonium Oxidation

Five genera of autotrophic nitrifying bacteria have been identified [77]. Four of these, *Nitrosomonas, Nitrosospira, Nitrosolobus,* and *Nitrosovibrio*, have been isolated from soil. Three species of the genus *Nitrosococcus* are found in marine waters. The most common and most intensively studied nitrifying microorganism is *Nitrosomonas europaea*, which is found in sewage and waters, as well as in soil.

The probable reaction sequence for the oxidation of ammonium to nitrite by Nitroso group bacteria is:

ammonia (NH_3/NH_4) $\rightarrow$ hydroxylamine (NH_2OH) $\xrightarrow{?}$ nitroxyl (NOH) $\xrightarrow{?}$ nitrohydroxylamine ($NO_2 \cdot NH_2OH$) $\rightarrow$ nitrite (NO_2^-).

The postulated intermediate compounds, NOH and $NO_2 \cdot NH_2OH$, have not been isolated, but their participation in the reaction sequence is consistent with the assumption that two electrons are transferred for each oxidation step between NH_4^+ and NO_2^-. Evidence for the reaction sequence is presented in several reviews (3, 53, 57, 58, 71, 83].

Although there is evidence that NH_3 and not NH_4^+ is the substrate for *Nitrosomonas* [81], the NH_4^+ oxidation will be used here. The oxidation of NH_4^+ to

NH_2OH probably requires

(1) copper as a component of cytochrome oxidase or an oxygenase;

(2) a cytochrome or peroxidase which binds carbon monoxide;

(3) molecular oxygen;

(4) a condition or substrate that activates NH_4^+;

(5) a catalase; and

(6) a free radical that is sensitive to nitrous oxide and methanol.

The oxidation of NH_4^+ to NH_2OH is thermodynamically unfavorable; this suggests the need for an energy-dependent activation of NH_4^+ before oxidation occurs. The mechanism by which this occurs is unknown but probably involves oxidative phosphorylation.

The oxidation of NH_2OH involves a dehydrogenation and electron transfer through a flavoprotein-cytochrome electron transport chain. Nitroxyl is believed to be the end product of this electron transfer. Little is known about the oxidation of the postulated intermediates, NOH and $NO_2 \cdot NH_2OH$, although some evidence indicates oxidation involves electron transfer through a flavin to a cytochrome.

Nitrous oxide can form during nitrification when O_2 concentrations are low, or where conditions resemble anaerobiosis, e.g., following cyanide addition [59]. The formation of N_2O apparently involves hydroxylamine oxidoreductase and nitrite reductase [68], in the presence of which at low O_2 tensions NOH or its dimer, hyponitrite ($H_2N_2O_2$) dismutates to form N_2O and H_2O. The possible practical significance of this reaction will be discussed later.

Nitroso group bacteria obtain carbon for synthesizing their cell constituents by assimilating CO_2. Fixation of CO_2 is by the Calvin reductive ribulose diphosphate (RUDP) cycle. The main reaction is the condensation of CO_2 with RUDP to form two molecules of phosphoglyceric acid, which undergo cyclic transformations to regenerate RUDP and yield one phosphoglyceric acid molecule for three molecules of CO_2 fixed. Details of the carbon metabolism of the nitrifiers are beyond the scope of this article and can be found elsewhere [68, 73, 77].

Nitrite Oxidation

Three genera of autotrophic bacteria have been found to oxidize NO_2^- to NO_3^- [77]. They are *Nitrobacter* (soil), *Nitrospira* (soil, fresh and marine waters), and *Nitrococcus* (marine waters). Nitrite oxidation by the Nitro group bacteria involves a two-electron transfer through a reverse flow cytochrome electron transport system [60]. It involves oxidative phosphorylation, yielding one ATP per NO_2^- [1]. The system, known as nitrite oxidase, functions as a respiratory complex and is closely associated with enzymes not directly involved in NO_2^- oxidation, such as nitrate reductase.

Carbon metabolism of the Nitro group bacteria is similar to that of the Nitroso group but differs in that, unlike the latter group, the nitrite oxidizers can be induced to function as heterotrophs. They can use organic carbon as an energy source (e.g., acetate, formate, peptone, and casein hydrolyzate) in the absence of NO_2^- [11, 78]. Because the organisms have an extremely sensitive energy balance, the ability to adapt to a heterotrophic mode may be a survival mechanism.

Much information about the biochemistry of autotrophic nitrification was obtained from studies in which the effect of chemicals on specific steps in the reaction sequence was observed. See [46] for a recent review of this subject.

Energy Relationships

As stated earlier, the oxidation of NH_4^+ to NO_2^- by Nitroso group bacteria and NO_2^- to NO_3^- by Nitro group bacteria produces the energy required for CO_2 fixation and growth. The reaction

$$NH_4^+ + 1.5\,O_2 \rightarrow NO_2^- + H_2O + 2H^+ \tag{31}$$

liberates energy estimated by various researchers to be between 58 and 84 kcal $mole^{-1}$ of NH_4^+ [70].

The reaction

$$NO_2^- + 0.5\,O_2 \rightarrow NO_3^- \tag{32}$$

liberates an estimated 15.4 to 20.9 kcal $mole^{-1}$ of NO_3^-. Assuming a cell content of 50–55% carbon, 25–30% oxygen, and 12–13% nitrogen, an empirical formula for nitrifier cells is $C_5H_7O_2N$. Using Eqs. (31) and (32) the relative amounts of microbial mass are as follows:

$$13\,NH_4^+ + 15\,CO_2 \rightarrow 10\,NO_2^- + \underset{\text{Nitroso cells}}{3\,C_5H_7O_2N} + 23\,H^+ + 4\,H_2O \tag{33}$$

$$10\,NO_2^- + 5\,CO_2 + NH_4OH + H_2O \rightarrow 10\,NO_3^- + \underset{\text{Nitro cells}}{C_5H_7O_2N} \tag{34}$$

The highly stylized Eqs. (33) and (34) do not reflect the large amount of NH_4^+ or NO_2^- that must be turned over to obtain the amount of energy required for growth. *Nitrosomonas* and *Nitrobacter* turn over about 35 and 100 moles of N respectively for every mole of CO_2 fixed [4]. However, the equations illustrate that nitroso group bacteria produce more energy and biomass than do Nitro group bacteria per unit of N oxidized. Also of practical importance is the production of hydrogen ions during NH_4^+ oxidation.

Heterotrophic Nitrification

Various actinomycetes, bacteria, and fungi oxidize NH_4^+ or other reduced N forms to NO_2^- or N compounds of a higher oxidation state. There is little evidence that the energy liberated during the oxidation is coupled with biosynthetic processes. However, at least one group, the methane oxidizing bacteria, which can oxidize NH_4^+ to NO_2^-, have some morphological and biochemical characteristics that are strikingly similar to those of the autotrophic nitrifiers [77]. This similarity notwithstanding, heterotrophic nitrification appears to be of minor importance to N transformations in soils.

Important Ecological Considerations

Nitrification has three general effects on the use of N by microorganisms and/or higher plants. First, it affects the N nutrition of lower and higher plants because the rates of uptake and use of the substrate (NH_4^+) and end product of nitrification (NO_3^-) are different. Under some conditions, it can result in the accumulation of NO_2^- in amounts toxic to lower and higher plants. Second, it promotes movement of N in the soil solution because it converts a relatively immobile cationic form of N, NH_4^+, into a relatively mobile anionic form, NO_3^-. Third, nitrification results in the formation of oxidized forms of N that can be volatilized from soils or waters by microbial and abiological chemical action – NO_2^- and NO_3^- by enzymatic denitrification and NO_2^- by nonenzymatic chemodenitrification.

Numerous studies with ^{15}N demonstrate that heterotrophic microorganisms use NH_4^+ in preference to NO_3^- [50]. Higher plants, in which NO_3^- reduction is coupled to photosynthesis readily assimilate both N forms, although the efficiency of using either form can be species dependent. Heterotrophic microbial use of NH_4^+ in soils may result in increased N immobilization and decreased higher (crop) plant use of ammonium- or ammonium-forming fertilizers, especially with the concommitant addition of crop residues.

As seen from Eqs. (31) and (33), considerable acidity is produced during the oxidation of NH_4^+ to NO_2^-. It follows that any material that neutralizes the acidity produced at the reaction site favors the rate of NH_4^+ oxidation. Conversely, addition of acidulating materials reduces the rate. The effect of reaction site pH on nitrification rate is readily observed during the nitrification of different N fertilizers. Fertilizer granules consisting of NH_4^+ salts of strong acids (e.g., hydrochloric, nitric, and sulfuric acids) dissolve and form an acid fertilizer-soil reaction site (e.g., for ammonium sulfate, about pH 5.5). Alkaline-hydrolyzing fertilizers include anhydrous NH_3, urea (after enzymatic hydrolysis to ammonium carbonate), and diammonium phosphate. In acid soils the alkaline-hydrolyzing fertilizers nitrify much faster than acid-hydrolyzing ones. Differences in nitrification rate between the two groups of materials tend to disappear in alkaline soils [51].

Although the optimum pH for nitrification in soil is reported to be in the range of 7.5 to 9.0 [1], NO_2^- and NO_3^- formation observed at pH 13.0 [62] indicates that the nitrifiers are relatively insensitive to high concentrations of OH^- *per se*. The combination of high NH_4^+ concentration at high pH to form unionized NH_3 represses nitrification at pH >8.5. Because *Nitrosomonas* has a higher pH tolerance and is less sensitive to NH_3 toxicity than is *Nitrobacter* [2], NO_2^- can accumulate in any microsite where NH_4^+ is present at a pH conducive for *Nitrosomonas* but not *Nitrobacter* activity [51].

The practical significance of these effects in regard to NH_3 volatilization, NO_2^- accumulation, leaching, denitrification, and immobilization have been outlined [47, 48], but additional documentation would be of value.

Two other direct effects of nitrification involve (1) the possible formation of nitrosamines, and (2) N_2O production.

Aromatic and aliphatic secondary amines can react with nitrous acid anhydride (N_2O_3) to form nitrosamines:

$$2\,HNO_2 \rightleftarrows N_2O_3 + H_2O \tag{35}$$

$$R_2NH + N_2O_3 \rightarrow R_2N{-}N{=}O \tag{36}$$

Nitrosation reactions (Eq. 35) occur only at low pH (optimum pH about 3.5–4.0).

Small quantities of nitrosamines have been detected in the atmosphere but their mode of formation is unknown [64]. Trace quantities have been detected in soils amended with amines (e.g., diethylamine and triethylamine) and NO_2^-, but none in soils outside of the laboratory [80]. Their possible formation in soils and manures merits concern because many nitrosamines are carcinogenic and mutagenic at low concentrations. If formed in soils, they could be a health hazard only if they were taken up by edible plant parts or leached into potable water supplies. Their possible formation in foods high in NO_3^- content (subsequently forming NO_2^-) has been suggested [88] but has not been demonstrated in food chains.

Nitrosamine synthesis in field soils in amounts that might pose a health hazard is unlikely because

(1) amines usually are present only in low concentrations and are rapidly mineralized in aerobic soils; they are present in higher concentrations in waterlogged soils, but there NO_2^- concentrations are very low;

(2) nitrite accumulates only in alkaline soils or alkaline microsites in acid soils; and

(3) nitrosation reactions occur only in acidic media.

Thus, the soil appears not to be a medium in which reactants are produced together at concentrations and under conditions that are conducive to nitrosation.

The possible significance of N_2O formation during nitrification is included with the discussion of denitrification.

Denitrification

Denitrification is most commonly defined as the biochemical reduction of NO_2^- or NO_3^- to N_2 or gaseous N oxides. However, the assimilatory reduction of NO_3^- to NH_4^+ and nitrification also produce N oxides (N_2O and/or NO), so that a more precise definition is desirable to keep pace with current knowledge. From a biochemical viewpoint, denitrification is a bacterial process in which N oxides (in ionic and gaseous forms) serve as terminal electron acceptors for respiratory electron transport. Electrons are carried from an electron donating substrate (usually, but not exclusively, organic compounds) through several carrier systems to a more oxidized N form. The resultant free energy is conserved in ATP, following phosphorylation, and is used by the denitrifying organism to support respiration.

Current knowledge indicates the following pathway involving the transfer of five electrons:

$$NO_3^- \rightarrow NO_2^- \rightarrow NO \rightarrow N_2O \rightarrow N_2 . \tag{37}$$

Although there is considerable evidence for this reaction sequence, the role of NO as an obligatory intermediate remains uncertain. Also, recent data from N isotope mixing experiments [17] raise the question of whether N_2O is an obligatory precursor of N_2 for all denitrifiers.

Denitrifying ability has been demonstrated in 17 genera of bacteria. Although more abundant than nitrifiers, the denitrifiers are not as diverse as bacteria capable of reducing NO_3^- to NO_2^- or the dissimilatory reduction of NO_3^- to NH_4^+, of which 73 genera have been identified [42].

Most denitrifying bacteria are chemoheterotrophs. They obtain energy solely through chemical reactions and use organic compounds as electron donors and as a source of cellular carbon. Among those of greater versatility are *Rhodopseudomonas sphaeroides*, which is photosynthetic but does not use light energy when in the denitrifying mode; *Paracoccus denitrificans* and several *Alcaligenes* species, which use H_2 as an electron donor; *Thiobacillus denitrificans*, which uses reduced sulfur and can obtain carbon (as does *Pseudomonas denitrificans*) autotrophically through CO_2 fixation; and *Azospririllum brasiliense* and *Rhizobium japonicum*, which fix N_2 as well as denitrify, but probably not simultaneously [33].

Most denitrifying bacteria are aerobic organisms which grow anaerobically only in the presence of oxidized N. Aerobic respiration using O_2 as an electron acceptor or anaerobic respiration using N oxides for this purpose is accomplished by the denitrifier with the same series of electron transport systems. This facility to function both as an aerobe and as an anaerobe is of great practical importance because it enables denitrification to proceed at a significant rate soon after the onset of reducing conditions (a redox potential of about 330 mv) without change in microbial population. Because denitrification is carried out almost exclusively by facultative anaerobic heterotrophs that substitute oxidized N forms for O_2 as electron acceptors in respiratory processes, and because these processes follow aerobic biochemical routes, it can be misleading to refer to denitrification as an anaerobic process. It is rather one that takes place under anaerobic (i.e., anoxic) conditions.

Detailed descriptions of the biochemistry of denitrification are given elsewhere [23, 33]. But two items of detail merit mention here because they are relevant to methods of measuring the extent of denitrification. Denitrifying bacteria use NO_2^- and NO_3^- in two ways. The first is as an electron acceptor for dissimilatory reduction to N_2 and intermediate N oxides. The second use is to satisfy cellular N requirements. Where NH_4^+ supply is limiting, the denitrifiers produce NH_4^+ from NO_2^- and/or NO_3^- by assimilatory reduction. Therefore, the rate of NO_3^- disappearance from soil under anaerobiosis is not always an accurate indication of N loss via denitrification.

A second technically important point is that nitrous acid reductase (which catalyzes the reduction of N_2O to N_2) is inhibited by acetylene. The degree of inhibition is independent of N_2O concentration and the inhibition is reversible. Methodology for making field measurements of N loss via denitrification has been developed that is based on the assumption that the quantity of N_2O that is produced in the presence of acetylene is a valid measure of the total gaseous N that would be produced in the absence of acetylene. The technique has several limitations and problems associated with its use, but appears to provide reliable measures of denitrifying activity over short time periods.

Reaction Conditions

Organic carbon supply, redox potential, pH, and temperature markedly affect denitrification rate. Organic carbon oxidation during aerobic and anaerobic respiration is illustrated by Eqs. (38) and (39), respectively:

$$C_6H_{12}O_6+6\,O_2\rightarrow 6CO_2+6H_2O \tag{38}$$

$$C_6H_{12}O_6+4NO_3^-\rightarrow 6\,CO_2+6H_2O+2\,N_2. \tag{39}$$

The energy yield per mole of glucose for reactions *38* and *39* is 570 and 686 kcal, respectively [31]. The greater amount of free energy liberated when O_2 is used as an electron acceptor favors O_2 over NO_3^- use when both are available.

Gaseous N production during denitrification also can be depicted by:

$$4\,CH_2O+4\,NO_3\rightarrow 4HCO_3^-+2\,N_2O+2\,H_2O \tag{40}$$

$$5\,CH_2O+4\,NO_3^-\rightarrow H_2CO_3+4\,HCO_3^-+2\,N_2+2\,H_2O. \tag{41}$$

Based on the stochiometry shown in Eqs. (40) and (41), 1.17 and 0.93 units of N as N_2O and N_2, respectively, are produced for each unit of carbon consumed [18]. Equations (40) and (41) do not consider the carbon requirement for bacterial cell synthesis. Biomass production can be described by:

$$14\,CH_2O+3\,NO_3^-+32\,H_2\,CO_3\rightarrow 3\,C_5H_7O_2N+31\,HCO_3^-+20\,H_2O. \tag{42}$$

Albeit highly stylized, Eq. (42) illustrates the importance of oxidizable substrate supply. Denitrification activity in soils correlates well with the supply of water-soluble carbon compounds and mineralizable carbon; the extent of activity under anoxic conditions is determined largely by the supply of readily decomposable organic matter [18]. Equation (42) also illustrates that some alkalinity is produced during denitrification, brought about by a decrease in carbonic acid (H_2CO_3) and an increase in bicarbonate (HCO_3^-) at the reaction site.

Denitrifier activity in soil is little affected by pH in the range between 6 and 8 [18] but is depressed in acid soils [33]. Low rates of denitrification can occur under extremely acid conditions (e.g., peat soils, pH about 3.5); raising the pH of such soils to 6.5 greatly stimulates denitrification.

The ratio of N_2O to N_2 produced during denitrification in soils can vary widely and is influenced by pH, temperature, and the ratio of NH_4^+ to NO_3^- in the denitrifying system. For example, in a laboratory study with 17 soils varying widely in chemical and physical properties, the ratio N_2O-N : N_2-N for soils in the pH range 5.8–6.6 (1.4:1) was markedly higher than for soils in the pH range 6.7–7.8 (0.38:1) [18].

Under environmental conditions that favor N_2O formation relative to N_2 (low pH, low temperature, and marginal anaerobiosis), denitrification rates are relatively low. Therefore, relatively small amounts of N_2O probably are formed under these conditions. Growing plants also appear to affect $N_2O:N_2$ ratios, tending to increase the amount of N_2 relative to N_2O produced [63, 87].

Minimum temperatures for denitrification in soils are in the range 2.7–10 °C; the maximum temperature appears to be about 75 °C [33]. For 95 denitrifiers isolated from temperate soils (mean annual temperature of about 20 °C), 68% could grow at 4 °C, 10% at 41 °C, and 22% only at ambient (about 25 °C) temperatures. Of 33 isolates from tropical soils none could grow at 4 °C, 67% could grow at 41 °C, and 33% only at ambient temperatures [34]. The optimum temperature for total denitrifying activity in soils appears to be 65 °C [14, 69]. One explanation for this surprisingly high optimum temperature for soil denitrification is that the activity of thermophilic bacteria is markedly increased in the temperature range 50–67 °C, resulting in the rapid reduction of NO_3^- to NO_2^-, which interacts with soil organic constituents to form gaseous N. The soil chemical reactions resulting in NO_2^- decomposition occur rapidly in the temperature range of 50 to 70 °C [55].

Freshwater and Marine Environments

Denitrification occurs in freshwater and sea sediments and in marine waters with low dissolved O_2 and NH_4^+ concentrations. The results of several studies [16, 22, 28, 56] suggest that the same sequence of ammonification, nitrification, and denitrification that occurs in waterlogged soils occurs in shallow and stratified lake and ocean sediments. Nitrogen cycle processes in aquatic environments appear to be similar qualitatively to those in soils, the main difference between the two environments being the species composition of their micropopulations and macropopulations and the mobility of N and other nutrients dissolved in seawater [82]. About 5% of marine bacteria are able to form N_2 from NO_3^- [92]. Additional groups of bacteria may reduce NO_3^- only to NO_2^-. The main process by which NO_2^- is formed in seas is not always apparent because NH_4^+ can be oxidized to NO_2^- in a given water at one time and NO_3^- can be reduced to NO_2^- by other microorganisms at another time [45].

Direct measurements of the gaseous products of denitrification are scarce. Where made, the product usually is found to be N_2 [35, 36]. There is evidence that nitrification and not denitrification is the main source of N_2O in the oceans. For example, N_2O has been found in excess of the atmosphere/water equilibrium level in ocean waters with O_2 cc•ncentrations too high to allow denitrification [74], implying its formation during nitrification, as reported for soils [59, 75, 89]. Also, it is thought that since N_2O appears in ocean waters low in dissolved O_2 (e.g., $<0.3\%$), denitrification may consume rather than produce N_2O [24]. The several extensive measurements of dissolved N_2O in surface and deep oceanic waters showing varying degrees of supersaturation [24, 39, 40, 90] lead to the conclusion that the oceans are a net producer of N_2O, but whether by nitrification or denitrification is unknown.

Chemodenitrification

Gaseous N losses that cannot be attributed to NH_3 volatilization or denitrification have been reported in numerous publications dating back to 1871 and have been the subject of several reviews [5–7, 15, 43, 65, 86]. These articles present evidence that reactions involving NO_2^- in soils lead directly to gaseous N loss.

Nitrous acid (HNO_2) is formed from NO_2^- added to or produced in acid soils; it then spontaneously decomposes as follows:

$$NO_2^- + H^+ \rightarrow HNO_2 \quad (43)$$

$$2\,HNO_2 \rightarrow NO + NO_2 + H_2O\,. \quad (44)$$

If NO_2 escape from the soil is impeded and the reaction site is anaerobic, the following reactions may occur:

$$2\,NO_2 \rightleftarrows N_2O_4 \quad (45)$$

$$N_2O_4 + H_2O \rightleftarrows HNO_2 + HNO_3 \quad (46)$$

$$3\,HNO_2 \rightleftarrows 2\,NO + HNO_3 + H_2O\,. \quad (47)$$

If NO and NO_2 escape from an aerobic soil is impeded, reactions *48* and *49* may occur:

$$2\,NO + O_2 \rightarrow 2\,NO_2 \xrightarrow{2H^+} 2\,HNO_2 \quad (48)$$

$$2\,HNO_2 + O_2 \rightarrow 2\,HNO_3\,. \quad (49)$$

Thus, the amount and composition of the N gases that may be liberated from acid soil via HNO_2 decomposition depend on the nature and geometry of the reaction site. The reaction rate is greatly influenced by pH. Based on a dissociation constant of 6×10^{-4} for HNO_2, the ratio of $NO_2^-:HNO_2$ decreases from about 61.5 to 6.1 to 1.7 at pH 5, 4, and 3, respectively (i.e., at pH 3, NO_2^- produces >39 times more HNO_2 than at pH 5). Extremely acid microsites in soils should not be uncommon, since H^+ concentrations 100 times greater than in the bulk soil solution have been observed in water films surrounding clay particles [44]. The existence of acid microsites in neutral or slightly alkaline soils would explain the liberation of NO and NO_2 from such soils following NO_2^- addition [66].

The decomposition of HNO_2 can be catalyzed by Cu^+, Fe^{2+}, and Sn^{2+} [67]:

$$Cu^+ + HNO_2 + H^+ \rightarrow Cu^{2+} + NO + H_2O. \quad (50)$$

Since aerobic soils contain negligible amounts of metallic ions in reduced form, reaction *50* and similar reactions probably do not occur to any significant extent. However, in waterlogged soils, it is possible that Fe^{2+}, which may be present in significant amounts, may catalyze the decomposition of NO_2^-. Formation of N_2 and N_2O has been observed after addition of NO_2^- to anaerobic soil containing large amounts of Fe^{2+} [20].

Numerous investigators have studied the reactions:

$$NH_4^+ + NO_2^- \rightleftarrows H_4NO_2 \rightarrow N_2 + 2H_2O. \quad (51)$$

There is evidence for and against the occurrence of this reaction in unamended acid or alkaline soils or soils amended with NH_4^+ and NO_2^- (for reviews see references [5, 65]).

Large amounts of N_2O evolution have been observed following the addition of NH_2OH to soils:

$$NH_2OH + HNO_2 \rightarrow N_2O + 2H_2O. \quad (52)$$

However, N_2O evolution has been observed after NH_2OH addition to soils in the presence or absence of added NO_2^- and there is little justification for including reaction *52* as a chemodenitrification process. A widely studied reaction with HNO_2 is the Van Slyke reaction, which is the basis for quantitative determination of amino-N:

$$R\text{-}NH_2 + HNO_2 \rightarrow ROH + N_2 + H_2O. \tag{53}$$

Whether reaction *53* occurs to a significant extent in soils is still open to question. However, because Van Slyke-type reactions occur only at pH <5, additional studies with ^{15}N-labeled reactants will be necessary to determine the extent to which HNO_2 spontaneously decomposes before reacting with amino groups, if it does so react, under prevailing conditions.

There is convincing evidence for the formation of N_2 and N_2O during the reaction of NO_2^- at pH 5 with lignin and humic acids [12] or with phenolic constituents of soil organic matter [13]. The reaction mechanism appears to involve the formation of a p-nitrosophenol, which tautomerizes to a quinone monoxime, followed by reaction of the monoxime with HNO_2 to form N_2 and N_2O.

Chemodenitrification reactions have been studied almost exclusively in the laboratory, often with soils amended with the reactants under study. The reactions have not been studied in field soils. There appears to be a problem of geometry relative to NO_2^- accumulation and its subsequent reaction to form gaseous N. Nitrite accumulates only in alkaline soils or alkaline microsites in acid soils, as indicated earlier. However, NO_2^- reactions leading to gaseous N formation require acidic media, usually pH <5. Therefore, NO_2^- must move from an alkaline microsite to an acid one without being biologically oxidized to NO_3^-. Many measurements made in fertilizer-soil reaction zones indicate that NO_2^- forming near the center of the reaction site is converted to NO_3^- near the periphery of the site [49]. However, data are available which show that NO_2^- may diffuse beyond the boundary of the reaction site into acid portions of the soil and may undergo reactions resulting in N loss [51]. The practical significance of chemodenitrification has yet to be established.

Sources and Sinks

Comment

Several quantitative aspects of the precipitation-volatilization and N_2 fixation-denitrification subcycles of the global N cycle are poorly understood. This is not surprising when one considers that:

(1) both subcycles, although relatively independent in the atmosphere, have components which interact in soils and waters;

(2) reaction products often originate from point sources while sinks for products may comprise vast areas; and

(3) the residence time of components in the subcycles may differ by many orders of magnitude.

Therefore, even where accurate data are obtained by direct measurement of subcycle components, incorporating these data into a valid quantitative concept of the N cycle requires one to make questionable assumptions and extrapolations of point source data to global dimensions and extended time periods. The resultant quantitative concept is of uncertain validity.

Precipitation-Volatilization Cycle

Models of NH_3 fluxes estimate annual NH_3/NH_4^+ circulations between the earth and its atmosphere as high as 1 179 Tg [76]. However, estimates based on other assumptions are much lower and usually show depositions to the earth to be greater than emissions to the atmosphere [52]. For example, a recent comprehensive analysis of global NH_3 fluxes gives values for annual emissions of NH_3 from known sources to be in the range of 25–53 Tg and total depositions (wet and dry) in the range of 91–186 Tg [79]. The latter figures suggest that a large source of NH_3 has not been considered in recent NH_3 flux estimates. The great lack of agreement among NH_3 flux estimates that are based on different interpretations of available data suggests that models differ greatly in characterizing the sizes of sinks and sources. One source of NH_3 that has not been considered in global estimates of NH_3 fluxes is plants. Considerable NH_3 can be evolved from senescing vegetation and the leaf canopies of healthy plants [61, 85]. Considering the vast amount of leaf surface that is available to act both as a source when internal NH_3 concentrations are higher than ambient NH_3 concentrations, or conversely as a sink [61], the NH_3 transfers to and from the phyllosphere may prove to greatly alter current estimates of global NH_3 fluxes.

Dinitrogen Fixation-Denitrification Cycle

Investigators who estimate N fluxes between the biosphere and atmosphere agree that values for the amount of gaseous N evolved during denitrification are the most uncertain. Estimates of annual emissions from soils and waters range between 107 and 390 Tg [26, 31, 52, 79, 80]. Sometimes estimates of the amount of N fixed by biological action and industrial activity are considered when estimating total gaseous N emissions via denitrification in order to adjust mass balances. Values for the relative amounts of N_2 and N_2O produced are even more uncertain. There is no reliable method for determining the $N_2:N_2O$ ratio in open field soils with growing plants, because the major product of denitrification, N_2, mixes with an enormous reservoir of atmospheric N_2. Distinguishing between evolved N_2 and atmospheric N_2 can be done only if the evolved N_2 is isotopically different in composition. Estimates of the $N_2:N_2O$ ratio to date have been made in systems in which the atmosphere above the soil is confined during all or part of the experimental period. Attempts have been made to calculate average $N_2:N_2O$ ratios and then use them to estimate the global emission of N_2O from croplands [26], but such estimates must be considered highly tentative.

Estimates of N_2 plus N_2O evolution from oceans are made by assuming that marine waters are in a steady state with respect to N; i.e., the N added annually that is not immobilized in sediments is presumably denitrified. Measurements of N_2 evolution have not been made in open waters. However, measurements made in marine waters supersaturated with N_2O are used to calculate N_2O flux from these waters. A stagnant film diffusion model usually is used as the basis for calculation [41]. Assuming no wind or water turbulence, the N_2O flux is proportional to the difference between the N_2O concentration measured in the supersaturated water and the theoretical concentration of N_2O in water in equilibrium with the atmosphere, as calculated from the measured N_2O concentration in air above the waters sampled. The N_2O flux is calculated from these relationships, and the validity of the model is corroborated by comparing the N_2O flux with the measured flux of radon from the ocean. The accuracy of the N_2O flux estimate therefore is determined in part by the degree to which waters sampled for their N_2O contents represent large oceanic regions. Values of annual total N_2O flux from marine waters range from 0 to 100 Tg [26]. If denitrification in soils produces annually about 70–100 Tg of gaseous N, about 100 Tg of gaseous N would need to be evolved from oceans to approximately balance annual gains of fixed N by the biosphere. For the oceans, information is needed on the total amount of gaseous N evolved, the proportion of N_2 to N_2O evolved, and the process by which N_2O is formed, i.e., *via* nitrification or denitrification, or both. The same kind of information is needed for soils, with special emphasis on N_2O formation during nitrification. The practical significance of chemodenitrification as a means of producing N_2, N_2O, NO, or NO_2 needs clarification. The possibility that significant amounts of N_2O can be produced by biological processes other than those discussed here needs continued investigation. For example, N_2O has been found in human breath and it is formed by several nitrate-respiring bacteria and assimilatory nitrate-reducing bacteria, yeasts, and fungi [10].

The nitrification-denitrification sequence in soils has long been of interest to ecologists interested in understanding the N economies of aquatic and terrestrial systems, to agronomists striving to improve the efficiency of N use by crop plants, and to those who model global N cycle processes. Knowledge of how N_2O diffusing from the troposphere into the stratosphere participates in an important ozone regulatory process is of recent origin. Therefore, it is an understandable concern that increases in the level of fixed N in the biosphere will lead, through the processes of nitrification and denitrification, to increased levels of tropospheric N_2O, which, in turn, may decrease the level of stratospheric ozone. Because future food needs can be met only with increased use of industrially fixed N and increases in grain legume production, important questions that need answers are:

(1) Are soils and waters both sources and sinks for N_2O, and, if so, how do amounts emitted compare with amounts absorbed?

(2) Do N_2O sinks exist that have not been discovered?

(3) What proportion of N_2O formation is induced by human activities such as crop production and fossil fuel burning compared to N_2O formed from natural processes over which humans have little or no control? and

(4) Does an increase in level of fixed N in the biosphere result in a permanent increase in the amount of tropospheric N_2O?

Concluding Remark

Increasingly, multidisciplinary effort is being directed toward answering these questions, bringing together the thinking and methodology of atmospheric chemists, ecologists, microbiologists, oceanographers, soil scientists, and others. The development of models of global N transfers and mass balances permits investigators to predict changes in N inputs and outputs as a function of time by comparing two or more time periods, or by taking the outputs of one time period as the inputs to a succeeding period. In the absence of a full understanding of the quantitative aspects of the N cycle and its subcycles, models can provide an interim understanding that becomes progressively more complete as the information base grows.

References

1. Aleem, M.I.H.: Ann. Rev. Plant Physiol. *21*, 67 (1970)
2. Aleem, M.I.H., Engel. M.S., Alexander, M.: Soc. Amer. Bact., Bact. Proc. *57*, 9 (1957)
3. Aleem, M.I.H., Nason, A.: Proc. Natl. Acad. Sci. USA *46*, 763 (1960)
4. Alexander, M.: Nitrification, in: Soil Nitrogen (Bartholomew, W.V., Clark, F.E., ed.), Agronomy *10*, 307 (1965)
5. Allison, F.E.: Evaluation of incoming and outgoing processes that affect soil nitrogen, in: Soil Nitrogen (Bartholomew, W.V., Clark, F.E., ed.), Agronomy *10*, 573 (1965)
6. Allison, F.E.: Adv. Agron. *18*, 219 (1966)
7. Allison, F.E.: Soil Organic Matter and Its Role in Crop Production, p. 639, Elsevier, New York, 1973
8. Bates, D.R., Hays, P.B.: Planet. Space Sci. *15*, 189 (1967)
9. Blackmer, A.M., Bremner, J.M.: Geophys. Res. Lett. *3*, 739 (1976)
10. Bleakley, B.H., Tiedje, J.M.: Appl. Environ. Microbiol. *44*, 1342 (1982)
11. Bock, E.: Lithoautotrophic and chemoautotrophic growth of nitrifying bacteria, in: Microbiology 1978 (Schlessinger, D., ed.), p. 310, Amer. Soc. Microbiology, Washington, D.C., 1978
12. Bremner, J.M.: J. Agric. Sci. *48*, 352 (1957)
13. Bremner, J.M., Nelson, D.W.: Int. Cong. Soil Sci. Trans. 9th (Adelaide) *11*, 495 (1968)
14. Bremner, J.M., Shaw, K.: J. Agric. Sci. *51*, 39 (1958)
15. Broadbent, F.E., Clark, F.E.: Denitrification, in: Soil Nitrogen (Bartholomew, W.V., Clark, F.E., ed.), Agronomy *10*, 344 (1965)
16. Brujewicz, S.V., Zaitseva, E.D.: Chemistry of sediments of the northwestern Pacific Ocean, in: Problems of Sea Chemistry, p. 1; translated from Russian, Akademiya Nauk SSSR, Trudy Inst. Okeanol., Vol. *42* (1960). Indian Natl. Sci. Doc. Cent., New Delhi, 1972
17. Bryan, B.A.: Cell yield and energy characteristics of denitrification with *Pseudomonas stutzeri* and *Pseudomonas aeruginosa*. Ph. D. Thesis, p. 127, Univ. California, Davis. University Microfilms, Ann Arbor, MI 1980
18. Burford, J.R., Bremner, J.M.: Soil Biol. Biochem. *7*, 389 (1975)
19. Burns, R.C., Hardy, R.W.F.: Nitrogen Fixation in Bacteria and Higher Plants, p. 189, Springer-Verlag, Berlin–Heidelberg–New York, 1975
20. Chalamet, A., Bardin, R.: Soil Biol. Biochem. *9*, 281 (1977)
21. Chemical Kinetic Data Needs for Modeling the Lower Troposphere; Workshop sponsored by EPA and NBS at the Sheraton-Reston Intern. Conf. Center, Reston, VA., 1978 (Herron, J.T., Huie, R.E., Hodgeson, J.A., ed.), Rep. No. PB-299439, p. 111, National Techn. Inform. Service, Springfield, VA, 1979
22. Chen, R.L., Keeney, D.R., Graetz, D.A., Holding, A.J.: J. Environ. Qual. *1*, 158 (1972)
23. Christensen, M.H., Harremoes, P.: Biological Denitrification in Wastewater, Rep. 2–72. Departm. Sanitary Engng., Technical Univ. Denmark, Copenhagen 1972
24. Cohen, Y., Gordon, L.I.: Deep-Sea Res. *25*. 509 (1977)
25. Cothern, C.R.: A Preliminary Analysis of Nitrous Oxide (N_2O) Including a Materials Balance, Rep. No. EPA-560/6-79-001, p. 84, Environmental Protection Agency, Office of Toxic Substances, Washington, DC, 1979
26. Council for Agricultural Science and Technology: Effect of Increased Nitrogen Fixation on Stratospheric Ozone, Rep. No. 53, p. 33, CAST, Ames, IA, 1976

27. Crutzen, P.J.: Quart. J. Royal Meterol. Soc. *96*, 320 (1970)
28. Crutzen, P.J.: Ambio *3*, 201 (1974)
29. Crutzen, P.J.: Atmospheric chemical processes of the oxides of nitrogen, including nitrous oxide, in: Denitrification, Nitrification, and Atmospheric Nitrous Oxide, p. 17 (Delwiche, C.C., ed.), John Wiley and Sons, New York, 1980)
30. Crutzen, P.J., Ehhalt, D.H.: Ambio *6*, 112 (1977)
31. Delwiche, C.C.: Sci. Am. *223*, 137 (1970)
32. Eriksson, E.: Tellus *4*, 215 (1952)
33. Firestone, M.K.: Biological denitrification, in: Nitrogen in Agricultural Soils (Stevenson, F.J., ed.), Agronomy *22*, 289 (1982)
34. Gamble, T.N., Betlach, N.R., Teidje, J.M.: Appl. Environ. Microbiol. *33*, 926 (1977)
35. Goering, J.J., Dugdale, R.C.: Science *154*, 505 (1966)
36. Goering, J.J. et al.: Nitrogen fixation and denitrification in the ocean: biogeochemical budgets, in: Proc. Symp. Hydrogeochemistry and Biogeochemistry, Vol. II, p. 12 (Ingerson, E., ed.), The Clarke Co., Washington, D.C., 1973
37. Goody, R.M., Walshaw, C.D.: Quart. J. Royal Meterol. Soc. *79*, 496 (1953)
38. Graetz, D.A., Keeney, D.R., Aspiras, R.B.: Limnol. Oceanogr. *18*, 908 (1973)
39. Hahn, J.: Nitrous oxide in air and seawater over the Atlantic Ocean, in: The Changing Chemistry of Oceans (Dryssen, D., Jagner, D., ed.), p. 53, John Wiley and Sons, New York, 1974
40. Hahn, J.: "Meteor"-Forschungsergeb., A *16*, 1 (1975)
41. Hahn, J., Junge, C.: Z. Naturforsch. *32A*, 190 (1977)
42. Hall, J.B.: Nitrate-reducing bacteria, in: Microbiology 1978 (Schlessinger, D., ed.), p. 296. Amer. Soc. Microbiology, Washington, D.C., 1978
43. Harmsen, G.W., Kolenbrander, G.J.: Soil inorganic nitrogen, in: Soil Nitrogen (Bartholomew, W.V., Clark, F.E., ed.), Agronomy *10*, 43 (1965)
44. Harter, R.D., Ahlrichs, J.L.: Soil Sci. Soc. Am. Proc. *31*, 30 (1967)
45. Hattori, A., Wada, E.: Biogeochemical cycling of inorganic nitrogen in marine environments with special reference to nitrite metabolism, in: Proc. Symp. Hydrogeochem. Biogeochem., Vol. II (Ingerson, E., ed.), p. 28, The Clarke Co., Washington, D.C., 1973
46. Hauck, R.D.: Mode of action of nitrification inhibitors, in: Nitrification Inhibitors-Potential and Limitations, p. 19, Amer. Soc. Agronomy, Madison, WI, 1980
47. Hauck, R.D.: Nitrogen fertilizer effects on nitrogen cycle processes, in: Terrestrial Nitrogen Cycles (Clark, F.E., Rosswall, T., ed.), Ecol. Bull. (Stockholm) *33*, 551 (1981)
48. Hauck, R.D.: Agronomic and technological approaches to minimizing gaseous nitrogen losses from croplands, in: Gaseous Nitrogen Emissions into the Global Atmosphere (Freney, J.R., Simpson, J.R., ed.), Martinus Nijhoff, The Hague, 1983
49. Hauck, R.D.: The significance of nitrogen fertilizer microsite reactions in soil, in: Nitrogen in Crop Production (Hauck, R.D., ed.), Amer. Soc. Agronomy, Madison, WI, (in press)
50. Hauck, R.D., Bystrom, M.: ^{15}N--A Selected Bibliography for Agricultural Scientists, p. 206, Iowa State University Press, Ames, IA, 1970
51. Hauck, R.D., Stephenson, H.F.: J. Agric. Food Chem. *13*, 486 (1965)
52. Hauck, R.D., Tanji, K.K.: Nitrogen transfers and mass balances, in: Nitrogen in Agricultural Soils (Stevenson, F.J., ed.), Agronomy *22*, 891 (1982)
53. Hofman, T., Lees, H.: Biochem. J. *54*, 579 (1953)
54. Johnston, H.: Proc. Natl. Acad. Sci. USA *69*, 2369 (1972)
55. Keeney, D.R., Fillery, I.R., Marx, G.P.: Soil Sci. Soc. Amer. J. *43*, 1124 (1979)
56. Kemp, A.L.W., Mudrochova, A.: Limnol. Oceanogr. *17*, 855 (1972)
57. Lees, H.: Ann. Rev. Microbiol. *14*, 83 (1960)
58. Lees, H.: Bacteriol. Rev. *26*, 165 (1962)
59. Lees, H.: Inhibitors of nitrification, in: Metabolic Inhibitors, Vol. 2 (Quastel, J.H., ed.), p. 615. Academic Press, New York, 1963
60. Lees, H., Simpson, J.R.: Biochem. J. *65*, 297 (1957)
61. Lemon, E., Van Houtte, R.: Agron. J. *72*, 876 (1980)
62. Meek, C.S., Lipman, C.B.: J. Gen. Physiol. *5*, 195 (1922)
63. Michoustine, E.N., et al.: Ann. Inst. Pasteur. Suppl. No. 3, 235 (1965)
64. Nat. Res. Council: Nitrates: An Environmental Assessment, p. 723, U.S. Nat. Acad. Sci., Washington, D.C., 1978

65. Nelson, D.W.: Gaseous losses of nitrogen other than through denitrification, in: Nitrogen in Agricultural Soils (Stevenson, F.J., ed.), Agronomy *22*, 327 (1982)
66. Nelson, D.W., Bremner, J.M.: Soil Biol. Biochem. *2*, 1 (1970)
67. Nelson, D.W., Bremner, J.M.: ibid. *2*, 203 (1970)
68. Nicholas, D.J.D.: Intermediary metabolism of nitrifying bacteria, with particular reference to nitrogen, carbon, and sulfur compounds, in: Microbiology 1978 (Schlessinger, D., ed.), p. 305, Amer. Soc. Microbiology, Washington, D.C., 1978
69. Nömmik, H.: Acta Agric. Scand. *6*, 195 (1956)
70. Painter, H.A.: Water Res. *4*, 393 (1970)
71. Peck, H.D., Jr.: Ann. Rev. Microbiol. *22*, 489 (1968)
72. Pierotti, D., Rasmussen, R.A.: J. Geophys. Res. *82*, 5823 (1977)
73. Quayle, J.R., Ferenci, T.: Microbiol. Rev. *42*, 251 (1978)
74. Richards, F.A.: La Mer *9*, 68 (1971)
75. Ritchie, G.A.F., Nicholas, D.J.D.: Biochem. J. *126*, 1181 (1972)
76. Robinson, E., Robbins, R.C.: J. Air Pollut. Control Assoc. *20*, 303 (1970)
77. Schmidt, E.L.: Nitrification in soil, in: Nitrogen in Agricultural Soils (Stevenson, F.J., ed.), Agronomy *22*, 253 (1982)
78. Smith, A.J., Hoare, D.S.: J. Bacteriol. *95*, 844 (1968)
79. Söderlund, R., Svensson, B.H.: The global nitrogen cycle, in: Nitrogen, Phosphorus and Sulfur – Global Cycles (Svensson, B.H., Söderlund, R., ed.), Ecol. Bull. (Stockholm) *22*, 23 (1976)
80. Stevenson, F.J.: Origin and distribution of nitrogen in soils, in: Nitrogen in Agricultural Soils (Stevenson, F.J., ed.), Agronomy *22*, 1 (1982)
81. Suzuki, I., Dular, U., Kwok, S.C.: J. Bacteriol. *120*, 556 (1974)
82. Vaccaro, R.F.: Inorganic nitrogen in sea water, in: Chemical Oceanography, Vol. I (Riley, J.P., Skirrowa, G., ed.), p. 365, Academic Press, New York, 1965
83. Wallace, W., Nicholas, D.J.D.: Biol. Rev. *44*, 359 (1969)
84. Warneck, P.: J. Geophys. Res. *77*, 6589 (1972)
85. Wetselaar, R., Farquhar, G.D.: Adv. Agron. *33*. 263 (1980)
86. Wilson, J.K.: Nitrous Acid and the Loss of Nitrogen, N.Y. (Cornell), Agr. Expt. Sta., Mem. No. 253, p. 25, 1943
87. Woldendorp, J.W.: Plant Soil *17*, 267 (1962)
88. Wolff, I.A., Wasserman, A.E.: Science *177*, 15 (1972)
89. Yoshida, T., Alexander, M.: Soil Sci. Soc. Amer. Proc. *34*, 880 (1970)
90. Yoshinari, T.: Mar. Chem. *4*, 189 (1976)
91. Zipf, E.C.: Nature *287*, 523 (1980)
92. ZoBell, C.E.: Marine Microbiology, p. 240, Chronica Botanica, Waltham, MA, 1946

Carbon Dioxide: A Biogeochemical Portrait

E. T. Degens, S. Kempe, A. Spitzy

Universität Hamburg
SCOPE/UNEP International Carbon Center
Bundesstraße 55
D-2000 Hamburg 13
Federal Republic of Germany

Introduction

Carbon dioxide, chemical formula CO_2, is a colorless and tasteless gas about 1.5 times as heavy as air. Its specific volume at atmospheric pressure (101,325 N m^{-2}) and room temperature (21 °C) is 0.546 m^3 kg^{-1}. Under natural atmospheric conditions it is a stable, inert, and nontoxic gas. When subjected to higher pressure and temperature, the gas can be liquefied and solidified. For instance at 21 °C and 838 psig (= pounds per square inch gage), CO_2 becomes a liquid. Further cooling will result in the formation of dry ice snow, which at atmospheric pressure has a sublimation temperature of −78 °C.

Carbon dioxide is found in all major compartments of our planet: air, water, life, and earth. It will be transferred from one compartment to another and from one generation to the next. During this transit through space and time, CO_2 will find itself moving through a variety of media and be subjected to different temperatures and pressures. It will chemically interact in the various bio- and geosystems and leave its imprint in a rather profound manner.

In this review we will draw a biogeochemical portrait of this intriguing molecule from its primordial point of origin to its various transitory resting places in atmosphere, hydrosphere, lithosphere, and biosphere.

History of CO_2

Cosmic Molecular Clouds

Comparison between stars has revealed that certain variations in element abundance do exist which are related to stellar age and position of a star in a galaxy. Yet, all in all it appears that the chemical composition of the Universe is remarkably uniform.

Table 1. Relative abundances of the more common elements in the solar atmosphere [1]

Element	Atomic number	Relative abundances (atoms per 10^6 atoms of Si)
H	1	3.18×10^{10}
He	2	2.21×10^9
C	6	1.18×10^7
N	7	3.74×10^6
O	8	2.15×10^7
Ne	10	3.44×10^6
Mg	12	1.06×10^6
Si	14	1.00×10^6
S	16	5.00×10^5
Ar	18	1.17×10^5
Fe	26	8.30×10^5

The relative abundance of the ten most common elements in the solar atmosphere is listed in Table 1 [1]. From this table we can deduce that (not considering helium because it is chemically inert) universal and living matter are principally composed of the same chemical elements:

hydrogen – oxygen – carbon – nitrogen

which in order of abundance are

H–O–C–N for universal matter vs. C–O–H–N for living matter.

The two other prominent biochemical elements, i.e. sulfur and phosphorus, rank also high in the cosmic abundance list [2]. In contrast, the Earth has an entirely different bulk composition:

iron – oxygen – silicon – magnesium.

It thus appears that organic matter is a more representative sample of the Universe than the bulk of our iron- and silica-dominated Earth.

No place in the Universe is empty. Even in the vast emptiness of the interstellar space there is still one lonely atom per cubic centimeter. For comparison, our atmosphere contains quadrillons of atoms per cm^3. Yet, interstellar matter is not distributed uniformly. There are high-density regions referred to as clouds and low-density regions known as intercloud gas.

Research on giant interstellar clouds was initiated in 1963 by a team of the Massachusetts Institute of Technology and the Lincoln Laboratory. The most important finding was the observation that such clouds were composed of molecules rather than atoms. Over the past 20 years more than 50 compounds have been identified [3, 4]. Adding to this number all isotopic species close to a hundred different types of molecules are presently known. Isotopic ratios are important indicators of interstellar chemistry; a certain isotopic version can locally be overabundant by orders of magnitude. Among the molecules discovered by radio telescopes and astrospectrographs are hydrogen, water, carbon monoxide, methane, formaldehyde, various alcohols, hydrogen cyanide, ammonia, and the hydroxyl radical. Tentative identification exists for the presence of protonated carbon dioxide [3].

The dark interstellar clouds are composed not only of gases but of solid dust grains in a mass ratio of about 100 to 1 [5]. The most dominant molecular species is hydrogen (99% of total), in a concentration of about 10^4 H_2 per cm^3. The bulk of the dust grains measuring less than 0.5 μ in diameter are in all probability crystalline carbon compounds (e.g. graphite), Fe-Mg silicates (e.g. olivine) and native iron. It is interesting to note that the grains might be covered by solid water, ammonia and methane, since these molecules freeze at space temperatures. The presence of solid particles has numerous consequences for the protection of organic molecules in the space environment. For example, the probability of collisions between atoms and molecules is increased, and three-body reactions become feasible. Furthermore, mineral surfaces may not only provide a convenient "resting place" for certain molecules, but by virtue of their crystalline order catalysis and epitaxis may proceed [6].

Under the condition of identical element distribution in the dark molecular clouds and in universal matter, the observational data suggest that in the space regions accessible for studies by radiotelescope all carbon is molecularly bound (that is, no atomic carbon remains). For oxygen, 30% can be accounted for, whereas only 1 ppm of all nitrogen theoretically present has entered a complex organic molecule [5].

In conclusion, the chemistry of interstellar clouds favors the formation of carbon-containing molecules. It is noteworthy that in the presence of a mineral catalyst, formaldehyde will readily yield complex biochemical molecules [7]. The same can be said for hydrogen cyanide [8] which is known to generate compounds such as amino acids, purines and pyrimidines. It is suggested that by such processes not only simple monomers are generated but a wide array of physiologically interesting macromolecules. In the line of this thought, carbon dioxide could be obtained via decarboxylation of, for instance, amino acids or fatty acids.

Meteorites

Molecular cloud complexes are considered to be the birthplace of stars and planets. Using carbon monoxide as fingerprint for aereal extension of such complexes, a typical cloud measures 45 parsecs in diameter (a parsec is 3.26 light years). One light year is equivalent to 9.461×10^{12} km; for comparison the distance from Sun to Pluto measures about 6×10^9 km. In turn, an enormous mass of pristine cosmic material is available, when it comes to the formation of solar systems such as ours [3, 9, 10].

On transit to Earth, carbon must have principally been present in a reduced state, due to the large excess of hydrogen molecules [11]. Consequently, carbon dioxide and carbonates should be late inventions in the evolution of matter. Let us now examine at what point CO_2 enters the carbon scene.

The first solid evidence for the type of mineral phases generated in our solar system is contained in meteorites which actually predate even the most ancient rock formation known from Earth. The age of meteorites is about 4.5 Ga (giga anni = 10^9 years). The wide range in mineralogical and chemical composition serves as a means to outline a sequence of events involved in the transformation of cosmic dust and gas into the present population of planetary bodies [12–14]. In principle, two schools of thought exist. According to the first one [15] planetary accretion is believed to have started from an intimate mixture of silicate and metal particles of the type found in chondritic meteorites. This chondritic material supposedly had formed prior to the formation of planets in the solar nebula. With respect to the proto-Earth, this mixture accreted over an extended period of time (10^7 to 10^8 years) and allowed gravitational potential energy to radiate away. The material, composed of various metal and silicate phases, remained undifferentiated and relatively "cool" (ca. 1,000 °C) during this stage. Heat released by the decay of long-lived radiogenic elements subsequently caused melting first of metals and then of silicates. In due course, a differentiated Earth developed. Some initial inconsistencies of the homogeneous accretion model were "ironed-out" in a revised version [16].

The second concept operates with a heterogeneous accretion. A chemically inhomogeneous solar nebula is assumed displaying a wide range in oxidation states and temperature progressions. Accordingly, planetesimals of various size, chemistry, and mineralogy were generated in different locations and at different times. Aggregation of such objects resulted in the present population of bodies in the solar system including the parent bodies of meteorites [17]. The thrust of evidence is in favor of the heterogeneous accretion model. This brief account should suffice and we will turn now again to carbon and the CO_2 question.

The forms of carbon present in different types of meteorites include: diamond [18], graphite [19], carbynes [20], carbide [21], extractable organics [22], polymeric organics [23], and carbonate [24].

The chemical state of carbon in a developing solar nebula at different pressure and temperature conditions has been outlined on the basis of direct measurements on various carbon species in meteorites [20]. The existence of carbynes earlier proposed on theoretical grounds [25] could be substantiated. It appears that carbynes not only form by condensation of carbon vapor above 2,600 K or by explosive shock of 600 kilobars but that in the temperature regime between 300° to 400 °C and in the presence of a suitable mineral catalyst it can form metastably by disproportionation of carbon monoxide [26]:

$$2\,CO \rightarrow CO_2 + C_{(solid)}. \qquad (1)$$

It is conceivable that carbynes formed in this manner are also responsible for the C_2 and C_3 band observed in comet spectra, whose parent molecules have thus far remained unidentified [27].

In conclusion, a developing solar nebula offers a series of possibilities for the generation of carbon dioxide. The presence of carboxylated organic matter and carbonates is direct proof for an early snythesis of CO_2. Furthermore, the wide abundance of carbynes in a number of meteorites suggests a route along which carbon dioxide could have been synthesized in the interstellar medium and in an evolving solar nebula.

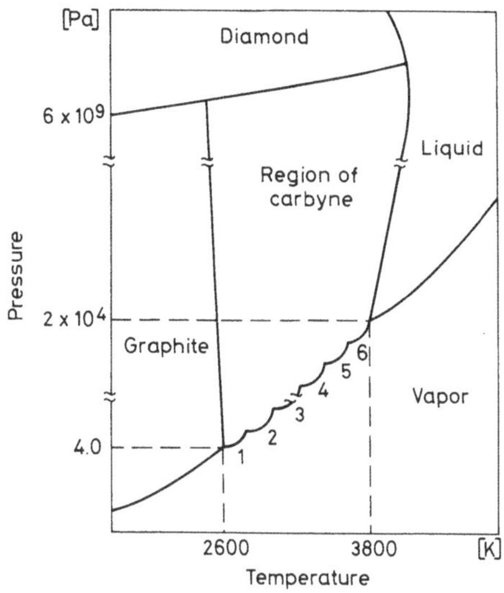

Fig. 1. Carbon phase diagram [28]

Solid Earth

Upon arrival on Earth, the reduced carbon pool became part of the differentiation process leading to: core, mantle, crust, hydrosphere, atmosphere, and biosphere. As a result, carbon distributed itself among the various compartments. New molecular carbon species developed in accordance with changing mineral equilibria, while other species survived molecularly intact the harsh treatment of the evolving Earth. Still other molecules known form outer space such as C_4H, HC_7N, and HC_9N have never been detected on the Earth.

Starting from a source of C_n of unspecified structure, the carbon phase diagram (Fig. 1) illustrates the suggested stability fields for the crystalline carbon structures graphite, carbynes, and diamond [28]. Carbon bonds itself in three different ways. By the sharing of a pair of electrons, single carbon–carbon (C–C) bonds develop. By the sharing of two pairs of electrons, double bonds are generated (C=C), and finally in triple bonds (C≡C) three pairs of electrons are shared. In diamond, each carbon makes a single bond with each of four neighbours in a three-dimensional array resulting in a crystal lattice schematically shown in Figure 2. In graphite, the carbon atoms lie in planes (Fig. 3) that consist of flat, six-carbon rings in which single and double bonds alternate. It is of note that between planes the opposing carbons do not share electrons and accordingly graphite has a weak crystal lattice and furthermore is soft. Carbyne is the name given to a series of crystalline carbon forms which are characterized by triple bonds.

At temperatures above 3,800 K and elevated pressures, carbon readily dissolves in melts of silicates and iron. In view of the lower melting point of iron, relative to silicates, carbon can be extracted from mantle material and carried via iron droplets to the central core. At temperatures exceeding the graphite stability, a single C–C bond can break and shift an electron into each of the adjacent double

Fig. 2. Point blank view down the axis of a channel in a diamond-like crystal lattice [29]. It is of note that liquid water is build according to the same structural principles [6], i.e. open circles represent the oxygen atoms and hydrogen atoms (for graphical reasons not shown) are located along the connecting lines

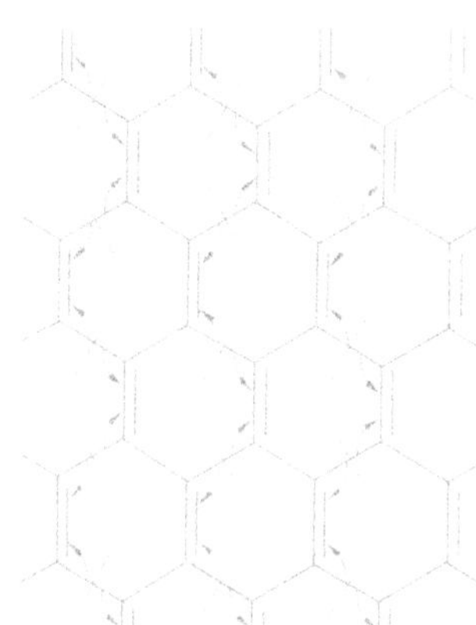

Fig. 3. Transformation of graphite into carbyne chains $(-C\equiv C-)_n$ [28]

bonds resulting in a kind of "chain reaction" until the graphite basal plane sheet is completely transformed into $(-C\equiv C-)_n$ chains (Fig. 3) [28].

Within the graphite stability field, graphite may also react with hydrogen gas:

$$C + 2H_2 \rightarrow CH_4 \tag{2}$$

or with water vapor:

$$C + H_2O \rightarrow CO + H_2. \tag{3}$$

The subsequent reaction of carbon monoxide with water vapor:

$$CO + H_2O \rightarrow CO_2 + H_2 \tag{4}$$

will yield CO_2. The data suggest that the formation of carbon dioxide and methane is thermodynamically favored.

Carbon dioxide can also be generated along a different route, i.e. from complex organic matter, which will be abbreviated CH_2. At depth a series of reactions may take place, for instance:

$$CH_2 \rightarrow C_{(diamond)} + H_2 \tag{5}$$

or

$$CH_2 + 3Fe_3O_4 \rightarrow CO_2 + H_2O + 9FeO. \tag{6}$$

An alternate pathway has recently been suggested [30] according to which atomic carbon present in solid solution in minerals such as olivine could give rise to CO_2 as well as to a series of hydrocarbon molecules. In principle, heating will cause atomic carbon to react with lattice oxygen yielding CO_2 whereas participation of co-dissolved hydrogen should produce hydrocarbons. In the light of this work the genesis of diamonds has been reconsidered (E. Galimov, pers. comm.) involving "collapsing" methane bubbles rather than native carbon as a source for isotopically light diamonds in a number of crystalline rocks.

In summary, carbon is an element which will readily combine with both hydrogen and oxygen; equilibrium should exist between these molecular bonding states:

$$(C\text{–}C):(C\text{–}OH):(C=O):(CO_2):(H_2O):(H_2).$$

It has been suggested [31] that the presence of CO_2 in an atmosphere such as on Venus implies that at least an equal number of moles of water must have been produced. However, since CO_2 production may start from C_n, CO, and CH_2 compounds and may even consume water, this statement is not valid.

Carbon dioxide and water are critical elements in the melting processes of mantle and crust. Crustal thickness or thickness of the lithosphere is not only a function of the geothermal regime, but it is also dependent on the tectonic processes taking place. Results in experimental petrology show that while CO_2 intensifies the polymerisation of a silicate melt, the presence of H_2O promotes its depolymerisation. The resulting lower viscosity of a melt containing traces of water relative to one that is water-free is reflected in its lower seismic velocities. In contrast, for a silicate melt containing CO_2 at the same temperature and pressure, higher seismic velocities can be expected [32].

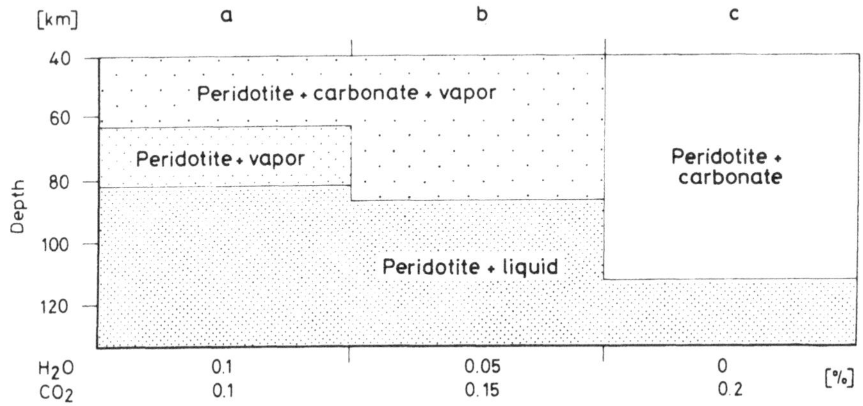

Fig. 4. Mantle cross-sections for the system: $CaO–MgO–SiO_2$ plus small quantities of H_2O and CO_2. Mantle cross-section with mixed volatiles (case a and b) retain excess vapor until melting begins, whereas if no H_2O is present (case c), all CO_2 reacts to produce calcic dolomite in the peridotite [32]

Analogously, in a peridotitic system such as the upper mantle, the process of partial melting is controlled not only by temperature and pressure, but also by the presence of trace constituents. Such a system may be schematically represented by $CaO–MgO–SiO_2$, to which traces of H_2O and CO_2 are introduced (Fig. 4). In the presence of $H_2O–CO_2$ fluids, the mantle peridotite with carbonate acts as a buffer, controlling the ratio of CO_2/H_2O as a function of temperature and pressure [33]. In the absence of H_2O, CO_2 reacts and carbonate is produced. A polymerisation of the magma along the lines of Figure 4c therefore results. In the presence of H_2O and CO_2, vapor is retained in the system until melting begins (Fig. 4a and b), but fractionation occurs depending on the molar ratio of the two volatiles. This results in a separation and stratification of H_2O and CO_2. H_2O-rich vapor migrates upwards and may even escape into the atmosphere, while CO_2-rich vapor is retained or fixed as carbonate in the mantle and lower crust.

It follows from the arguments given above that the upward or downward displacement of the liquidus/solidus of a peridotitic melt does not necessarily imply an alteration in the geothermal regime but may simply be the result of the addition of CO_2, H_2O or other trace constituents into the melt. When H_2O in the form of interstitial water, hydrous minerals or hydrothermal fluids along fractures is introduced, melting may be initiated and a batholith can rise, leading to the familiar granitic intrusions. H_2O and other trace constituents may subsequently be removed from the system along pneumatolithic or hydrothermal channels. Synorogenic and postorogenic events could represent such a development. In contrast, if carbonate is the principal additional material involved in a remelting process, the probability of granite formation would be greatly reduced [34].

Juvenile CO_2

Carbon dioxide has been released from the interior of the Earth practically from start. Most of it has come from the upper mantle and lower parts of the crust. Mechanisms leading to its release could be catastrophic (e.g. meteoritic impact), episodic (e.g. rifting) or continuous (e.g. slow decay of organic matter). The rate of discharge at any one time might vary considerably. In all probability, the CO_2 streaming-off from the crust and mantle is tuned to the pulses of rock-forming and tectonic events that shaped the Earth [35].

As will be shown later (see p. 176) the weathering of silicate minerals furnishes an effective way to remove gaseous CO_2 into solution as bicarbonate or carbonate ions. Continental weathering today is capable to sequester one volume of atmospheric CO_2 within 7,000 years. The products are dissolved Ca, Mg, K, and Na carbonates and dissolved silica. Closed basins within volcanic areas where volcanic CO_2 and unaltered igneous silicates are available develop terminal lakes of a characteristic soda chemistry. Na, K, and carbonates collect in dissolved form, while Ca and Mg are precipitated as calcite, aragonite or even dolomite. Typical soda lakes are e.g. Lake Van in eastern Anatolia, the majority of the East African rift lakes and many lakes in the volcanic belt of the western Rocky Mountains.

By analogy with these lakes one has to conclude that the primordial ocean had a chemistry dominated by sodium and carbonate [378]. The crust, oceans, biosphere, and atmosphere together contain some 65.5×10^{21} g carbon (see Table 2). The present ocean has a mass of $1{,}350 \times 10^{21}$ g. If all carbon would have been released at once during early degassing of the crust, the concentration of carbon in the ocean would have been 48.5 gC kg^{-1} H_2O. This is in fact close to the solubility of the system $NaHCO_3$–Na_2CO_3–H_2O at 20 °C (32.7 gC kg^{-1}). Thus the early oceans must have been quite capable to keep any amount of sodium carbonates in solution which originated from the reaction of water, carbon dioxide and silicates. Such an ocean must have had a high pH-value (between 9.5 and 10.5) and should have lead to a low atmospheric PCO_2 very early in history in accordance with chemical and kinetical equilibria established in the sea.

The hypothesis of an "early soda ocean" has interesting geological ramifications and may explain several geochemical problems associated with the Precambrian environment:

1. Precambrian carbonate rocks are dominated by dolomite. Dolomite can not form from present sea water. The high pH, the large Mg/Ca ratio and the high sulfate content of a soda ocean would, however, favor dolomite formation.
2. Precambrian rocks contain massive chert formations. These precipitated most likely from an alkaline environment by small pH variations or upon evaporation due to high solubility of SiO_2 at high pH.
3. Early rocks display a predominance of illites and chlorites over kaolinite and montmorillonites.
4. Development of life could have been forstered by the soda chemistry: Much higher inorganic carbon and phosphate (due to the low Ca-content) concentrations prevailed. Humic substances would stay dissolved and Na_2O, which would be formed in the eruption clouds of planetesimal impacts, is a strong catalyst for organic reactions. Soda lakes of today flourish with endemic life and are by no means hostile environments.

The present sodium chloride ocean would have evolved over three billion years by slow addition of chlorine degassing from the mantle (even today the ratio of total chlorine to carbon in crust and oceans would favor soda chemistry). This is in accordance with the late appearance of halite as a major evaporite. Carbon and

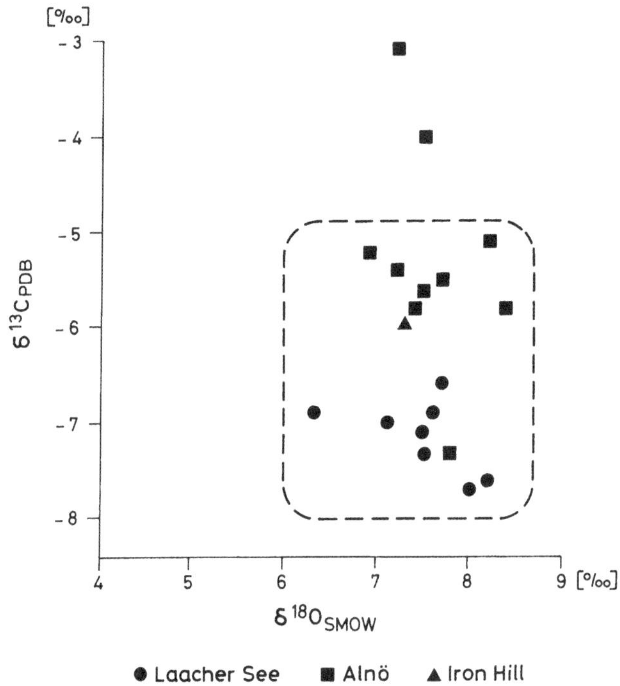

Fig. 5. Probable field of primary igneous carbonatites [37]

sodium on the other hand by being subducted with pore waters and sodium-rich oceanic crust became selectively removed from the sea and in large parts retained in the continental crust.

One way to trace the flow pattern of juvenile carbon from the mantle is by means of stable isotopes. Figure 5 depicts the ^{13}C and ^{18}O contents of Late Pleistocene carbonatites which were explosively transported upwards from depth through the entire continental crust. Point of origin is the upper mantle in the vicinity of the Moho and mafic or ultramafic material is their starting material. The vents or pipes that formerly were generated by carbonatite projectiles presently serve as escape routes of carbon dioxide gases released from the mantle or lower crust in the aftermath of the explosive carbonatite event. The CO_2 is about 2 per mil richer in ^{13}C than the cogenetic carbonate. This is the expected fractionation between CO_2 and calcite in the temperature range 500 to 700 °C. At temperatures prevailing in the upper mantle carbon isotope fractionation between graphite, carbonate and diamond are insignificant and the magmatic carbon pool should have a mean $\delta\ ^{13}C$ close to -6.5 per mil. However, the CO_2 released at depth is slightly enriched in ^{13}C. Thus juvenile CO_2 should have a $\delta\ ^{13}C$ value close to -5 per mil, whereas juvenile carbon has one of about -6.5 per mil. Terrestrial magmatic carbon has a mean $\delta\ ^{13}C$ that corresponds with carbonaceous chondrites (Type 1 or C1 chondrite). This implies that the bulk of the carbon present in our lithosphere has C1 chondritic material as its most likely source [36] (Fig. 6).

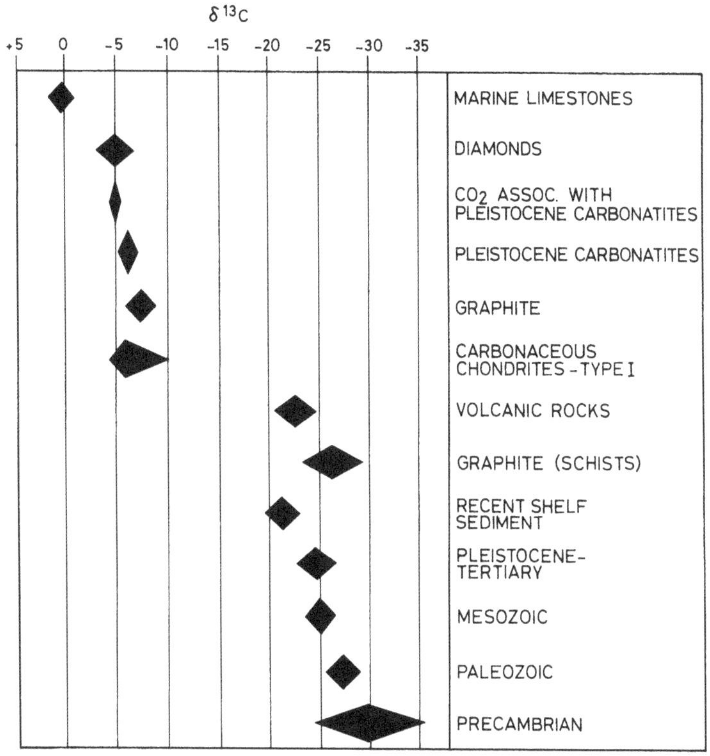

Fig. 6. $\delta\ ^{13}C$ in various geological materials. Diamond-shaped figures represent 2 σ ranges; the center of each diamond refers to the mean $\delta\ ^{13}C$ value. Marine limestones (biologic origin) have a uniform $\delta\ ^{13}C$ independent of geological age. Reduced and oxidized magmatic carbon is similar in ^{13}C content and has a mean $\delta\ ^{13}C$ of about -6 per mil. In contrast, $\delta\ ^{13}C$ in C_{org} of basalts, metamorphic schists and sediments exhibits a wider range with a weighted mean of -25 per mil. Of note is the apparent age effect for organic carbon in sediments. This is principally related to differences in carbon isotope fractionation characteristics of organisms. For instance, marine chemoautotrophic bacteria produce a $\delta\ ^{13}C$ in their organic matter which is depleted by 7 per mil relative to the $\delta\ ^{13}C$ of phytoplankton grown under the same circumstances. The data bank used for the construction of this figure is based on more than 5,000 analyses [38–40]

The partitioning of carbon in the course of its subsequent biochemical cycling can be documented and summarized by means of $\delta\ ^{13}C$ ranges for various carbon-bearing materials (Fig. 7). The two most outstanding endmembers of carbon isotopic fractionation are CO_2 and CH_4, the first one being generally depleted and the second one enriched in ^{12}C. Photosynthesis shifts $\delta\ ^{13}C$ towards more negative values by about 15 to 25 per mil relative to its CO_2 precursor. In contrast, chemical carbonate precipitation moves $\delta\ ^{13}C$ into the more positive range by 4 per mil relative to the starting bicarbonate ion. However, calcification in biological systems (see p. 190) involves no isotope fractionation between solid carbonate and bicarbonate ion. In case methane becomes oxidized to CO_2 or a contribution of biologically generated CO_2 is indicated, carbonates derived from such sources will be isotopically light.

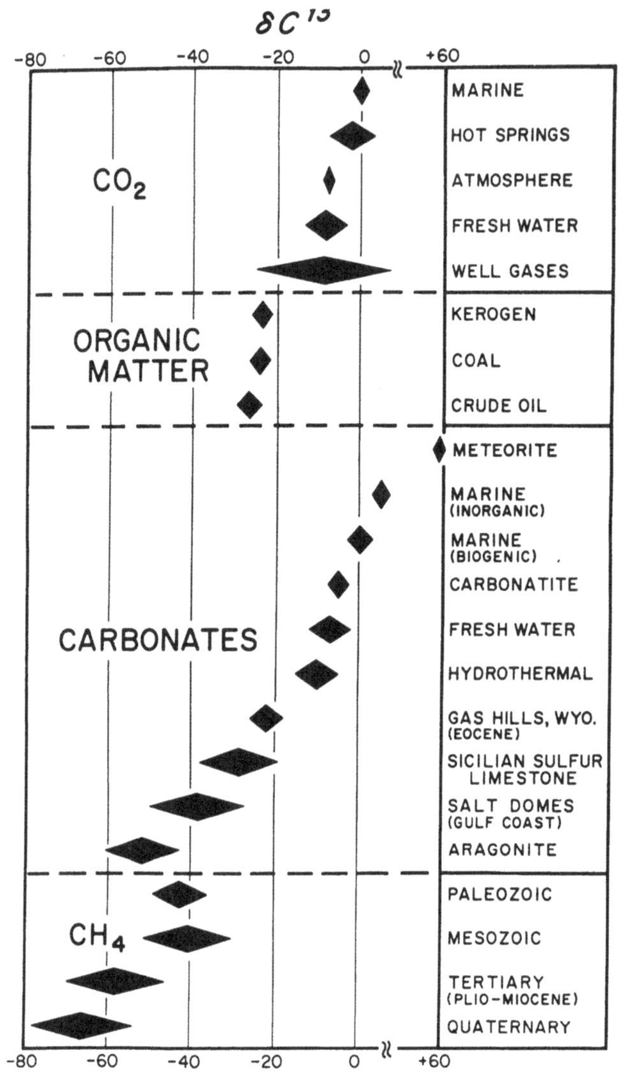

Fig. 7. δ ^{13}C ranges for various carbon-bearing materials. Width of diamond-shaped symbols equals 2 standard deviations for the data [38–40]. The individual samples have been grouped into systematic classes of compounds. (After [41])

At present, the bulk of carbon dioxide released from the lithosphere either through volcanism or hydrothermal vents its secondary in nature, i.e. recycled. Carbonate, sedimentary organic matter, and sea water bicarbonate is the principal progenitor of carbon dioxide emitted by volcanism and hydrothermal activities (Fig. 8). Judged from helium-3 anomalies [43] mantle-derived and thus juvenile CO_2 is today only a minor contributor of carbon to the biogeochemical cycle of air, water, life, and crust. Estimates on total CO_2 released from the solid Earth to the surface environment run in the order of 200 million tons of carbon per year [44].

Fig. 8. Hydrothermal vent on top of East Pacific Rise discharging heavy-metal solutions at temperatures of a few hundred degrees centigrades. Water depth close to 2,500 m (redrawn after photograph in [42])

It is this amount which drives the exogenic carbon cycle. Carbon will return, after many transmutations, to the solid Earth in different organic and inorganic forms. From this complex mixture CO_2 will arise over and over again in a never ending cycle.

Today's Carbon Cycle

Sinks, Sources and Fluxes

The wide attention presently paid to CO_2 has, on the one hand, to do with man's activity in the area of fossil fuel combustion and deforestation of tropical forests. On the other hand, there are still many unresolved questions concerning the interplay of CO_2 within and between the four major terrestrial carbon pools: air, water, life, and earth. Consequently, we are dealing with two major issues, one that is focussed on anthropogenic events and the other that pertains to the pristine global carbon cycle.

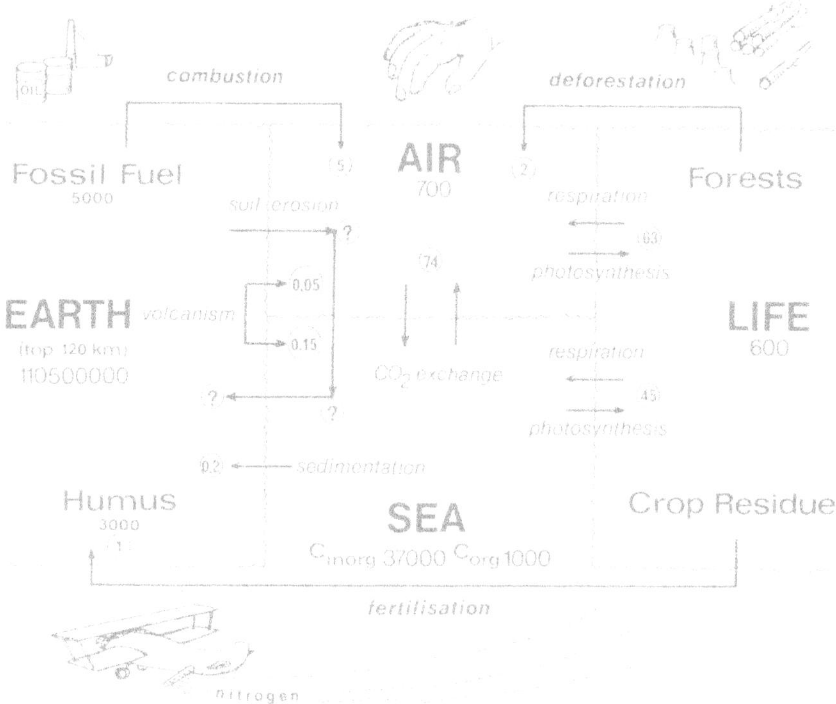

Fig. 9. Simplified scheme of principal reservoirs and fluxes in the natural carbon cycle (after [44]); man's input is depicted as combustion, deforestation and fertilisation

When the general CO_2 rise in the atmosphere, due to man's impact, became substantiated by the famous Mauna Loa curve (see p. 157), tree-ring records [45, 46] and related observations [47], it was felt that this trend would lead to a global warming of 2–3 °C in a matter of decades. Climate watchers became alerted, research on CO_2 prospered and many models on environmental consequences were proposed (see p. 203). The models should be viewed with some skepsis, since key processes like e.g. the dynamics of the ocean have not yet been properly accounted for (see p. 200). Yet, in fairness and in spite of all the shortcomings, many of the proposed models can be viewed as a kind of stepping stone in finding the true answer to the problem at hand.

From the beginning we wanted to elucidate the principles of the pristine global cycle in terms of sinks, sources and fluxes before even attempting to deal with man-made CO_2 in relation to climatic change. From a more elaborate scheme presented elsewhere, we extracted the critical parameters of the global cycle and differentiated between natural and man-imposed entries (Fig. 9).

Close examination of this simplified scheme reveals that carbon dioxide is the principal driving force of the global carbon cycle. The amount of CO_2 in the four major terrestrial reservoirs depends on a variety of factors most notably the exchange kinetics through boundaries such as the air-sea interface, biological membranes (photosynthesis-respiration), plus a series of carbonate equilibria displayed in marine and fresh water systems. The turnover rate of the exogenic carbon cycle falls in the range of about 180 Gt C yr^{-1} (gigatons carbon per year; 1 Gt $=10^9$ metric tons). The corresponding figure for the endogenic carbon cycle is lower by several orders of magnitude [48, 49].

In the following, we will briefly discuss the major sinks, sources and fluxes of carbon in the present-day exogenic cycle and focus on those points where CO_2 intersects. The data base is given in Table 2.

Volcanic Emanations

CO_2 has been released from mantle and crust practically from start of terrestrial accretion. It is reasonable to assume that in primordial time the streaming-off of CO_2 was quite substantial largely due to the more energetic state of the Earth and events such as the "terminal cataclysm" that is the period of major meteorite bombardment about 4 billion years ago [50]. The majority of impact craters on Moon and other planetary bodies were created at that time. Volcanism on the Moon is linked to this event.

Much speculation has been given to the PCO_2 in the early atmosphere with values as high as that of Venus today; the subject matter has been reviewed elsewhere [31, 51–54]. At what time PCO_2 in air has acquired its present level is still unresolved. However, in all probability a PCO_2 value ranging between 0.02 and 0.04 vol.-% has been with us for at least the past 1 billion years [36] (see p. 135).

We now make the heuristic assumption that over geologic times the global carbon cycle has developed into a steady-state system where input is matched by output. It follows that the amount of carbon released by the geosphere (largely as CO_2) should be matched by an equal return of carbon in the form of sediment (largely carbonates and organic matter) to the geological column. To what extent sea water is involved in this scheme is presently undecided for the following reasons. Cold sea water can enter the oceanic crust at discrete points and circulate in convection cells down to depths of several kilometers [55–61]. The idea has been advanced that the Mohorovičić discontinuity (abbr. Moho) may represent the level to which intracrustal water penetrates [62]. Along submarine ridge crests or vents generated in the vicinity of subduction zones, sea water may emerge as hot springs and release excess CO_2. The relationships are depicted in Figure 10.

It is too early to judge how fast sea water becomes recycled through the oceanic crust. One volume of sea water in 5 to 10 million years is a rough approximation. Only so much can be said for sure that the fluxes of various elements passing through the oceanic crust and entering the ocean do exceed the fluxes of the same elements moving from land to sea in the course of riverine erosion [55].

Measurements made on a number of active land volcanoes over the globe place an upper limit on the volume of carbon dioxide released through these channels at 0.002 Gt C yr^{-1} (see Table 2). Less certain is the rate of CO_2 emanations along active submarine rifts and subduction zones. In using global spreading rates as a means to calculate the volume of basalt, the CO_2 contribution from these sources should fall in the range of 0.1 to 0.2 Gt C yr^{-1}. Thus, the total volume of volcanic CO_2 comes out to be not more than ca. 0.2 Gt C yr^{-1}. But it is clear from the pre-

Table 2. Carbon cycle

Reservoirs sizes	Authors	Concentration	10^{15} g C
Atmosphere (total mass 5.14×10^{21} g)			
1982	(a)	340 ppmv	717
Preindustrial	(b)	260 ppmv	550
Ocean (total mass 1.384×10^{24} g)			
Inorganic	(c)	29 g m^{-3}	39,000
Organic	(d)	0.7 g m^{-3}	1,000
Particulate org.	(e)		30
Biosphere			
Living biomass	(f)		560
Standing dead	(f)		30
Litter	(f)		60
Humus	(f)		1,600–2,000
Marine plankton	(e)		3
Lithosphere (total mass of crust 24×10^{24} g)			10^{21} g C
Inorganic	(g)	0.20%	48
Organic	(g)	0.07%	17.5

Of this fossil fuels in 10^{15} g C		(h)	
	Best current estimate	Upper limit	Consumption until 1978
Oil	230	380	35
Gas	143	230	15
Coal	3,510	6,315	100
Oil shales	173	9,530	–
Tar sands	75	200	–
Total fuels	4,131	16,655	150
Methane clathrate resources		(i)	
Below permafrost		2,000	
In marine sediments		100,000	

References:
(a) Fraser, P.J. et al.: Proc. WMO/ICSU/UNEP Sci. Conf. Anal. Interpr. Atmosph. CO_2 Data, p. 179, Bern 1981; (b) Barnola, J.M. et al.: Nature *303*, 410 (1983); (c) Bolin, B. et al. (eds.): The Global Carbon Cycle, p. 491, SCOPE Rep. *13*, J. Wiley & Sons, Chichester-New York-Brisbane-Toronto 1979; (d) Mycke, B., Kempe, S.: In: RV. SONNE Cruise May/June 78, Bremerhaven-Panama-Hawaii, Final Report Germ. Res. Council (unpublished); (e) Mopper, K., Degens, E.T.: In: (c) Chapter 11; (f) Ajtay, G.L. et al.: In: (c) Chapter 5; (g) Kempe, S.: In: (c) Chapter 12; (h) Laurmann, J.A., Rotty, R.M.: J. Geophys. Res. *88*, 1295 (1983); (i) Bell, P.R.: In: Carbon Dioxide Review: 1982 [Clark, W.C. (ed.)], pp. 401–406, Oxford Univ. Press, New York 1982; (j) Bolin, B. et al.: Carbon Cycle Modelling [Bolin, B. (ed.)], pp. 1–28, SCOPE Rep. *16*, J. Wiley & Sons, Chichester-New York-Brisbane-Toronto 1981; (k) Aselman, I., Lieth, H.: In: Transport of Carbon and Minerals in Major World Rivers, Part II [Degens, E.T. et al. (eds.)], Mitt. Geol.-Paläont. Inst. Univ. Hamburg *54* (in press); (l) de Vooys, C.G.: In: (c) Chapter 10; (m) Leavitt, S.W.: Environm. Geol. *4*, 15 (1982); (n) Kempe, S.: In: (c) Chapter 13; (o) Kempe, S.: (unpublished results); (p) Meybeck, M.: Amer. J. Sci. *282*, 401 (1982); (q) Schlesinger, W.H., Melack, J.M.: Tellus *33*, 172 (1981); (r) Rotty, R.M.: J. Geophys. Res. *88*, 1301 (1983); (s) Moore, B. et al.: In: Carbon Cycle Modelling [Bolin, B. (ed.)], pp. 365–385, SCOPE Rep. *16*, J. Wiley & Sons, Chichester-New York-Brisbane-Toronto 1981; (t) Houghton, R.A. et al.: Ecol. Monogr. *53*, 235 (1983); (u) Olson, J.: In: Carbon Dioxide Review: 1982 [Clark, W.C. (ed.)], pp. 388–398, Oxford Univ. Press, New York 1982; (v) Bacastow, R., Keeling, C.D.: In: Carbon Cycle Modelling [Bolin, B. (ed.)], pp. 103–112, SCOPE Rep. *16*, J. Wiley & Sons, Chichester-New York-Brisbane-Toronto 1981

Table 2 (continued)

Exchange and transport rates	Authors	Rate (10^{15} g C yr^{-1})
Gross rate air-sea exchange	(j)	≈ 100
Gross primary productivity	(k)	110–120
Net primary productivity	(k)	57
Marine primary productivity	(l)	43.5
Fresh water primary productivity	(l)	2.3
Volcanism	(m)	0.002
Rivers inorganic dissolved	(n)	0.45
Rivers organic	(o, p, q)	0.2–0.8
Net silicate weathering	(g)	0.01–0.02
Marine sedimentation	(g)	0.12–0.26
Anthropogenic disturbances:		
Fossil fuels and cement (1979, 1980)	(r)	5.3
Destruction of biosphere	(s, t)	1.8–4.7
	(n)	1.0
Total anthropogenic releases		6.3–10
Net increases:		
Atmospheric increase (57% air-borne-fract.)	(v)	3.0 ± 0.2
Ocean uptake	(j)	1.9 ± 0.4
Disturbances of fresh water system	(o)	0.8 ± 0.5
Total:		5.7 ± 0.6
Biospheric sink (CO_2-fertilisation, deposition on land, reforestation):		0.6–4.3

vious argumentation that – with respect to its absolute value – this figure is beset by many uncertainties.

In relation to the ongoing discussion it is not essential whether the proposed value of 0.2 Gt C yr^{-1} for volcanic CO_2 is off by plus-minus 50% in the light of the 182 Gt C yr^{-1}, which are presently reshuffled between air, water and organisms. Furthermore, the majority of mid-oceanic rifts and subduction zones lie below the CCD (carbonate compensation depth) and any volcanic or hydrothermal CO_2 discharge will be readily picked up at depth by $CaCO_3$-unsaturated sea water, thus causing at most a minute rise of the CCD.

Carbon in the Sea

The ocean is the principal carbon reservoir at the surface of the Earth. Marine carbon is partitioned into an oxidized and a reduced pool. Bicarbonate (HCO_3^-) is the principal oxidized species and dissolved organic carbon (DOC) the major reduced form. The sizes of the two reservoirs are 39,000 and 1,000 Gt C, respectively. These figures are substantial considering that the estimated total biomass on Earth is only 560 Gt C.

Inorganic Carbon

The concentrations of inorganic carbon compounds CO_{2aq}, HCO_3^-, and CO_3^{2-} in the ocean arise from thermodynamic and kinetic balances (for details see p. 169). To avoid the cumbersome calculation of ionpair influence in sea water, equilibrium

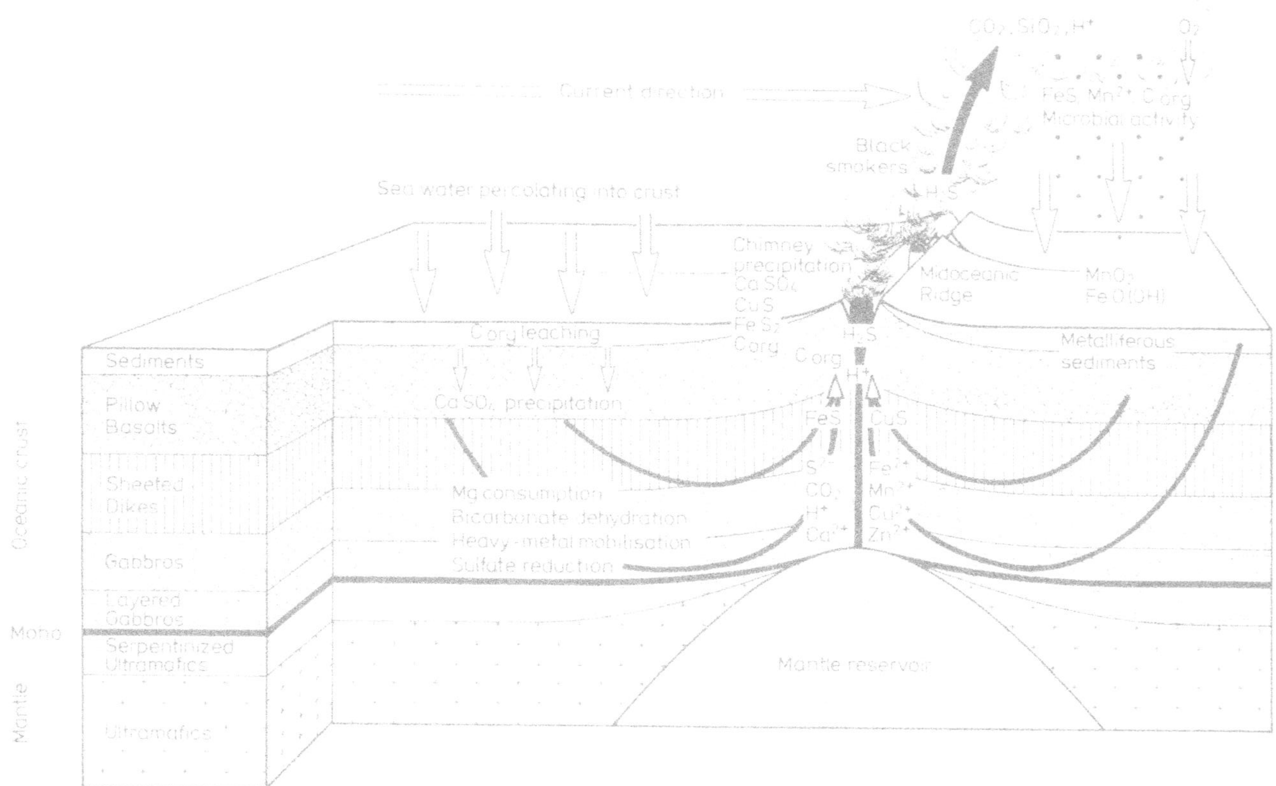

Fig. 10. Cycling of sea water through the oceanic crust. A simplified cross section exhibiting the main petrographical features of crust and mantle are shown on the left side of the graph. The Moho (named after its discoverer Mohorovičić) represents the crustal base about 8–11 km below sea surface. The schematic section is principally derived from ophiolite sequences exposed on land; ophiolites are thought to be former oceanic crust. The serpentinized ultramafics are in most observed cases products of hydrothermal alteration after emplacement on land. For details see [62]. The arrows indicate the general direction of water movement through the "porous" basaltic crust. Zones of mineral precipitations, metal leachings, ion extractions, redox and pH changes, or biological activities both in crust and sea water are roughly sketched. The information is principally drawn from [55]

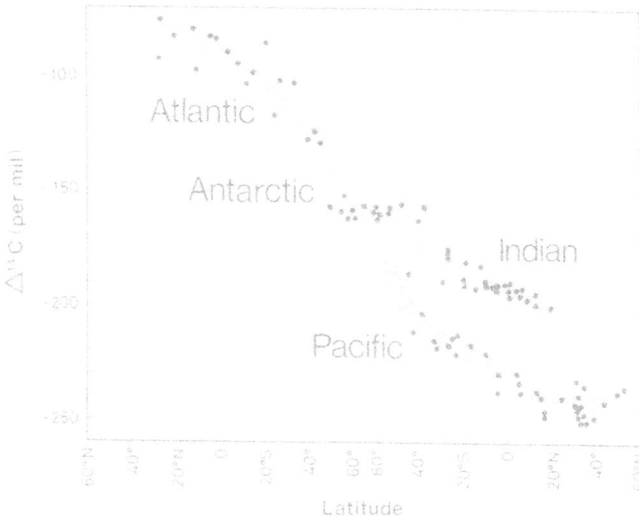

Fig. 11. ^{14}C (per mil) variations in water samples below a depth of 1,500 m from various regions of world ocean [74]

constants have been measured in sea water separately [63–68]. Determination of total alkalinity and total CO_2 can be done with potentiometry at high precision [69]. The largest set of measurement was obtained in the "Geochemical Ocean Section Study" (GEOSECS) [70, 71] and its successor project "Transient Tracers in the Ocean" (TTO) [72]. Results of titrations are accurate within a few micromoles [73].

The ocean is virtually saturated with respect to $CaCO_3$ and cannot enlarge its reservoir size significantly. River input is therefore balanced by output from the sea. River input results from the weathering of limestone during which one mole of biospheric CO_2 combines with one mole CO_2 from the rock. Thus there are two net outputs from the oceans: a constant (very small) net loss of CO_2 to air to close the weathering cycle and an output of organic and inorganic carbon to marine sediments. The average carbonate ion would have to wait about 165,000 years before it will come to rest in the sediment. During that time interval carbonates play an active role in the biogeochemistry of the ocean and will travel around the globe and from the warm surface layer to the deep sea many times over and over again.

During the GEOSECS program samples from 122 stations in all oceans have been analysed for ^{14}C content. Radiocarbon has a half life of 5,740 years, and is thus a suitable tracer to measure time scales up to a few thousand years. Furthermore, it is generated only in the atmosphere by cosmic rays and enters the ocean through its surface which is in isotopic equilibrium with air. Stuiver et al. [74] compiled measurements for deep ocean water (below 1,500 m) as shown in Figure 11. The pattern of ocean circulation becomes apparent from this graph: recent water sinks at the northern end of the Atlantic and travels south. The uniform ^{14}C-age of Antarctic circumpolar waters indicates rapid mixing of the Atlantic input with the cool, south polar surface waters. These sources then feed the abyssal Indian and

Pacific oceans, which contain the "oldest" waters. Eventually, the water upwells slowly over the entire ocean surface as it becomes displaced by new and colder water masses.

Applying a box model to the ^{14}C data the average replacement times of ocean waters read: Atlantic, Indic and Pacific waters change every 275, 210, and 510 years. Mean replacement time for the whole ocean is about 500 years [74]. The rate of upwelling inferred from these data is 4, 10, and 5 m yr^{-1} for Atlantic, Indic and Pacific, respectively. Should our carbonate ion choose to move downward with the North Atlantic polar water, and to continue its travel through the circumpolar Southern Ocean into the Pacific, it would stay below surface between 500 to 1,000 years. The more rapid surface currents would eventually return it to the North Atlantic.

While residing in the surface water dissolved CO_2 can be sequestered by phytoplankton and reduced to organic matter. In the marine environment, the atom ratio of dissolved phosphorus to dissolved carbon is about 1 to 800 [75]. As surface waters are generally low in phosphorus, it is commonly assumed, that all phosphorus entering the euphotic zone is incorporated by plants. There should be a note of caution, since mineral detritus (e.g. clays) can become covered by phosphate ions upon entering the marine habitat [6]. The C/P ratio in phytoplankton is about 1/100 (the so-called Redfield-ratio). Thus about one eighth of the carbon is sequestered as organic tissue. Also shells of $CaCO_3$ are formed consuming another 5% of the carbon. The phytoplankton is consumed by zooplankton and – after some recycling – carbon settles to the sea floor as particulate debris in the form of fecal pellets or macro flocs [76]. At depth most of the infalling C_{org} is remineralised and $CaCO_3$ dissolves thereby recharging the dissolved inorganic carbon pool.

Nitrate is another limiting nutrient. However, many algae can fix nitrate from dissolved nitrogen and thus ease nutrient shortage in surface waters [77]. Inasmuch as phosphorus can not be supplied biologically, phosphate is thought to be the essential factor limiting primary productivity in the sea.

These processes cause the ocean to differentiate into carbonate-rich bottom waters and carbonate-poor surface waters [75]:

	Warm	Cold	Deep Atlantic	Deep Pacific
	Surface			
Gaseous CO_2 (mmol l^{-1})	0.01	0.01	0.015	0.02
HCO_3^- (mmol l^{-1})	1.65	1.95	2.10	2.35
CO_3^{2-} (mmol l^{-1})	0.35	0.20	0.15	0.10
Total CO_2 (mmol l^{-1})	2.01	2.16	2.26	2.47
Alkalinity ($HCO_3^- + 2*CO_3^{2-}$) (meq l^{-1})	2.35	2.35	2.40	2.55

Again, we see that deep Pacific water is much older than Atlantic water, since the former had more time to accumulate dissolved carbonates from the infalling detritus. Dissolved phosphate concentration is also much higher in the deep Pacific than in the deep Atlantic. Mixing of oceans tends to undo this differentiation, yet at the same time it counteracts this "homogenization" process by upwelling which recharges phosphorus to the surface layer. This implies that a change in the rate of ocean circulation would not seriously effect the equilibrium. Less phosphorus offered per time unit would reduce remineralisation and the difference in inorganic carbon concentration between surface and bottom waters would remain constant. Only a removal of phosphorus from the exogenic system could lower the rate of photosynthesis and keep the rate of carbon loss smaller.

Thermodynamics predicts that at lower extraction rates for inorganic carbon the concentrations of free dissolved CO_2 should rise. As the concentration of free CO_2 in the sea determines PCO_2 in the air, the ocean can – via its phosphate content – exert influence on the Earth's climate. Measurements on air trapped in ice cores from the Antartic and Greenland [78, 79] indicate that PCO_2 increased from about 200 ppm during glacial times to 260 ppm in the postglacial. Broecker [80, 81] concludes that this increase is linked to the cycling of phosphorus. At times of sea level rises following the melting of polar ice, the continental shelfs become inundated and marine life can fix mineral nutrients in shelf sediments. Upon retreat of the ocean during a glacial stage, shelf sediments are exposed to erosion and release their nutrient content to the sea (fertilisation effect) and enhance primary productivity.

In the past, as well as in the near future ocean chemistry could influence climate. Even though the rate of photosynthesis in the sea is climatically controlled, the ocean is sluggish in accomodating sudden increases of CO_2 in the air. The homogeneous buffer factor, which excludes any reaction which could change the alkalinity (as, for example, calcite dissolution) [82] determines how much additional CO_2 can be dissolved in the surface layer following a PCO_2 increase:

$$B = \frac{\Delta PCO_2/PCO_2}{\Delta \sum CO_2/\sum CO_2}. \tag{7}$$

In essence the buffering capacity of the water stems from the presence of the CO_3^{2-} ion which can combine with CO_2 to form HCO_3^-. This reaction increases total carbon in the water, but will not change its charge balance nor alkalinity. The buffer factor may be calculated from standard mass balance data and varies between 14 near 0 °C and about 8 at 30 °C. Ocean surface water follows this slope closely, indicating that the PCO_2 of the ocean surface is quickly adjusted to temperature changes. Lowering the surface temperature by 1 °C will cause a drop in the $CO_{2aq.}$ content by 4% due to recombination with CO_3^{2-}. In consequence, the water comes out of equilibrium with the well-mixed atmosphere and will – on its poleward course – dissolve additional CO_2 from the air. The half-life of this process is about 0.8 years [75]. Cooler water will gain more total CO_2 by uptake of gaseous CO_2 from the air. There is a constant net flux of CO_2 stripped from warm tropical surface currents which redissolves in polar waters. It is concluded that the major spatial difference in surface PCO_2 of the ocean is temperature controlled. This is also true for the seasonal PCO_2 variation which appears to be largely induced by changing temperatures and salinities rather than biological activities [83]. Thus, removal of future atmospheric CO_2 increases will depend less on air-sea exchange

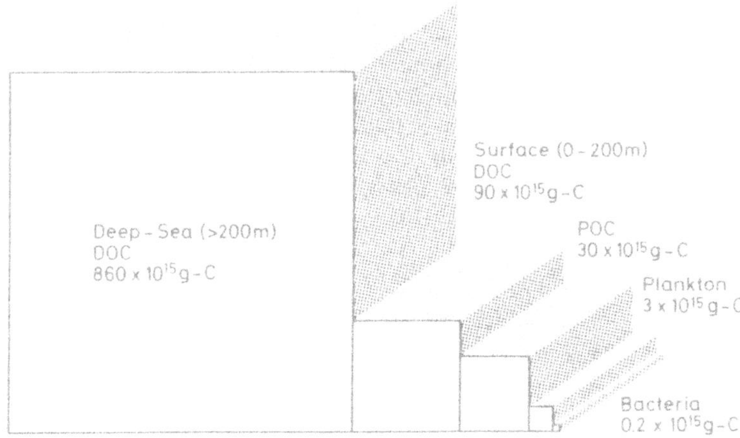

Fig. 12. Distribution of organic carbon pools in the ocean [84]

and buffering but on the formation of polar deep water. The impact of calcite dissolution on carbonate buffering will be discussed elsewhere (see p. 196).

Organic Carbon

The distribution pattern for the various forms of marine organic carbon is illustrated in Fig. 12. Living matter in the sea represents just a minute fraction, i.e. 0.3% of total marine organic carbon or 0.5% of total biomass. Nevertheless, primary productivity in the ocean is only one third lower than on land, i.e. 43.5 versus 57.2 Gt C yr^{-1}.

This number of 43.5 Gt C yr^{-1} has been derived at by comparing and weighting the annual primary production figures in the world's ocean reported by various authors [86] (Table 3). Critique has recently been expressed as to the reliability of the weighted figure which is based principally on the radiocarbon method. In measuring the observed distribution of dissolved oxygen below the euphotic zone as production indicator and using the transient tracers ^3H and ^3He, it was found [87] that the oxygen utilization rates in a test region: the oligotrophic waters of the subtropical Atlantic, were more in line with Riley's 1946 estimates [88] given in Table 3. This viewpoint also agrees with excess oxygen data observed in the Pacific shallow oxygen maximum [89] which indicate plankton production rates for oligotrophic waters several times those estimated previously. Since picoplankton contributions [90], rate of DOC excretion during plankton blooms, cell rupturing during filtration, or biomineralization effects ($CaCO_3$ deposition) are not or cannot be duly considered in the majority of cases, the 43.5 Gt C yr^{-1} value is at best a minimum figure. Furthermore, a plankton bloom is an episodic event lasting for a few weeks or months only. In consequence, a proper sampling device for primary production representative for the whole ocean and throughout the year has still to be invented.

The DOC pool is principally aliphatic in nature and of low molecular weight (5,000 daltons) [91]. Recent data suggest [92] that the bulk of DOC is of non-ma-

Table 3. Annual primary production in the world's oceans

Production in 10^{15} g C	Source
126	Riley (1946)
15	Steemann Nielsen (1953)
20	Ryther (1969)
23	Koblentz-Mishke et al. (1970)
44	Bruevich and Ivanenkov (1971)
60–80	Sorokin (1973)
31	Platt and Subba Rao (1975)

Riley, G.A.: J. Mar. Res. *6*, 54 (1946); Steemann, Nielsen, E.: J. Cons. Perm. Int. Explor. Mer *19*, 309 (1953); Ryther, J.H.: Science *166*, 72 (1969); Koblentz-Mishke, O.J. et al.: Nat. Acad. Sci., Washington (1970); Bruevich, S.V., Ivanenkov, V.N.: Okeanologiya *11*, 835 (1971); Sorokin, Y.I.: In: General Ecology, Biocenology, Hydrobiology, Vol. 1 (Biology Series), G.K. Hall & Co, Boston, Mass. 1973/74; Platt, T., Subba Rao, D.V.: In: Photosynthesis and Productivity in Different Environments [Cooper, J.P. (ed.)], pp. 249–280, Cambridge Univ. Press, Cambridge-London-New York-Melbourne 1975

rine origin and derived from riverine DOC. Low molecular weight organic matter generated in the sea by planktonic organisms either as excretion product or through cell lysis becomes principally harvested by microbial grazers or particularizes in the euphotic zone. It is only the inert riverine DOC which temporarily survives in the sea to slowly oxidize at a rate of about 12×10^{-4} mg C 1^{-1} yr^{-1}. Expressed differently, all DOC in the ocean would be gone in a few thousand years, unless the DOC pool becomes recharged by riverine DOC at about its annual oxidation rate [93]. In short, a kind of steady state exists between DOC oxidation and riverine DOC supply and may explain why DOC in the ocean is so uniformly distributed.

Carbon on Land

Fixation of CO_2 into organic matter and its subsequent decomposition are the main features of the biospheric carbon cycle. Green plants utilize CO_2 from the atmosphere for their gross primary production (GPP) (Fig. 13). Part of the production is used by the plant to meet its metabolic requirements and is respired through leafs and the root system (respiration, R). Only the net primary production (NNP) is available for heterotrophic organisms which either consume the plant matter directly or decompose the plant litter gradually. Residence times of assimilated carbon ranges from a few hours for immediately respired carbon to several thousand years for stable humic substances accumulating in soils and peat. Only fractions of the NPP persist through geological times as lignites and organic matter in sediments.

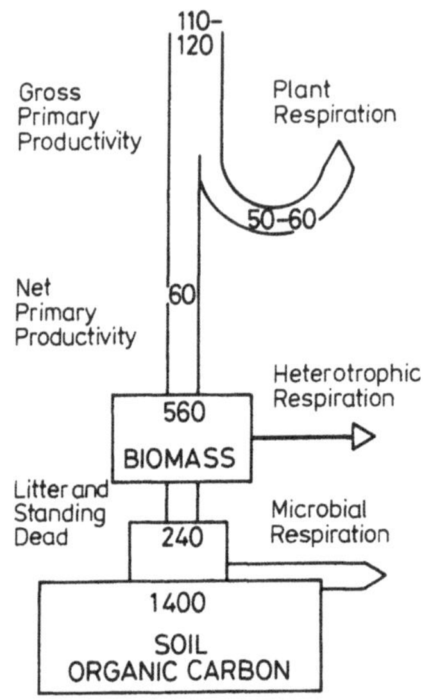

Fig. 13. Pool sizes and annual fluxes of terrestrial biospheric carbon cycle

CO_2-Fixation

Life depends on its ability to fix gaseous or dissolved CO_2. Evolution has created two general pathways: (i) oxidation of energy-rich compounds such as H_2S, CH_4, CO, NH_3, H_2, Fe^{2+}, S_0, NO_2, and (ii) the photolysis of water.

The chemosynthetic route is restricted to bacteria:

$$mAH_2 + nH_2O \xrightarrow{\text{Dehydrogenase}} nAO + 4n(H^+ + e^-) . \tag{8}$$

Faunal communities established near deep sea hydrothermal vents solely utilize H_2S emanating from below [55, 56]. Symbiosis of chemosynthetic bacteria, e.g. *Desulfovibrio* or *Thiobacillus* and higher organisms such as worms is observed; it may have played a larger role in past anaerobic oceans. Most chemosynthetic bacteria need O_2 as a final proton and electron acceptor, but some anaerobic bacteria can use oxygen bound in nitrate or sulfate.

In photo-autotrophic bacteria and plants light is captured within the thylakoid membrane of the chloroplasts resulting in the oxidation of water:

$$2H_2O \xrightarrow[\text{chloroplasts}]{h * \nu} O_2 + 4(H^+ + e^-) . \tag{9}$$

Two chlorophyll systems (I and II) are linked by various redoxsystems (cytochrome b, ferredoxin) to produce NADP-H as a stable proton-transport metabolite and to combine ADP and phosphate to ATP. The energy efficiency of this process is 30% of the light quanta captured by the chlorophyll system.

The biochemical reduction of CO_2 (carbon oxidation stage +4) to hydrocarbons (−1) is a dark reaction involving the oxidation of two NADP-H$^+$ and the consumption of three ATP's:

$$CO_2 \xrightarrow[3ATP \;\to\; 3ADP + 3P_i]{2NADP\text{-}H^+ \;\to\; 2NADP} CH_2O\,. \tag{10}$$

Plants have developed three modes of CO_2-assimilation [94]. The most common one is the Calvin-Benson or C_3-cycle [95]. Plants of the Hatch-Slack or C_4-cycle first produce oxaloacetic acid, a 4-C atom compound [96].

The third photosynthetic scheme is used by succulent plants such as Crassulacea. This Crassulacean acid metabolism (CAM) absorbs CO_2 during the night as organic acids. During daytime the stored CO_2 is reduced photochemically either by the C_3- or the C_4-pathways. Leaf stomata stay closed during the heat of the day, thus effectively saving water. The CAM-plants are common in deserts and savannas: the only cultivated species is the pineapple.

Due to the different metabolism C_3- and C_4-plants show marked differences in their assimilitation performance and their isotopic fractionation [97]:

	C_3	C_4
CO_2 assimilation rate in high light	2–4 g CO_2/m^2h	4–7 g CO_2/m^2h
Temperature optimum	20–25 °C	30–35 °C
CO_2 compensation point in high light	50 ppm	10 ppm
Photorespiration	Present	Not present
Typical water respiration per unit dry matter	400	200
Average $\delta^{13}C$	−27‰	−13‰

C_4-plants have a much higher optimum temperature and are therefore more common in tropical areas. Many tropical grasses, corn (maize), sorghum, and sugar cane use the C_4-metabolism. C_4-plants can operate at lower PCO_2 than C_3-plants before the compensation point is reached where the rate of photosynthesis is matched by the rate of respiration. This is because C_3-plants experience photorespiration, an additional respiration processes invoked by light and oxygen.

Water and CO_2 stresses regulate the opening of the stomata. When water is abundant (like in greenhouse cultures) stomata may not close at all and assimilation can be fertilised by artificially higher PCO_2 (lettuce, cucumber, tomato cultures). Under field conditions and under water stress stomata open to maintain a PCO_2 of 120 ppm in the substomatal cavity in C_4- and of about 220 ppm in C_3-plants. Hence C_4-plants do not need to open their stomata as often as C_3-plants and are much better adopted to water stress in the tropics. However, potential net primary production is – due to the much higher plant respiration at elevated temperature – similar for C_3-plants in temperate regions and for C_4-plants in tropical regions [98] and amount to about 20 g dry matter per m^2 and day. Due to longer growing seasons, tropical plants still achieve a higher annual potential productivity of about 7,000 g DM m^{-2} compared to 2,000 g DM m^{-2} in temperate regions. These productivities are normally not achieved because of water, light and nutrients stresses.

Decomposition of Organic Matter

Heterotrophs recover solar energy sequestered by photosynthesis by means of biological oxidation. The biochemical respiration chain serves to produce water from hydrogen of the reduced organic carbon and from free oxygen. The formation of water is highly exergonic. One pathway of the oxidation starts at the level of the $NADH\text{-}H^+$ so that a total of 52 kcal per mol of water formed is available. In the oxidation process three ATP's are generated saving a total of 21 kcal or 40% of the available energy for the metabolism of the heterotrophic organism:

$$NADH\text{-}H^+ + \tfrac{1}{2}O_2 + 3ADP + 3P_i \rightarrow H_2O + 3ATP + NAD^+. \tag{11}$$

Amino acids, fats, and carbohydrates are transformed into activated acetic acid bound to coenzyme-a. In the citric acid cycle this molecule is either oxidized to CO_2 to recover energy or is used to produce new organic matter for the metabolism of the heterotroph. Details are given in biochemical textbooks (e.g. [99, 100]).

In the absence of free oxygen, certain bacteria can use nitrate, sulfate or carbonate as an oxygen source:

$$NO_3^- + 2e^- + 2H^+ \rightarrow NO_2^- + H_2O. \tag{12a}$$

This reaction occurs in many aerobic bacteria which contain nitrate reductase. *Pseudomonas* and *Bacillus* species can reduce nitrate to nitrogen (denitrification reaction):

$$2NO_3 + 10e^{-+}12H^+ \rightarrow N_2 + 6H_2O. \tag{12b}$$

This reaction could play an increasing role in the respirative processes in rivers where much organic substances and high concentrations of nitrate are present and low oxygen concentrations prevent rapid oxidation. By comparing the PCO_2 and the oxygen deficit, one can show for the Rhine (NO_3 about 15 mg l^{-1}) that the river generates more dissolved CO_2 than can be explained by its oxygen deficit. The reason for this behavior is probably related to denitrification [101].

Sulfate reduction is – contrary to denitrification – not a facultative way to produce substrate oxidation:

$$SO_4^{2-} + 8e^- + 8H^+ \rightarrow S^{2-} + 4H_2O. \tag{13a}$$

A classical environment for sulfate reducers is the deep water of the Black Sea. Skopintsev et al. [in 102] have shown that concentrations of dissolved sulfide and carbonate increase linearly with depth at a rate almost expected by the loss of organic matter through sulfate reduction. Sulfate reduction is also common in marine and fresh water sediments as well as in wet soils.

Methane producers – a group of anaerobic microbes – use CO_2:

$$4H_2 + CO_2 \rightarrow CH_4 + 2H_2O. \tag{13b}$$

Also methanol or acetic acid may function as substrates for this reaction (fermentation).

A typical summation of respiration of organic matter with the composition of marine plankton (lacking cellulose and lignins) is:

$$C_{106}H_{263}O_{110}N_{16}P + 138\,O_2 \rightarrow$$

$$106\,CO_2 + 16\,NO_3^- + HPO_4^{2-} + 122\,H_2O + 18\,H^+. \tag{14}$$

Thus heterotrophic organisms return CO_2 to the atmosphere or the aquatic environment and remineralize nitrates and phosphates.

Reservoir Size

The size of the continental biosphere may be estimated by multiplying the areas of certain vegetation types by the biomass per unit area measured at sites of that vegetation type. This procedure has been followed for example by Ajtay et al. [103], Duvigneaud [in 104], and Olson [105]. These authors derived at estimates of 560, 592, and 560 GTC, respectively (Fig. 13). However, maps of biome distribution are often outdated. Even the most modern compilations of vegetation types (e.g. [106]) can rely only partly on satellite images and have to use older vegetation maps. Most carbon is stored in forests (428 Gt or 76% [103]), especially tropical rainforests (189 Gt C or 34%), although these forests cover only 21% and 7%, respectively, of the continental surface. The area of agriculture is 17.6×10^6 km^2 or 11.8% of the 149.3×10^6 km^2 of total continental surface [106].

According to one source [105] carbon density is still overestimated for certain vegetation types and total biomass has to be corrected towards lower values. Dead organic material also belongs to the biosphere. Estimates for "standing dead", i.e. old trees still above ground, may add-up to 30 Gt C [103] and for litter above soil to 210 Gt C with a turnover time of about 3 years [107]. Published figures for soil organic carbon differ even more than for biomass running from 2,950 Gt C [108] to 1,460 Gt C [109]. Recent data from 2,700 soil profiles (down to 1 m depth) gave 1,400 Gt C of soil organic carbon [110]. Most soil carbon is contained in wetlands (14.5%), tundra (13.7%), agricultural areas (12%), wet boreal forests (9.5%), tropical woodland and savanna (9.3%), and cool, temperate steppe (8.6%). Wet and moist tropical forests contribute only 9.9% to total soil carbon illustrating that such soils are poor in humus.

Rates of Primary Productivity

Net primary productivity on land is estimated at about 50 to 60 Gt C yr^{-1} [103, 104, 111]. Gross primary productivity about doubles these figures and dry matter fixation is calculated by using a factor 2.2 [103]. Climate data of 1,141 weather stations and regressions of NPP with precipitation and temperature serve as basis to derive a net primary productivity figure of 57.2 Gt C yr^{-1} [112]. This value is not the potential but the actual NPP (areas as of 1969) including the NPP (4.2 Gt C yr^{-1}) of agriculture. Agricultural areas have 40% less NPP than the natural potential NPP. Globally the NPP diminished by 2.7 Gt C yr^{-1} or 4.5% due to agricultural activities.

These figures suggest, that the gross rate of photosynthesis (110–120 Gt C yr^{-1}) could deplete the atmosphere in less than 10 years of its CO_2, if plant respiration and heterotrophs would not return carbon to the air.

All animals together have a biomass of only 0.45 Gt C and annually they consume 3.5 Gt of the NPP only [103]. Thus bacteria and fungi play the major role in decomposing the net primary production. CO_2 liberated in soils can diffuse only slowly upwards thus increasing the PCO_2 of soil air well above that of the free atmosphere. PCO_2 values up to several 10,000 ppm have been recorded in soils.

Biospheric changes due to man on a global scale are difficult to quantify for the past as well as for the present. For a full explanation of observed atmospheric

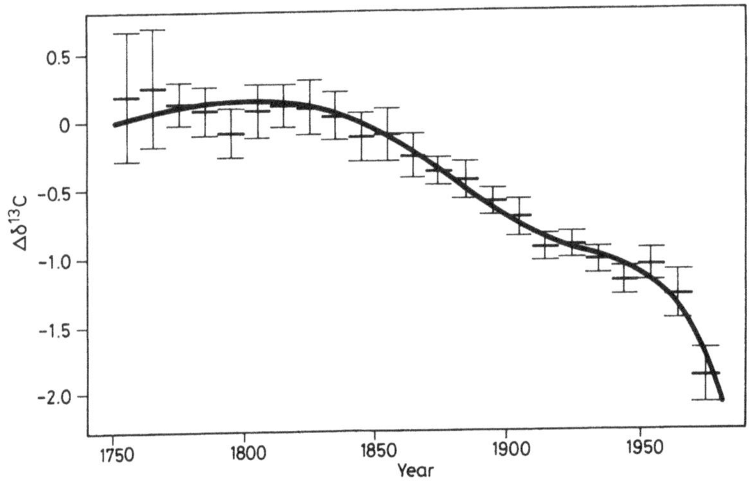

Fig. 14. Mean δ ^{13}C variations in Northern Hemisphere trees. From Freyer and Belacy [121]

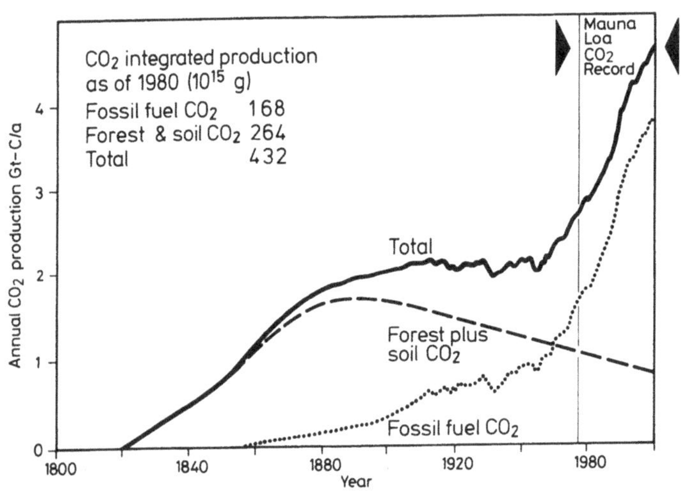

Fig. 15. Evolution of carbon releases by fossil fuel combustion and destruction of biosphere. Biospheric release is calculated based on the δ ^{13}C measurements of Freyer and Belacy [121] by model deconvolution by Peng et al. [122]

CO_2-trends, however, such data are needed (see p. 203). For the present, estimates of the net CO_2-release associated with those changes (e.g. deforestation, soil oxidation), range from -1 Gt C yr^{-1} to $+7$ Gt C yr^{-1} [103, 113–117]. Estimates for the past 150 years have been obtained by two different methods. One is, to assemble historic deforestation and agricultural expansion data and to calculate the tons of carbon involved [118, 119]. According to this approach, the biosphere has lost between 150 and 1,280 Gt C since around 1850, and continues to lose between 2 and 5 Gt C yr^{-1}. The other method is, to match a carbon cycle model with stable carbon isotope records. Application of a box-diffusion model [120] to tree-ring

$\delta^{13}C$ records (Fig. 14) [121] led to the conclusion [122] that about 260 Gt C have been liberated since the beginning of large scale land conversion, and the present release is estimated close to 1 Gt C yr^{-1}. The results are shown in Figure 15. Note the maximum release at the turn of the century ("pioneer effect"). The advantage of this approach is that the isotope record mirrors the carbon cycle's integral response as it actually has taken place. The lower CO_2-release estimates for the present may be due to the fact that part of the burned vegetation is transferred to unreactive charcoal, hence representing a net sink [117]. Moreover, large parts of destroyed soils are not oxidized but rather eroded and deposited in e.g. stream gulleys, alluvial fans and behind dams. Finally, deforestation and soil destruction may be counterbalanced not only by reforestation, but also by regrowth on cleared land. Stimulated plant growth by increased PCO_2 ("CO_2-fertilization") might be yet another effect, but does not seem likely when regarding nutrients, water and light as growth-limiting factors. In controlled experiments with higher PCO_2 it turns out that C_3-plants increase photosynthesis, grow larger leafs and acquire more dry matter [123]. Wheat closes the stomata and thus improves water-use-efficiency [124]. Larger yields therefore seem possible even under water stress. Possibly plants also react to higher PCO_2 even if no additional nutrients are available by simply decreasing the percentage of proteins produced. Controlled experiments so far deal mainly with agricultural plants. Open field communities have to be studied to obtain data for a meaningful assessment of the global biosphere's response.

A continuous and high resolution time series of erosion rates during the last 2,000 years is recorded in Black Sea sediments [125]. The record, as shown in (Fig. 16), provides geological evidence of a well known historical fact: large scale deforestation and agricultural expansion in Eastern Europe in the early middle ages and a corresponding accelerated rate of soil erosion. We strongly feel that soil erosion will become one of the most serious environmental problems in the decades immediately lying before us. The Black Sea record should be a warning signal.

Carbon in the Air

The masses and mean global mixing ratios of carbon gases in the atmosphere are [126]: 720 Gt C (=340 ppmv) carbon dioxide, 3.0 Gt C (=1.41 ppmv) methane, 0.227 Gt C (=0.11 ppmv) carbon monoxide. The methane cycle contributes 1–2% to the atmospheric carbon cycle. Because of the photochemical coupling of the methane and carbon monoxide cycles, this figure also applies to carbon monoxide [127, 128].

In the following section only CO_2 will be discussed, which is by far the most abundant carbon gas in air. For the atmospheric chemistry of the non-CO_2 carbon gases we refer to: [129–132]. For a review of organic carbon in the troposphere see: [133].

The atmosphere contains no significant internal sources/sinks of CO_2, because CO_2 is a photochemically non-reactive gas and hence rather stable [131]. However, it exchanges large amounts of CO_2 with the biosphere (see p. 151) and with the ocean (see p. 182).

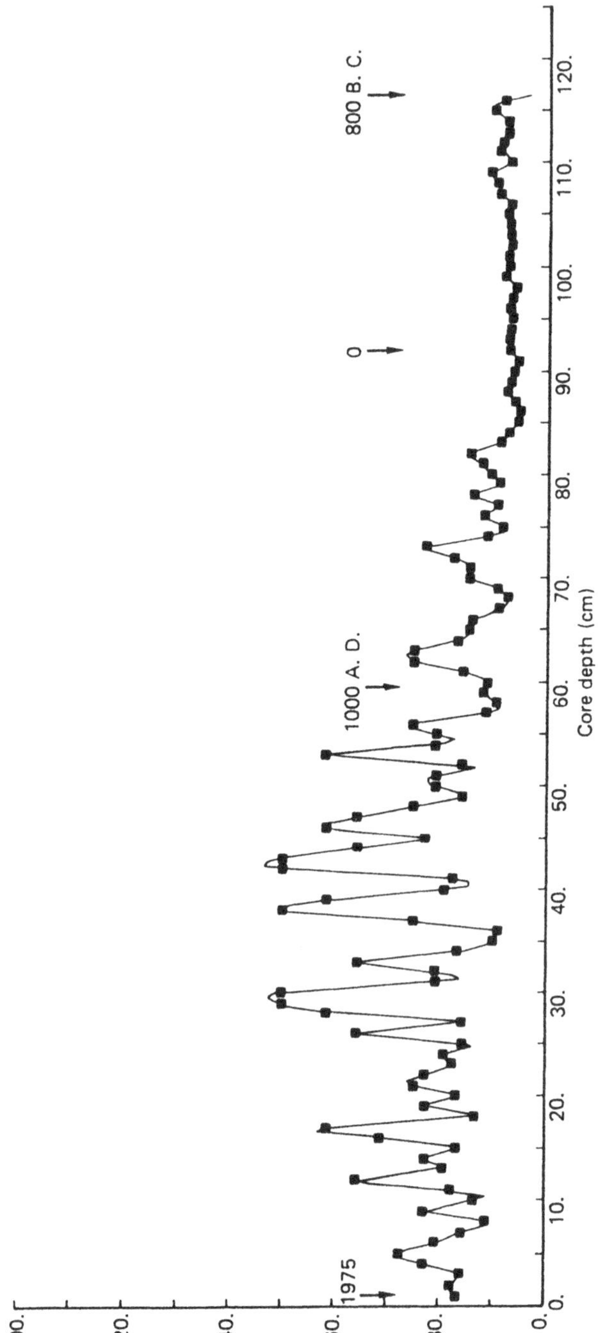

Fig. 16. Rates of sedimentation in a sediment core from the Black Sea apron at a water depth of 470 m. The age assignment is based on varve counts (1 light and 1 dark layer = 1 year), and stratigraphic cross-correlation with 7 other Black Sea cores. Between 800 B.C. and 1,000 A.D. repeated varve counts check within 1%. An uncertainty of about 20% exists for the upper 1,000-year-section. It is noteworthy that varve dating of a nearby core which goes back to 4,100 B.C. shows no apparent change in sedimentation rate below 800 B.C. This is the first long-term record which quantitatively shows man's impact on soil erosion which has increased 3-fold over the past 2,000 years [125]

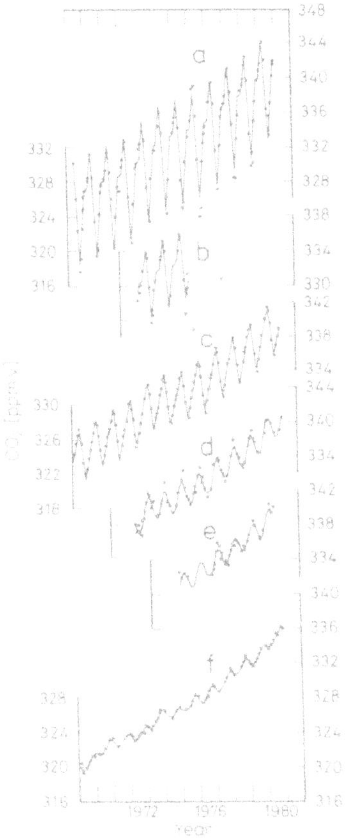

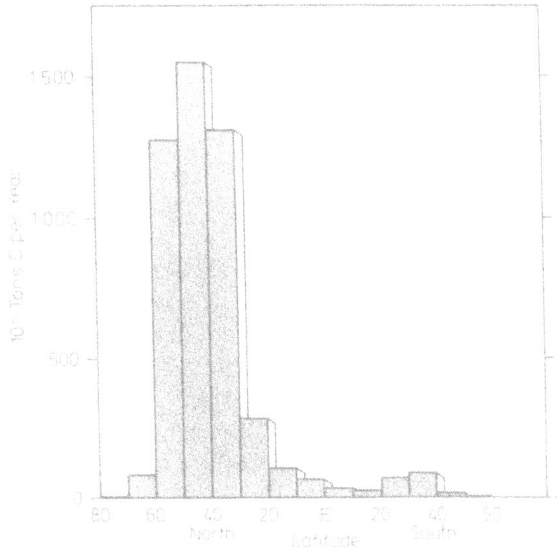

Fig. 17. **Fig. 18.**

Fig. 17. Secular trend and seasonal amplitudes of atmospheric CO_2 at various latitudes from north to south (from [134]): a) Weather ship P, 50.5° N; b) La Jolla, California, 32.9° N; c) Mauna Loa observatory, Hawaii, 19.5° N; d) Fanning Island, 3.9° N; e) Christmas Island, 2.0° N; f) South Pole

Fig. 18. Latitudinal distribution of annual fossil-fuel-CO_2 input to the atmosphere [321]

Less than 0.1% of the annually grown organic carbon on land and in the sea remains unoxidized and is deposited to form new fossil carbon. This deposition is roughly compensated by lithospheric sources of CO_2 to the atmosphere such as volcanoes, fumaroles and hot springs. Superimposed on this natural cycling is the anthropogenic CO_2 release (see Fig. 18).

On time scales longer than a few years the atmosphere may be considered homogenously mixed. Therefore, long-term monitoring of global trends in atmospheric CO_2 can be achieved at a single station. In 1958 accurate measurements started at the Mauna Loa Observatory in Hawaii, latitude 20°N [367]. The mean annual concentration of CO_2 in the atmosphere (PCO_2) has increased since then by 25 ppmv to a value of 340 ppmv as of 1982 [134]. This secular trend is confirmed by records from other monitoring stations spread over the globe (Fig. 17). Superimposed on the secular trend is a seasonal cycle. It results from photoysynthesis by land plants in summer and respiration by land plants and soils throughout the year. It's amplitude varies with latitude and altitude. The maximum occurs around late spring/early summer, the minimum occurs in autumn. The peak-to-peak amplitude is about 15 ppmv at Barrow, Alaska, about 6 ppmv at Mauna Loa and about 1 ppmv at the south pole (see Fig. 17).

When the annual cycle and secular trends are removed from the Mauna Loa record, the PCO_2 appears to fluctuate on time scales from about 2 to 10 years with an amplitude in the order of about 1 ppm [134]. Equatorial sea surface temperature changes have been proposed as explanation [135–139].

Over the last twenty years the amplitude of seasonal variation has increased by about 0.66% yr^{-1} [134]. If the change is interpreted as reflecting enhanced biospheric growth, the effect is equivalent to a 8% change in net summer uptake of carbon over the years 1959–1978 and to a growth of the northern hemisphere seasonal biosphere of 0.5 Gt C yr^{-1} [140].

In a synoptic view the 1979 records of monthly mean CO_2-concentrations of 14 globally distributed locations were analysed by eigenvector analysis [141]. Three independent patterns of CO_2-variations emerge. The first (69% of the variance) is interpreted as a globally averaged pattern caused primarily by fluxes into and from the biosphere. The second (13% of the variance) suggests a phase change of the annual CO_2 fluctuation due to oceanic temperature fluctuations and appears to be strongest near the poles. Pattern 3 (7.5% of the variance) suggests a biospheric effect that deepens the minimum CO_2 concentrations in northern latitudes.

Diurnal variations of atmospheric CO_2 content occur at ground level because at night no CO_2 is removed from the air by photosynthetic uptake. The CO_2 respiration rate from the soil and from plant leaves was estimated in the Heidelberg region, Germany [142]. A total CO_2 respiration flux to the atmosphere of 21 mmol CO_2 m^{-2} h^{-1} for a short time period in August was observed. Direct measurements derived from CO_2 concentration profiles in the soil yield an average regional CO_2 flux from the soil alone of 12 mmol m^{-2} hr^{-1}, with a maximum of 20 mmol m^{-2} hr^{-1} in June and July and a minimum of 3 mmol m^{-2} hr^{-1} or less in winter.

The range of diurnal variation can be as great as 100 ppm [126]. Only slight variations occur in winter and during spring. The effect may be recognizable, under certain conditions in temperate climates as high as 150 m above ground. In the tropical atmosphere over Kenya, Africa, where convection is the dominant meteorological phenomenon, the diurnal signal has been found up to 4,000 m above ground on sunny days [143]. Over savanna regions day time drawdowns were to 322 ppm and night time buildups to more than 400 ppm. In and around tropical rain forests, drawdowns were to 310 ppm and buildups to more than 400 ppm. On the higher reaches of Mount Kenya the diurnal CO_2 cycle was considerably reduced in amplitude, with variations in the range of 2–6 ppm during the 16 months of observation. Measurements over the Indian Ocean to distances 450 km upwind of the coast showed no diurnal fluctuations. (This different signature of continental and marine air was also observed, for example, in the South East Australian atmosphere: seasonal variations in the order of 5–10 ppm decreased to near zero as the coast was approached [144].)

Form an urban atmosphere hourly average maxima as high as 580 ppm and minima as low as 285 ppm are reported [145], the diurnal pattern being strongly influenced by meteorological conditions and by photosynthesis. Seasonally, maxima occured in winter due to stationary combustion sources. The winter-summer difference was 30 ppm. The long-term trend of the base line CO_2 concentration shows a time integrated increase of 0.6 ppm yr^{-1}, consistent with observations at global background sites.

Analyses of troposphere and lower stratosphere air samples from the northern hemisphere revealed seasonal variations of about 6.5 ppm at 2 km altitude and 3.5 ppm in the upper most part of the troposphere [146]. The phase shift of the seasonal variation between these two levels is 25–30 days. The amplitude of seasonal variation in the lower stratosphere at 11–12 km is less than 1 ppm, and the phase is delayed 1 ½ months compared to the upper troposphere. Hence, vertical CO_2-exchange is damped by the tropopause. The gradient of CO_2 mixing ratios across the tropopause has been confirmed by measurements [147]. Anomalously large and yet unexplained variations of CO_2 in the stratosphere between 25 and about 40 km were reported from balloon flights over Texas, USA [148]: at one particular night flight PCO_2 was 400 ppm. These data contradict the small variations reported earlier [146, 149, 150]. Recently, a decrease of CO_2 by about 7 ppm between the tropopause and 33 km was observed in air samples collected during balloon flights and during commercial airliner flights over Europe [151]. The interpretation given is, that increased tropospheric CO_2 concentrations reach an altitude of 25 km with a delay of 5–6 years.

Very detailed observations of the variability of CO_2 in a polar atmosphere from the northern coast of Alaska (Pt. Barrow) during sommer are available [152], with day to day and within-day variations of up to 50% of the annual range (14 ppm). It turns out that the tundra of the Alaskan North Slope is a net source of CO_2 and the ice-free areas of the sea bordering Alaska are a significant sink for CO_2.

Measurements of atmospheric CO_2 in the 19th century have been compiled by Callendar [153] (the measurements of Lecher (1881) may be added to the list [154]), whose analysis of the 19th century data suggests a PCO_2 content near 290 ppm. Tree-ring stable and radioactive carbon isotope data suggest a pre-industrial atmospheric PCO_2 in the range between 245 and 270 ppm [122, 155]. The mean CO_2-level recorded by Antarctic ice for the period 800–2,500 years before present (B.P.) is about 260 ppmv [156]. This value is slightly, but not significantly lower than the 271 ppmv suggested by measurements performed on a ice core section from Greenland for the period 1650–1800 [79].

Ice core records extending back to the last ice age are available for Greenland and Antarctica. Measurements of CO_2 have been performed on two Greenland cores [79] and on two Antarctica cores [157]. For a detailed discussion of the four records see Lorius and Raynaud [158].

For the four records a similar general trend is indicated, with a mean of 270 ppm for the Holocene and, as the main feature, a marked depletion in CO_2 corresponding to the latest part of the last ice age (around 18,000 yrs B.P.). Concentrations for this period are of the order of 200 ppmv (Fig. 19). Earlier during the ice age the atmospheric CO_2 concentrations seem to have been higher and of the same magnitude as the Holocene values. Various hypotheses have been proposed to explain these surprising and important findings. Broecker [80] suggested, as one scenario, that phosphate-fertilization of the ocean during the glacial was involved:

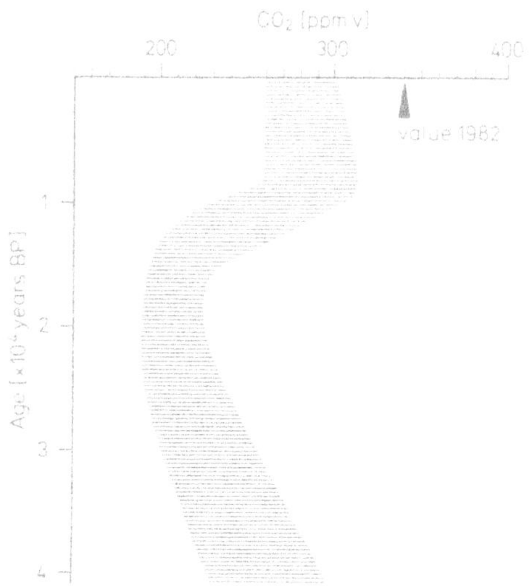

Fig. 19. Variation of CO_2 in polar ice cores: estimated range of atmospheric CO_2 during the last 40,000 years, as deduced from the records of the stations "Byrd" (West Antarctica; 80° 01′ S, 110° 31′ W) and "Camp Century" (Greenland; 77° 10′ N, 61° 08′ W). From Neftel et al. [79]

the drop in sea level would expose the phosphate-rich continental shelfs to erosion. The fertilized ocean would, as a consequence, extract more CO_2 from the atmosphere. As an alternative scenario he proposes [81] that during the glacial there was a change in the phosphorus to carbon ratio of the organic debris falling to the deep ocean. The potential role of nitrogen in this context was discussed by McElroy [159]. Alternately, Berger [160] advocates increased deposition of shelf carbonates during sea-level-rise as responsible for rising PCO_2 after a glacial. This contrasts the findings (see p. 190) that calcification in biological systems will not release free CO_2 except in a few species. For empirical evidence supporting the idea of a more fertile glacial ocean see: [161–163]. Rather than sea-level-changes, Müller and Suess [163] favor climate related changes in wind and ocean current patterns, which affect the rates of nutrient-rich upwelling, as fertility controlling mechanism. This view is supported by the recent record of PCO_2 excursions in the time between 20,000 and 30,000 years B.P. reported by Neftel et al. [164]. The amplitude of PCO_2 variations is comparable in magnitude to the glacial-interglacial difference, but the time scale involved is much shorter: the fluctuations correlate with the stable oxygen isotope record in the same core, which reflects concurrent climatic and PCO_2 fluctuations on time scales of centuries. Broecker's sedimentation-erosion-hypothesis, though plausibel for the glacial-interglacial cycle, cannot account for such short term fluctuations. In the light of these new data Siegenthaler [165] suggested that oceanic nutrient chemistry is still involved, but now responding to short term changes in oceanic circulation, which modify in particular the equatorial Pacific upwelling. Meltwater stratification during deglaciation is also involved in this scenario. How precisely physical, chemical and biological changes in the ocean interacted to change atmospheric PCO_2-levels remains to be explained. As a further process we suggest that during glacial times episodic dust storms (loess) which covered the ocean may have reduced the depth of the euphotic zone by increasing turbidity. Thus primary productivity might have been lowered for short time periods increasing atmospheric PCO_2. At the termination of the glaciation fine debris carried to sea by meltwater rivers may have induced a permanent rise in PCO_2.

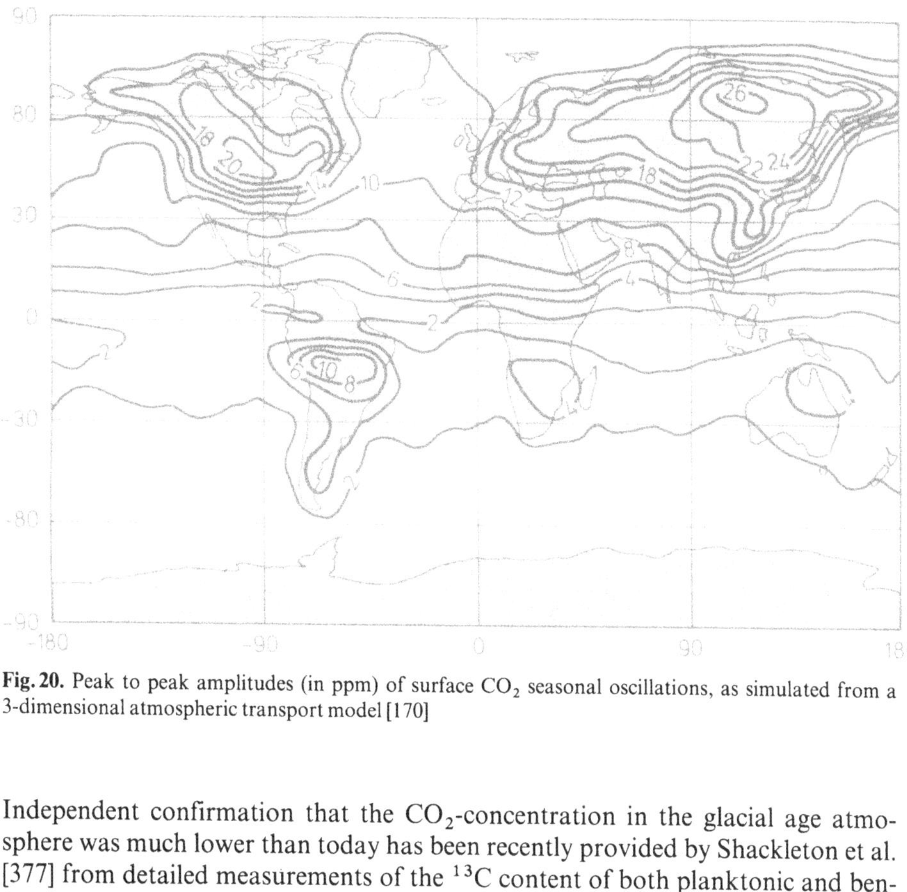

Fig. 20. Peak to peak amplitudes (in ppm) of surface CO_2 seasonal oscillations, as simulated from a 3-dimensional atmospheric transport model [170]

Independent confirmation that the CO_2-concentration in the glacial age atmosphere was much lower than today has been recently provided by Shackleton et al. [377] from detailed measurements of the ^{13}C content of both planktonic and benthic foraminifera in a deep sea core.

Modelling of atmospheric CO_2 transports has mainly been done with diffusive or advective-diffusive models [166–169]. Pearman et al. [169] related estimated air to surface exchanges of CO_2 to spatial and temporal variations of atmospheric concentrations of CO_2. They assumed that, at present, the equatorial oceans release a net total of 1.3 Gt C yr^{-1} into the atmosphere, while high latitude oceans take up a net total of 4.4 Gt C yr^{-1}. Associated with the increasing release of fossil-fuel CO_2 into the atmosphere of the northern hemisphere the model suggests that the interhemispheric difference in concentration has changed from -1 ppmv preindustrially to $+1$ ppmv in 1960 and to 4–5 ppmv at present. From the model results it seems unlikely that any equatorial deforestation source, combined with a similar northern hemisphere sink (temperate forest uptake) would yield more than about 2 Gt C yr^{-1}. A three-dimensional atmospheric tracer transport model, based on a wind field provided by a general circulation model of the atmosphere was used by Fung et al. [170] to investigate the annual cycle of atmospheric CO_2 concentration produced by seasonal exchanges with the terrestrial biosphere. The geographical distribution of the seasonal amplitude is shown in Figure 20. It turns out that zonal homogeneity in surface CO_2 concentrations can never be achieved at mid-latitudes because the time required for zonal mixing exceeds the time scale of biospheric exchanges by 50% or more. Year-to-year variations of the annual

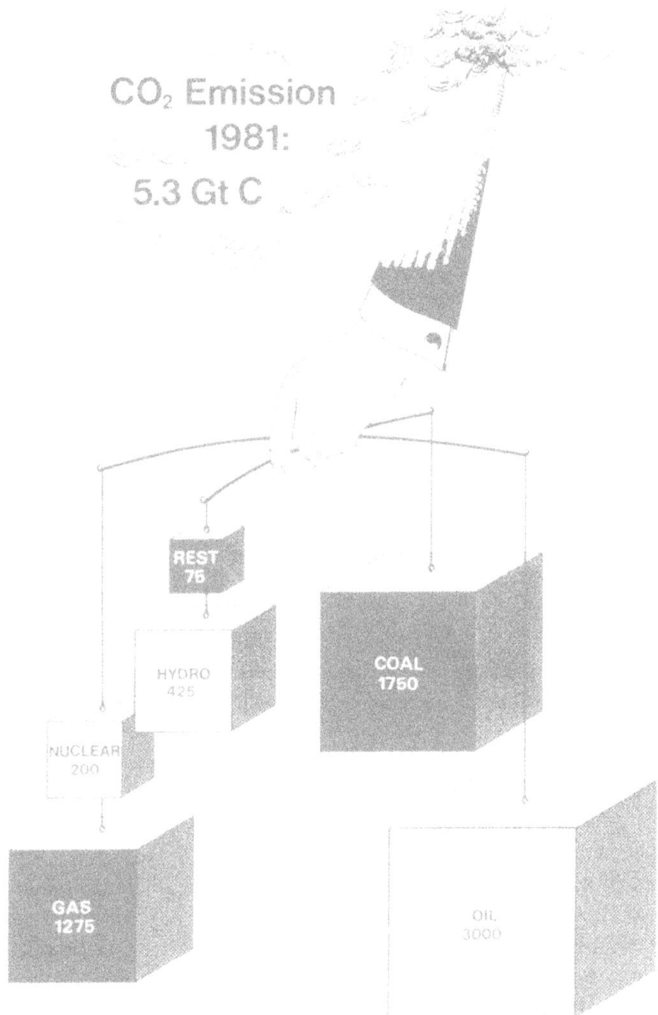

Fig. 21. World energy consumption figures for the year 1981 in millions of metric tons oil equivalent [172]. Burning of fossil fuel has caused an emission of 5.3 Gt C in the form of CO_2

CO_2-cycle may therefore result not only from changes in biospheric sources and sinks, but also from the natural variability of the atmospheric circulation.

Man-Made Carbon Dioxide

Combustion, Deforestation, Fertilisation

Combustion of oil, coal and gas and profound changes in land use patterns have not remained unnoticed by the natural cycle of carbon. An expression of these activities is, among others, a systematic rise in atmospheric CO_2. Direct measure-

Table 4. World energy consumption[a]. (After "1983 Shell Briefing Service")

	1961	1971	1981	Reserves[b]
Oil	1,175	2,530	3,000	32
Coal	1,200	1,430	1,750	248
Gas	400	910	1,275	59
Hydro	215	305	425	
Nuclear	tr.	40	200	
Others	65	65	75	
Total	3,055	5,280	6,725	

[a] In millions of metric tons oil equivalent p.a.
1 barrel oil per day is about 50 t per year
1 barrel = 159 l or 0.136 t

[b] Known reserves (in years) at 1981 consumption figures [172]

ments on the CO_2 trend exist for the past 25 years [134] but inferences drawn from carbon isotope distribution in tree rings [45] permit an extension of the CO_2 record into past centuries. Carbon isotopes signal an about 2 per mil decrease in $\delta\ ^{13}C$ and an increase of approximately 2 per cent in the ^{14}C content of wood ("Suess-effect") [171, 368]. This phenomenon is caused by contributions of light, i.e. ^{12}C-rich, and dead, i.e. ^{14}C-free, carbon dioxide derived from the burning of fossil fuel and the manufacturing of cement. Tree ring records calibrated against the Mauna Loa curve [134] and extended to the year 1750 [46] suggest a CO_2 level according to one source of 265 ppm [47] and to another of 290 ppm [44].

The extent of fossil fuel burning and changing land use on the pristine carbon cycle is shown at the upper and lower margins of Figure 9. Carbon dioxide emissions into the air are indicated by the terms: combustion and deforestation. In 1981, a total of about 5.3 Gt C were emitted into the air by oil, coal and gas burning and cement fabrication. To illustrate how much the various fossil carbon sources contributed individually, Figure 21 has been prepared from production figures given in [172]. For comparison, data on world energy consumption for 1961–1981 from all energy sources are given in Table 4. Added to the about 5.3 Gt C yr^{-1} value are ca. 2 Gt C yr^{-1} from deforestation activities largely in tropical forests [44]. There is consensus with respect to the number for combustion but the deforestation value is much debated (see p. 154). Of the total man-made CO_2, 3.0 Gt C yr^{-1} remain air-borne and are the cause of the present rise in atmospheric CO_2 at an annual rate of about 1 ppm [134].

Initially, there had been much speculation where the remaining two thirds of man-induced CO_2 could hide. There is indication (see p. 200) that about 1.9 Gt C yr^{-1} of human induced CO_2 is removed by the sea which leaves a missing CO_2 of ca. 2 Gt C yr^{-1} to be accounted for.

A brief glance at Figure 9 will reveal that there are only a few options left for finding a sink for the final one third of man-made CO_2. It has been proposed that perhaps an increase in land biomass in the virgin forests of the world has and still is occurring, but proof is difficult to obtain [44]. Another line of thought is a general increase in rate of photosynthesis in a CO_2-enriched atmosphere, an idea which has frequently been voiced. Contrary to these commonly held opinions, the photosynthetic ability of plants is rarely a factor that determines rates of growth; mineral nutrients, moisture and temperature are the determining elements which limit growth of plants to only a fraction of their photosynthetic potential [103]. One possible net sink for atmospheric CO_2 is elemental carbon, due to incomplete combustion of biomass to charcoal [117]. This sink was certainly more effective during the last centuries than today but still has to be considered in the final analysis. What other sinks are left?

World food production has doubled over the past 30 years [173, 174]. This impressive growth rate has been achieved by a combination of factors: (1) green revolution, (ii) putting more land under the plow, and (iii) profound changes in cultivation techniques. The idea has been advanced that humus production could intensify along with food production. Humus or soil organic matter is the largest organic carbon reservoir at the surface of the Earth.

Of the about 2,000 Gt C stored in humus about two thirds are in the form of peat and the rest is distributed among ordinary soils. There is only little data on how much humus becomes annually generated, oxidized and eroded. Only one fact is certain that farming activities have accelerated the pace of humus generation, oxidation and erosion [175].

Humic substances of all sorts are quite resistant to oxidation and can be transported over wide distances in the particulate or dissolved form. They will accumulate in soils or when eroded enter marine sediments or the DOC pool in the ocean. Soils and the DOC pool may thus act as temporary carbon sinks. Eventually, however, the sea will be the final burying ground for the human-induced CO_2.

At present rates of fossil fuel burning (Table 2), the known reserves will last for at most a few hundred years. The present CO_2 problem is thus – geologically speaking – only a short-term event. On the basis of carbon, all fossil fuel depots contain about ten times the amount of C_{org} fixed in modern biota and fifty times the quantity of organic matter annually generated by photosynthesis. Expressed differently, it took Earth about 500 million years to come up with all the oil, gas and coal we presently know of (Fig. 22). In one year man burns as much carbon as Earth has laboriously collected during roughly 1 million years of its Phanerozoic history.

Energy Future

The issue of "Energy Future" is pressing because there is a foreseeable end to easy oil and gas and environmentally benign energy resources are hard to come by. Temporary relief is to be expected due to adaptation and conservation measures. Yet, all efforts combined are not adequate to meet the energy demands of the next century.

In 1981, renewable energy technologies of various provenance contributed about 8% to the total world energy consumption figures [172]. Whether this percentage will increase in the years to come is questionable because even "safe" energy such as wind and water may environmentally be hazardous when used on a much wider scale than at present. A wealth of data exists on future energy planning and the various options have been discussed at length (e.g. [176–192]).

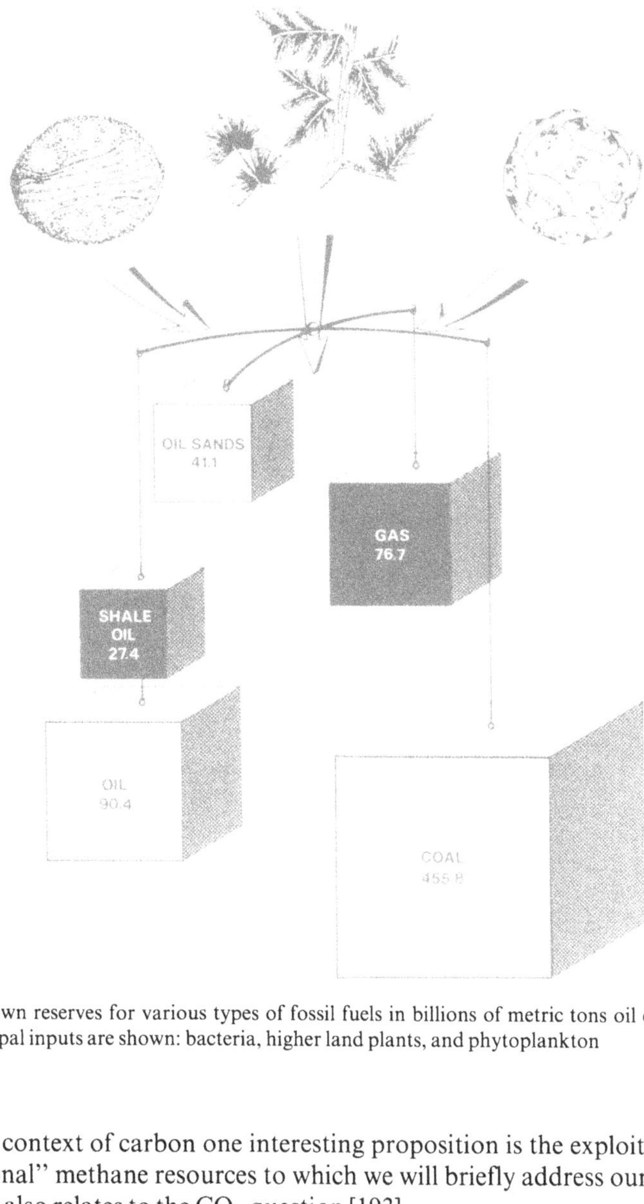

Fig. 22. Known reserves for various types of fossil fuels in billions of metric tons oil equivalent [172]. Three principal inputs are shown: bacteria, higher land plants, and phytoplankton

In the context of carbon one interesting proposition is the exploitation of "unconventional" methane resources to which we will briefly address ourselves below, because it also relates to the CO_2 question [193].

Most of the conventional methane deposits come from rock formations which are burried not more than two kilometers. Only a few natural-gas wells exist which go deeper than that. However, there is evidence for the presence of "non-associated" gas at greater depth [194]. Here the so-called "geopressured" methane is dissolved in brines which – when brought to the surface – will release gas and heat energy. First calculations show that this methane reservoir would more than double the supply of natural gas at the present rate of use [195] but doubt exists, whether this resource can be tapped profitably and environmentally safe [193].

In permafrost regions and below the sea floor, massive sections of gas hydrates (clathrates) are encountered. Seismic records allow to trace the distribution of gas hydrate layers in marine sediments. Inasmuch as the pressure and temperature relationships for the formation of gas hydrate are known, such records also reveal information on the geothermal regime. Presence and actual recovery of gas hydrates has been documented at Site 565 of Glomar Challenger operations (Leg 84) at a water depth of about 3 km and 300 m below the sea floor [196]. Other sites produced white ice-like substances with a gas composition containing mainly C_1, C_2, C_3, and i-C_4. The gases found in the gas hydrates of Leg 84 are considered to be partly of microbial origin (early diagenesis) and partly of thermally degraded organic matter (late diagenesis or catagenesis). It is too early to estimate the total volume of gas hydrates in the marine sedimentary column and reported figures should be used cautiously (Table 2). Yet, in view of difficult logistics, the future energy potential of marine clathrates is hard to judge. In permafrost regions, the economic outlook stands on better grounds. In taking all numerical uncertainties into account, unconventional methane could about triple the reservoir capacity of natural gas in the United States alone [193]. It is questionable, however, whether these resources can be extracted at a reasonable price.

At the beginning of this article we stated that in the Earth crust the formation of carbon dioxide and methane from some unspecified carbon source is thermodynamically favored. Even though primordial methane is known to be released in trace quantities from mantle and crust and should occur in high abundance below the lithosphere, there is virtually no chance to tap this huge reservoir.

Unconventional methane could be generated from volcanic carbon emissions via an alternate route as exemplified in Lake Kivu, an African rift lake with a water volume of 580 km^3 and surrounded by active volcanoes. In its deep water (250 to 500 m) it contains about 50 km^3 of methane and 150 km^3 of CO_2 (STP) [197]. Carbon isotope data indicate that while the dissolved CO_2 is volcanic in origin, the associated methane has been formed by bacteria utilizing free hydrogen derived from volcanic gases emanating from hydrothermal vents into the deep water. Such vents release hydrothermal jets up to 80 m high into the open water column. Since the lake is anoxic below about 100 m water depth, some of the hydrogen has also been liberated by the formation of heavy metal sulfides. The reservoir is confined to the deep water because the lake is permanently stratified; its size increases at a rate of about 1% per year [198].

Complex communities of thermophilic bacteria which produce methane, hydrogen gas, and carbon monoxide have been cultured from superheated vent water ("black smokers") emanating from the sea floor in the vicinity of rifts and submarine volcanoes [199]. The gases will rapidly disperse, but in Lake Kivu they conservatively accumulate. It is conceivable that the two observations namely high chemosynthetic primary production along vents and excellent reservoir potential of stratified water bodies may open up new vistas for the generation and accumulation of renewable methane deposits. At the same time, methane production technologies using organic substrates as an energy source for microbial populations could accompany these developments. The advantage is that in such a case methane is produced from recently photosynthesized carbon thus lowering the amount of respiration CO_2 [200].

In conclusion, unconventional methane is a promising future energy source especially at the expected escalating costs. Because of its high solubility in water it can dissolve and readily be extracted. Many stratified water bodies exist worldwide that are suitable to accomodate methane at depth.

The discussion on "Energy Future" would not be complete if there were no mentioning on some new promising technologies involving supercritical fluid CO_2 as a means to enhance oil recovery [201, 202]. In principle, CO_2 above its critical temperature of 31 °C becomes a dense phase at pressures encountered in many oil-producing rock formation. There, it can extract hydrocarbons and act as displacing fluid. The necessary CO_2 can be supplied from natural wells. Alternately, CO_2 generated in power plants could be collected and injected into oil-bearing strata. In this manner not only the life-time of petroleum as a source of energy is extended, but some of man's CO_2 would be safely returned to its source.

Boundary Phenomena

Air-sea-life-earth interactions are driven by the radiation from the sun ("fire"), the heat derived from the decay of radiogenic elements within the earth, and the energy released from phosphate bonds in the living cell. The alchemistic symbol for fire △ can be likened to the four triangular faces of a tetrahedron with fire at its center coordinating the flow of matter, be it gaseous, liquid, solid, or living (Fig. 23).

Carbon is a unique element in that it exists in all four compartments of Planet Earth. It is not only the key element of life, but occurs as rock (limestone), dissolved mineral (bicarbonate in the sea) or as a gas (CO_2 in the air). In all instances, CO_2 is the starting point of matter and the chief agent driving the carbon cycle.

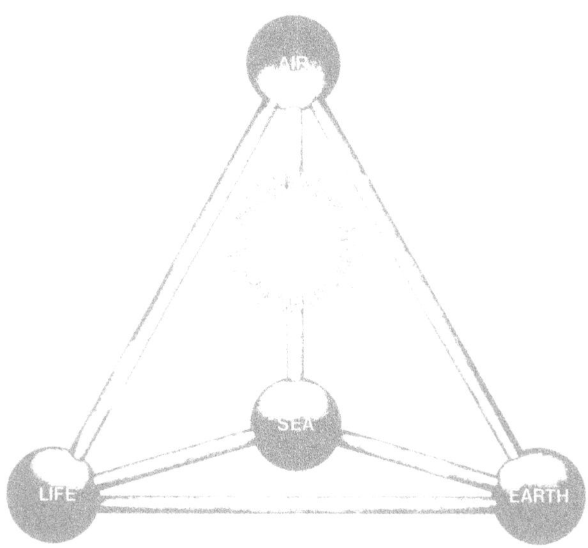

Fig. 23. Air-sea-life-earth interactions. Note that all spheres interconnect with one another [236]

The individual spheres depicted in Figure 23 or shown as compartments in Figure 9 are very well separated from one another. It is a fact that the most interesting reactions will not proceed within the spheres but at or close to their boundaries. At the contact air-solid earth we observe weathering and at the contact water-solid earth there is erosion on the continents and sediment deposition in the open sea. Air-sea interactions are responsible for the exchange of water, gases, heat, just to name a few. Biological membranes which can be considered phase boundaries between the organic cell and the outside environment, be it air or water, are chemically and structurally involved in channelling photosynthesis, osmosis, charge transfer (=electron donor – acceptor reactions) and hydration or dehydration processes. As is the case for the sciences, where the most interesting problems may no longer be found in the centers of physics, chemistry, biology or geology but at the peripheries of these sciences (e.g. biogeochemistry), we have to examine closely the phase boundaries between spheres, where the key for the understanding of systems such as the global carbon cycle is to be found. The forthcoming discussion focusses attention on some strategic boundaries where CO_2 is a critical element.

Weathering and Erosion

Carbon in the Rock Cycle

Carbon is a versatile agent of the rock cycle. The total carbon content of the Earth's crust amounts to 65,500,000 Gt, that is 0.27% of all rocks [203]. One quarter of this is organic carbon with an average isotopic composition of −25‰ on the $\delta\,^{13}C$ PDB scale. Carbonate carbon has a $\delta\,^{13}C$ of 0‰. The carbon streaming-off from crust and mantle in the form of CO_2 has a $\delta\,^{13}C$ of −5‰ and is thus almost identical to the $\delta\,^{13}C$ of atmospheric CO_2 in pre-industrial times.

The formation of oceanic and continental crust is a continous process. At the oceanic spreading centers ultramafic magma of the upper mantle differentiates and forms a several km thick crust of basalts and gabbros with an accretion rate of about 10 $km^3\ yr^{-1}$ (see also Fig. 10). The oceanic crust is in turn subducted and remelted at converging crustal boundaries. Thereby a second differentiation takes place and andesitic extrusiva and felsic intrusiva produce continental crust material of granitic composition. This material becomes aggregated onto the slowly growing continents. Due to the lesser density of granite – compared to basalt and gabbro – the continents can be much thicker – on average 40 km – than the oceanic crust and continental crust will rise above sea level.

Crystalline rocks will subsequently be subjected to weathering and erosion yielding sediments of various distinctions. They can be folded, subducted or undergo metamorphosis. In the process of mountain-building former sediments can rise and be subjected to weathering. Thus the exogenic rock cycle is closed [204]. Table 5 lists the carbon contents in the major crustal compartments. Sediments are particulary rich in carbon compared to crystalline rocks. The relationships are depicted in Fig. 24. Numerical models of the exogenic rock cycle reveal that the residence time of crustal carbon is close to 400 million years [205]. In short the rock cycle is the slowest "wheel" of the biogeochemical carbon cycle.

Table 5. Carbon contents of crustal compartments. (After [203])

	Total rock 10^{24} g	Total C 10^{21} g	Carbonate C 10^{21} g	Non-Carbonate C 10^{21} g
Basaltic oceanic crust	3.22	0.66	0.34	0.32
Oceanic sediments	1.20	20.16	14.16	6.0
Continental granitic crust	17.58	9.05	7.91	1.14
Continental sediments	2.0	35.64	25.64	10.0
Total crust	24.0	65.51	48.05	17.46
Average C-content in %		0.27	0.20	0.07

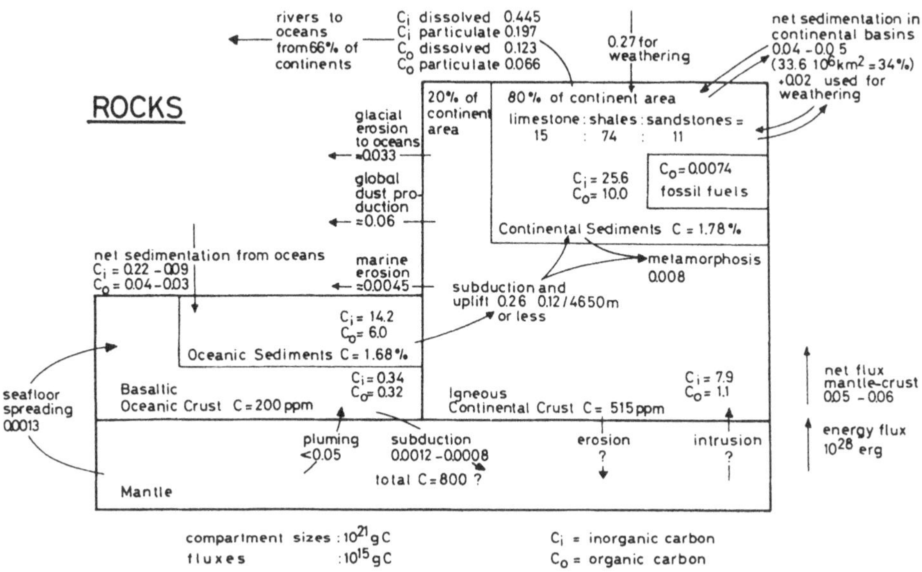

Fig. 24. The unbalanced rock cycle, compartment sizes and flux rates [203]

The Chemistry of CO_2 Induced Weathering

CO_2 is the most common acid anhydride on the Earth's surface and carbonic acid is the main agent of chemical weathering. Free CO_2 for weathering can be supplied by one or more of the five listed processes:

(1) atmospheric transport supplies CO_2 to exposed rock surfaces,

(2) root respiration and microbial degradation of organic substances supply large quantities of CO_2 to soil air,

(3) volcanic emanations release recycled or juvenile CO_2,

(4) oxidation of certain carbonate minerals (e.g. $FeCO_3$) may liberate CO_2, and

(5) CO_2 may be expelled from water bodies for equilibrium reasons.

Water charged with CO_2 and carbonic acid dissolves carbonates and silicate minerals in the following simplified scheme:

$$CaCO_3 + CO_2 + H_2O \rightarrow Ca^{2+} 2HCO_3^- \text{ (calcite dissolution)}, \quad (15)$$

$$FeSiO_3 + 2CO_2 + 3H_2O \rightarrow Fe^{2+} + H_4SiO_{4aq.} + 2HCO_3^- \quad (16)$$
$$\text{(iron silicate weathering)}.$$

The reaction involves a number of separate equilibria:

CO_2 dissolves in water and produces carbonic acid (H_2CO_3). H_2CO_3 in turn dissociates in two steps producing bicarbonate (HCO_3^-), carbonate (CO_3^{2-}) and hydrogen (H^+) ions. The process is described by a set of mass balance equations, the equilibrium constants can be obtained from Table 6 (as p-values, i.e. as negative decadic logs). For details see [205, 206]:

$$(H_2CO_3^*)/PCO_{2gas} = K_{CO_2} = K_H, \quad (17)$$

$$(H^+) * (HCO_3^-)/(H_2CO_3^*) = K_1, \quad (18)$$

$$(H^+) * (CO_3^{2-})/(HCO_3^-) = K_2, \quad (19)$$

$$(H^+) * (OH^-)/(H_2O) = K_W. \quad (20)$$

The brackets denote activities (X_i), which are related to total concentrations of the ion in question (mX_i in mol/kg = molality) by:

$$(X_i) = \gamma_i \, mX_i. \quad (21)$$

γ_i is the activity coefficient which may be approximated in fresh waters (I below 0.1) by the extended Debye-Hückel equation:

$$-\log \gamma_i = (A * z_i^2 * \sqrt{I})/(1 + a_i * B * \sqrt{I}). \quad (22)$$

A and B are tabulated constants related to temperature (compare Table 6); a_i symbolizes the ionic radii (a_i in 10^{-8}): $OH^- = 3.5$, $HCO_3^- = 4.0$–4.5, $CO_3^{2-} = 4.5$, $Ca^{2+} = 6$, $Mg^{2+} = 8$, and $H^+ = 9$. I denotes the ionic strength, i.e. half of the sum of the concentrations of all ions multiplied by the square of the absolute values of their charges (z_i):

$$I = 0.5 * \Sigma mX_i * |z_i|^2. \quad (23)$$

The activity coefficient of a strong electrolyte is identical in aqueous solutions of the same ionic strength.

Concentrations of the free ion in an aqueous solution is, however, not identical with the total concentration of that element as determined by chemical analysis. Alkali earth metals, sulfate and to a certain extent also carbonate ions form neutral or charged ion pairs such as $CaHCO_3^+$, $CaCO_3^0$, $CaSO_4^0$, $CaOH^+$, $MgHCO_3^+$, $MgCO_3^0$, $MgSO_4^0$, and $MgOH^+$. Chloride does not form such pairs but sodium does at higher concentrations, e.g. in sea water. Wigley [207] suggested a procedure to account for these ion pairs in equilibrium calculations of natural waters. First one calculates the ionic strength of the water, assuming that the analysed concen-

Table 6. Mass balance constants of carbonate system

Temperature °C:		0	5	10	15	20	25	30	35
pK_H	(a)	1.11	1.19	1.27	1.32	1.41	1.47	1.53	
pK_1	(a)	6.579	6.517	6.464	6.419	6.381	6.352	6.327	6.309
pK_2	(a)	10.625	10.557	10.490	10.430	10.377	10.377	10.290	10.250
pK_W	(a)	14.93	14.73	14.53	14.35	14.17	14.00	13.83	
$pK_{Calcite}$	(b)	8.350	8.351	8.359	8.373	8.393	8.420	8.452	8.491
$pK_{Calcite}$	(c)	8.390	8.396	8.406	8.423	8.446	8.475	8.509	8.549
$pK_{Dolomite}$	(d)	16.56	16.63	16.71	16.79	16.89	17.00	17.12	17.25
$pK_{Magnesite}$	(e)						4.9		
$pK_{Aragonite}$	(e)						8.22		
$pK_{Siderite}$	(e)						10.40		
Constants for sea water at 35‰									
pK_H	(f)	1.19	1.27	1.34	1.41	1.47	1.53	1.58	
pK_1	(g)	6.20	6.15	6.10	6.06	6.03	6.00	5.98	5.96
pK_2	(g)	9.45	9.39	9.33	9.25	9.18	9.11	9.06	9.01
Constants for extended Debeye-Hückel equation (h)									
A		0.4883	0.4921	0.4960	0.5000	0.5042	0.5085	0.5130	0.5175
$B * 10^{-8}$		0.3249	0.3249	0.3258	0.3262	0.3273	0.3281	0.3290	0.3297

(a) Stumm and Morgan [102], Tables 4.7, 4.8, 4.9, 3.1, at infinite dilution
(b) Jacobson and Langmuir [212], ignoring ion pairs
(c) Jacobson and Langmuir [212], assuming ion pairs
(d) Langmuir [216]
(e) Stumm and Morgan [102], Tables 5.1, at infinite dilution
(f) Buch [63] (19‰ Cl)
(g) Mehrbach et al. [65]
(h) Garrels and Christ [205]

trations are free ions. Then the activity coefficients and the activities of the ions are calculated. Activity coefficients of uncharged species are normally taken as unity, but experiments [208, 209] reveal, that activity coefficients for the pairs in question are significantly less than unity. Next the concentration of the ion pairs are evaluated by using their mass balance constants (e.g. [207, 210–213]). Finally the concentration of the free ions is calculated by subtracting the concentrations of the ion pairs – recalculated from their activities – from the total concentration of the element. Now the free ion concentration is used in a new estimate of the ionic strength and the process is repeated. Normally no more than five iterations give a sufficient accuracy.

It is found, that even in dilute solutions not more than 85% of the sulfate is present as a free ion. In solutions rich in sulfate only 70% of the sulfate is present as its free ion, 70–72% of the Ca and Mg are free ions, and 80–85% of the bicarbonate. Even in sulfate-free solutions of low ionic strength up to 5% of the bicarbonate exists in pairs with alkali earth metals (e.g. [214, 215]).

The mass balance constant K_H (17) describes the relation between PCO_2 and total free CO_2, i.e.:

$$(CO_{2aq.}) + (H_2CO_3) = (H_2CO_3^*). \tag{24}$$

The constant is the Henry law constant which brings the partial pressure of a gas (mol per atm) in relation with a concentration (activity) of the same molecule in water (see p. 183). The partial pressure may also be noted as its negative decadic log abbreviated as $pPCO_2$:

$$pPCO_2 = -pK_H - \log(H_2CO_3^*). \quad (17b)$$

Values for pK_H are given by Table 6 and can be approximated by:

$$pK_H = 1.12 + 0.014T \quad (25)$$

with T as temperature in °C [210].

K_1 is the "first acidity constant" which is a composite constant for the protolysis of $H_2CO_3^*$ and includes the hydration reaction:

$$(H_2CO_3)/(CO_{2aq.}) * (H_2O) = K_0 \quad (18b)$$

and the protolysis of the true H_2CO_3:

$$(H^+) * (HCO_3^-)/(H_2CO_3) = K_{H_2CO_3} \quad (18c)$$

The equilibrium of the hydration reaction(K_0)lies far to the $CO_{2aq.}$ side, i.e. most of the $H_2CO_3^*$ is free CO_2, at 25 °C less than 0.3% is true H_2CO_3. K_0 and $K_{H_2CO_3}$ may be replaced by

$$(H^+) * (HCO_3^-)/[(H_2CO_3) + (CO_{2aq.})] = K_{H_2CO_3}/(1 + K_0) = K_1. \quad (18d)$$

At 25 °C K_0 ist about 650, thus K_1 is about $= K_{H_2O_3}/K_0$ and the concentration of $CO_{2aq.}$ is almost identical with the $H_2CO_3^*$ concentration. K_1 can accurately be determinated experimentally [102].

The charge balance of water equilibrated with air PCO_2 ($pPCO_2 = 3.5$):

$$(H^+) = (HCO_3^-) + 2(CO_3^{2-}) + (OH^-) \quad (26)$$

will determine the pH of the solution. Water will yield a pH of 5.65 should the hydrogen concentration be balanced by the charge of the bicarbonate. The carbonate and hydroxide ions are one and a half orders of magnitude lower in concentrations and do not influence the charge balance. Addition of strong acids and bases will shift the system according to the sloping lines in Figure 25.

Total carbon in solution is defined by:

$$C_T \text{ (often } \Sigma\, CO_2) = mH_2CO_3^* + mHCO_3^- + mCO_3^{2-}. \quad (27)$$

The alkalinity of a solution is the sum of all anions of weak acids:

$$\text{alkalinity} = mHCO_3^- + 2mCO_3^{2-} + mOH^- - mH^+ + \text{others} \quad (28)$$

and the acidity is defined as:

$$\text{acidity} = mHCO_3^- + 2mCO_3^{2-} - mOH^- + mH^+. \quad (29)$$

From Figure 25 ist is apparent, that at high pH values alkalinity and total carbon increase, while at lower pH values the water will show high concentrations of H^+ and an increasing mineral acidity.

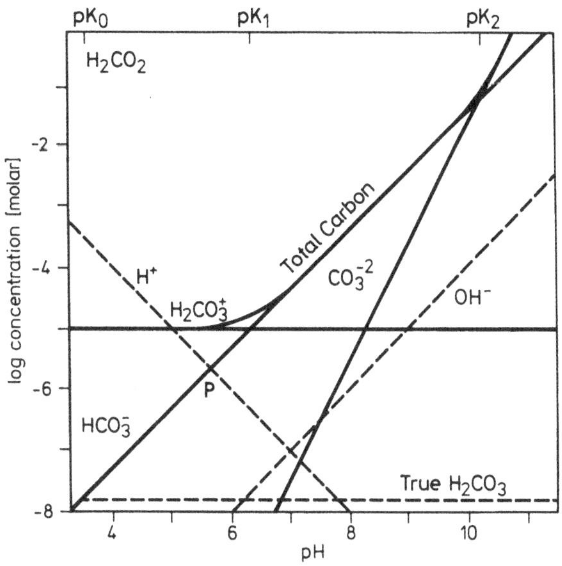

Fig. 25. Aqueous carbonate equilibrium at a constant PCO_2 (3.5 $pPCO_2$). Rainwater devoid of other bases or acids would equilibrate at point P (about 5.65 pH) where H^+ is balanced by HCO_3^-, other species playing a negligible role. Addition of acids or bases would shift carbon according to the upper, heavy line. Graph is valid for 25 °C with constants $pK_H = 1.5$, $pK_1 = 6.3$, $pK_2 = 10.25$, and $pK_0 = -2.8$ been used. Equilibration concentration of $-\log(CO_{2aq.}) = -\log(H_2CO_3^*) = 5.0 \times 10^{-5.0}$ mol/l about = 0.44 ppm CO_2 [102]

Carbonate Dissolution

The solubility of carbonate minerals is described by the equations:

$$(Ca^{2+}) * (CO_3^{2-})/(CaCO_3)_s = K_{Calcite}, \quad (30)$$

$$(Ca^{2+}) * (Mg^{2+}) * (CO_3^{2-})^2/(CaMg(CO_3)_2)_s = K_{Dolomite}, \quad (31)$$

$$(Mg^{2+}) * (CO_3^{2-})/(MgCO_3)_s = K_{Magnesite}, \quad (32)$$

$$(Ca^{2+}) * (CO_3^{2-})/(CaCO_3)_s = K_{Aragonite}, \quad (33)$$

$$(Fe^{2+}) * (CO_3^{2-})/(FeCO_3)_s = K_{Siderite}. \quad (34)$$

The respective pK-values are listed in Table 6. Much work exists on the solubility of calcite [212], but questions remain as to the precise constant of dolomite [216]. A perfect dolomite is rarely encountered in nature; most have slightly higher Ca than Mg concentrations which explains that dolomites dissolve incongruently. At dissolving a Ca-rich dolomite, first the saturation of calcite is reached. Further dissolution of dolomite will cause precipitation of calcite. The ratio of Ca/Mg in a solution equilibrated with respect to calcite and dolomite is more related to temperature, than to PCO_2. The mol-ratio is 0.59 for 10 °C and 1.1 for 25 °C. In turn dissolution of dolomite can lead to a range of Ca/Mg ratios. Upon evaporation of Mg-rich solutions first calcite ($CaCO_3$), then Mg-calcite ($Ca_{0.96-0.5}Mg_{0.04-0.5}CO_3$), aragonite ($CaCO_3$) together with monohydrocalcite ($CaCO_3 * H_2O$), and hydromagnesite [$Mg_4(OH)_2(CO_3)_3$] jointly with nesquehonite ($MgCO_3 * H_2O$)

precipitate in nature [217]. Synthetic protodolomite ($Ca_{0.65-0.5}Mg_{0.35-0.5}CO_3$) too can form in pure Ca-Mg solutions.

In discussing saturation of waters with respect to certain minerals the concept of the saturation index (SI) is helpful. SI can be definded as:

$$SI = \log(IAP/K_{mineral}) \tag{35}$$

i.e. the SI compares the ion activity product (IAP) of all ions of a certain mineral with its equilibrium constant K. The log scale results in zero-values at equilibrium, positive values at supersaturation, negative values at undersaturation. In detail the SI's for calcite and dolomite are given by:

$$SI_{calcite} = \log[(Ca^{2+}) * (HCO_3^-) * K_2] - \log[(H^+) * K_{Cal}] \tag{36}$$

and

$$SI_{dolomite} = \log[(Ca^{2+}) * (Mg^{2+}) * (HCO_3^-) * (HCO_3^-) * K_2 * K_2] - \log[(H^+) * (H^+) * K_{Dol}]. \tag{37}$$

Calculation of SI without considering ion pairs leads commonly to an overestimate of SI.

Calcite reacting with water consumes CO_2 and yields bicarbonate. If no CO_2 is recharged from ambient air, we speak of a closed system. In such a case PCO_2 of the solution drops to a point where calcite and PCO_2 are equilibrated. In an open system calcite dissolution will continue until equilibrium with ambient PCO_2 is established.

The evolution of the open system is exemplified by:

$$CO_2—CaCO_3—H_2O\,.$$

Dividing K_1 and K_2 yields:

$$K_1/K_2 = (HCO_3^-)^2/(CO_3^{2-}) * (H_2CO_3^*)\,. \tag{38}$$

Substituting $PCO_2 * K_H$ for $(H_2CO_3^*)$ and $K_{Cal}/(Ca^{2+})$ for (CO_3^{2-}) gives:

$$K_1/K_2 = (HCO_3^-)^2 * (Ca^{2+})/PCO_2 * K_H * K_{Cal}\,. \tag{39}$$

Due to the charge balance of $mHCO_3^-/2 = mCa^{2+}$, we can write for solutions containing mostly bicarbonate and calcium ions:

$$(HCO_3^-) = 2mCa^{2+} * \gamma_{H_2CO_3^-}$$

and

$$(Ca^{2+}) = mCa^{2+} * \gamma_{Ca^{2+}}\,. \tag{40}$$

Substituting this we obtain:

$$K_1/K_2 = 4(mCa^{2+})^3 * \gamma_{Ca^{2+}} * (\gamma_{HCO_3^-})^2/K_{Cal} * K_H * PCO_2 \tag{41}$$

and:

$$K_1 * K_{Cal} * K_H * PCO_2/4K_2 * \gamma_{Ca^{2+}} * (\gamma_{HCO_3^-})^2 = (mCa^{-+})^3 \tag{42}$$

Evaluating this equation for various PCO_2's results in curves as depicted in Fig. 26.

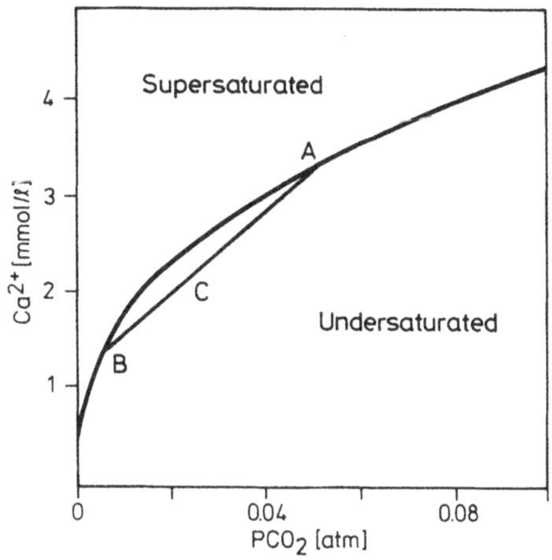

Fig. 26. The relation between PCO_2 and the equilibrium $CaCO_3$ concentration

The Ca ion concentration – and similarly the alkalinity – increases non-linearily with PCO_2. At atmospheric PCO_2 (340 ppm) only about 0.4 mmol Ca/l or about 40 mg $CaCO_3$/l dissolve. At soil PCO_2 of 100,000 ppm (0.1 atm) ten times as much calcite will dissolve. In fact such high values are often found in rivers, springs and ground waters from limestone regions indicating the importance of soil PCO_2 on weathering whereas CO_2 dissolved in rain water plays only a minor role in mineral weathering.

Karstification

Areas dominated by limestone develop special morphology and hydrology, called karst. Karst lowering and erosion is considerably faster if outcrops are covered by vegetation. Any pocket of soil in barren limestone will attract plants. Punctually the effective PCO_2 is increased and the rock below the soil dissolves rapidly forming a doline that is a surface dissolution pit.

Figure 26 illustrates another essential feature of carbonate dissolution and hence of karst development. This concerns the mixing of two waters (A and B) of different PCO_2 and $CaCO_3$ concentration. Their equal-volumen conservative mixture (C) will have an indermediate Ca-concentration not in equilibrium with PCO_2 at the point of mixture. The mixture appears undersaturated with calcite, even though the endmembers had been saturated.

The mixture is capable of dissolving additional $CaCO_3$. This phenomenon is called mixing-corrosion [218, 219]. In detail the mixing-corrosion is a composite of an algebraic, the Δ PCO_2 and an ionic strength effect [220]. The mixing of carbonate waters has been thermodynamically evaluated and a computer program (WATMIX) was designed that allows theoretical mixing of various waters, including sea water [220]. Redistribution of ionic species in the mixture often leads to lower pH-values of the mixture than of both endmembers, an effect sometimes measured in estuaries at the mixing front of fresh and salt waters.

In limestone mixing-corrosion may occur deep below the surface. At intersections of e.g. a bedding plane and a joint, water different in PCO_2 may cause corrosion, and create small tubes far below the water table. Such tubes enlarge gradually and finally interconnect to form an integrated cave where water can drain quickly. Later intruding surface water not in equilibrium with calcite can continue to enlarge the cave. In the final stage of the karst development, all surface water is completely drained to underground rivers.

The kinetics of calcite dissolution are described at fixed temperature and PCO_2 by:

$$R = k_1(H^+) + k_2(H_2CO_3^*) + k_3(H_2O) - k_4(Ca^{2+})(HCO_3^-). \tag{43}$$

k_1, k_2, and k_3 are first order rate constants, depending on temperature. k_4 depends on both temperature and PCO_2. Three mechanisms seem to occur simultaneously:

$$CaCO_3 + H^+ \rightleftharpoons Ca^{2+} + HCO_3^-, \tag{44}$$

$$CaCO_3 + H_2CO_3^0 \rightleftharpoons Ca^{2+} + 2HCO_3^-, \tag{45}$$

$$CaCO_3 + H_2O \rightleftharpoons Ca^{2+} + HCO_3^- + OH^-. \tag{46}$$

These reactions are counterbalanced by the backward interaction of calcium and bicarbonate ions with the solid surface [221, 222]. Reaction (44) controlls the dissolution in low pH-ranges. Reaction (46) controlls dissolution in absence of hydrogen ions and carbonic acid, i.e. at high pH and low PCO_2. Reaction (44) is also the fastest when compared to (45) and (46).

These reaction rates were applied to PCO_2 and carbonate concentrations found in natural waters and rates of cave enlargement were calculated using transport models [223, 224]. Here are some results: CO_2-charged water entering joints (20 μ wide) will already be saturated after 23 cm of travel. Saturation in 200 μ wide joints is established within one meter suggesting that dissolution will not proceed within the limestone strata, except for mixing-corrosion. Calculations show, that within a few thousand years mixing-corrosion tubes can develop. If no lateral transport along the tube occurs, the dissolution would have created a tube 1 m wide within 600,000 years. The diameter of this tube would grow according to the square root of time. However, tubes more than 5 mm wide will sustain turbulent flow. A characteristic dissolution rate for turbulent flow is 0.0005 μg $CaCO_3$ $cm^{-2}sec^{-1}$. Cave conduits can be enlarged from a few millimeter to 50 cm wide within a few 10,000 years. The diameter increases linearily with time. Mixing-corrosion occurring within turbulent cave conduits have high saturation lengths, i.e. $^2/_3$ of their corrosion potential is often consumed only kilometers below the point of confluence of the two original endmembers and very long cave channels usually form.

Reactions of Silicates with CO_2

When reacting carbonates with carbonic acid, one mole CO_2 is consumed from the water and one is provided by the carbonate. Upon recrystallisation of calcium carbonate, one mole of CO_2 is returned to the environment. Thus, consumption of CO_2 by carbonate dissolution is not a long-term sink for CO_2. This does not apply to silicates. The Mg-olivine forsterite disintegrates in the presence of carbonic acid:

$$Mg_2SiO_4 + 4CO_2 + 4H_2O \rightarrow 2Mg^{2+} + 4HCO_3^- + H_4SiO_4. \tag{47}$$

4 moles CO_2 are taken from soil air, of which two return when magnesium is precipitated as magnesite. With every mole of forsterite weathered, 2 moles of CO_2 are fixed from the gaseous pool in the form of carbonate.

The weathering of the Na-feldspar albite will produce an alkaline solution plus the clay mineral kaolinite:

$$2\,NaAlSi_3O_8 + 2CO_2 + 11\,H_2O \rightarrow$$
$$Al_2Si_2O_5(OH) + 2\,Na^+ + 2\,HCO_3^- + 4\,H_4SiO_4. \quad (48)$$

The consumed CO_2 will not return to the atmosphere. Weathering of anorthite (Ca-feldspar) ($CaAl_2Si_2O_8$) will produce kaolinite, 1 mole Ca and 2 moles bicarbonate but no dissolved silica. Upon precipitation of calcite, only one mole of CO_2 is returned to the gaseous pool. In consequence, the weathering of silicates forms an effective sink for CO_2.

Global Rates

Export of dissolved products of chemical weathering provide an efficient way to establish global weathering rates. Recalculating river quality data [225] and using newer discharge values for the Amazon, rivers carry an average salinity of 122.4 ppm, a HCO_3^- content of 59.6 ppm and a silica concentration of 13.1 ppm [203, 226]. More recent data [227] gave average concentration of 52 ppm bicarbonate and 10.8 ppm for silica. Rivers discharge annually 37.7×10^{18} g of water [228].

From these figures we can calculate the contribution of silicate weathering to the transport of carbonate carbon. The most common silicate mineral weathered is the feldspar plagioclase (average composition $Na_{0.62}Ca_{0.38}Al_{1.38}Si_{2.62}O_8$) which disintegrates into one mole of HCO_3^- and SiO_2 each. The mole ratio of SiO_2 to HCO_3^- in rivers is $8.2 \times 10^{12}/37 \times 10^{12}$ moles. Thus 0.22 or one quarter of the bicarbonate discharge could be induced by silicate weathering. Silicate weathering on the continents could deplete the atmosphere of its CO_2 within 7,000 years.

One third of the silicate-carbonate is balanced by Ca derived from plagioclase and can form carbonates, i.e. one sixth returns as CO_2. Thus the annual production of carbonates from silicates amounts to 0.016 Gt C.

Twenty percent of the continents are covered by limestone or calcareous rocks of various compositions [204]. Four percent of total continental area show distinct karst features [229]. The bulk of karst occurs in Europe and Asia between 60° and 20 °N with smaller areas in North America and marginal regions in the southern hemisphere. This is because marine limestones occur mostly in young, active mountain belts or as uplifted deposits of former epicontinental seas. This irregular distribution of limestones explains the large differences in continental mean carbonate concentrations of rivers as given in Table 7 [226].

Figure 27 compares long-term means of calcite saturations and PCO_2 of several rivers. Clearly, South-American and African streams have the lowest saturation of calcite. Other rivers from Europe and Asia are highly supersaturated because they are fed by groundwaters from densely vegetated limestone regions. In case of the Huanghe, the high calcite supersaturation paired with very high total dissolved calcite favors precipitation of calcite in the river and leads to a cementation of the river bed [230].

Table 7. Discharge of total dissolved matter and inorganic carbon of the continents. (After Baumgartner and Reichel [228], Balazs [229], and Livingstone [225]

Continents	Europe	Asia	N. Amer.	S. Amer.	Africa	Australia
Data of Livingstone:						
Discharge $km^3\ yr^{-1}$	2,498	11,108	4,557	8,008	5,901	316
Salinity ppm	182	142	142	69	121	59
HCO_3^- ppm	95	79	68	31	43	32
Inorg. C ppm	18.7	15.5	13.3	6.1	8.5	6.2
Data of Baumgartner and Reichel:						
Discharge $km^3\ yr^{-1}$	2,800	12,200	5,900	11,100	3,400	2,400
Data of Balazs:						
Area $10^6\ km^2$	10.5	43.9	30.3	24.2	17.9	8.5
Karst $10^6\ km^2$	1.42	1.60	0.99	0.86	0.09	0.38
Karst %	13.5	3.6	3.2	3.5	0.5	4.5

Calculated sums and weighted means:

Livingstone:	Total discharge $km^3\ yr^{-1}$	32,390
	Salinity average ppm	122.4
	HCO_3^- average ppm	59.8
	Diss. inorg. C ppm	11.8
Baumgartner and Reichel:	Total discharge $km^3\ yr^{-1}$	37,700
	Total diss. load Gt yr^{-1}	4.20
	Total HCO_3^- Gt yr^{-1}	2.31
	Total diss. inorg. C Gt yr^{-1}	0.45
Balazs:	Total continents (without Antarctica) $10^6\ km^2$	135.3
	Total karst area $10^6\ km^2$	5.34
	Total karst percentage	4.0

Erosion of Organic Carbon

Little is known on global erosion rates for organic carbon, for the following reasons:

1. Techniques to measure dissolved organic carbon (DOC) and particulate organic carbon (POC) have only recently been developed. Formerly, chemical and biological oxygen demands were used as indirect measurements of organic carbon loads.

2. The DOC concentration of a river often undergoes marked changes during the year, depending on seasonal biological activities and on erosion from floodplains. Thus single measurements often do not reveal the total organic carbon transported and long-term records of the type shown for the Niger River (Fig. 28) are needed [231].

3. Measurements for particulate carbon require new collecting techniques to obtain representative water samples [232].

4. Much carbon is lost by sedimentation on alluvial fans, in deltas, on flood plains, in natural lakes, and behind dams [233].

5. Organic carbon undergoes alterations during its travel from land to sea. In lakes with stable stratification photosynthesis will decrease the PCO_2 of the water producing organic matter (Fig. 29). This is greatly enhanced by pollution with nu-

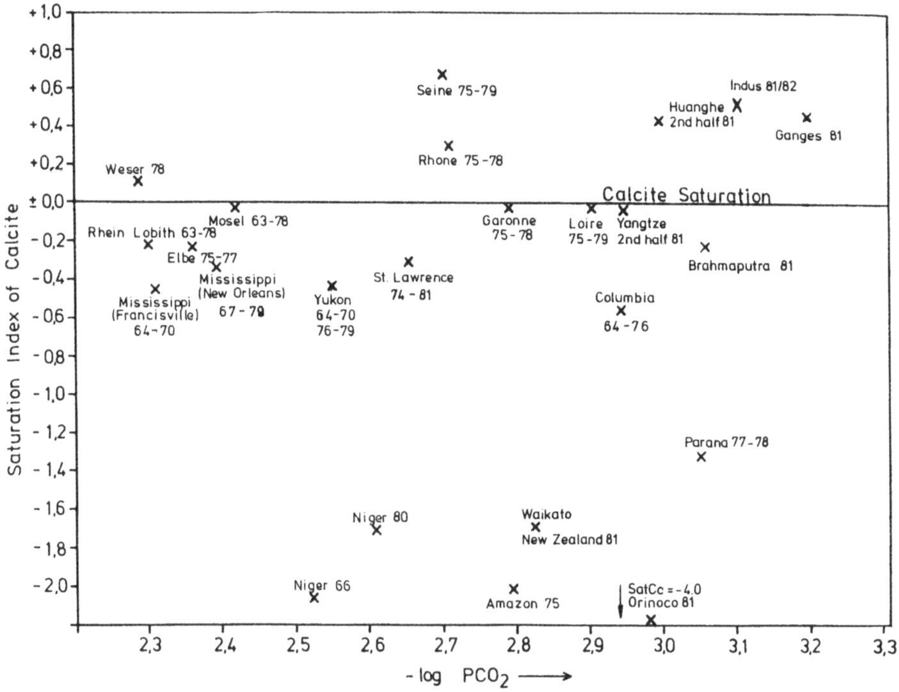

Fig. 27. Annual or long-term means of PCO_2 and calcite saturation index of various rivers. Note $-\log$ scale for $PCO_2 = pPCO_2$, $pPCO_2$ 3.5 equals 330 ppm PCO_2 and $pPCO_2$ 2.5 equals 3,300 ppm PCO_2. Saturation index is defined in Eq. (35). Its dimensionless value is positive at supersaturation, zero at saturation and negative at undersaturation [101]

trients. The long-term trend of the PCO_2 of the river Rhine from Lake Constance to the North Sea is illustrated in Figure 30 [101]. Clearly the build up of very high PCO_2 values below Lake Constance can be followed. Rivers polluted with organic wastes gain a PCO_2 of more than 30 times the equilibrium value with ambient air. This artificial increase of PCO_2 due to pollution with organic waste is clearly seen in many rivers from industrialized regions. The Mississippi, Rhine, Weser, and Elbe Rivers group at much higher PCO_2 values than less polluted rivers (Fig. 27).

Estimates of annual organic carbon transport to the sea ranges from 0.1 to 1.0 Gt C yr^{-1} [230]. A widely accepted value is 0.4 Gt C yr^{-1} [227, 234, 235]. The SCOPE/UNEP project "Transport of Carbon and Minerals in Major World Rivers" is currently studying many important rivers for their carbon budgets. Measurements will eventually include several annual cycles [236, 237].

Even though annual transport values of combined inorganic and organic carbon amount to only one fifth of the annual fossil carbon releases, there might be a series of sinks associated with the processes and transformations of carbon in the freshwater system equalizing the total export to the sea:

1. Destruction of natural vegetation cover and transformation of land for agriculture increase the rate of soil erosion several fold [238]. Much of this material is burried inland and thus spared from rapid oxidation. No estimate for the amount of this carbon is currently available.

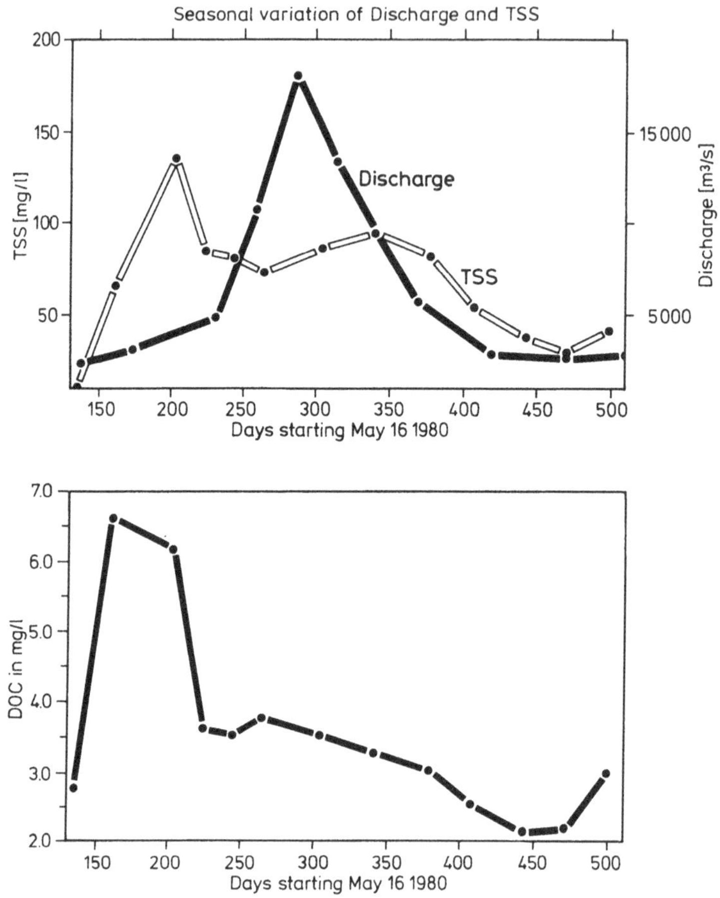

Fig. 28. Seasonal cycle of DOC concentration in relation to the annual discharge cycle. Note the maximum in DOC at initial overbank of the river and first inundation of the floodplains [231]

2. The inaccuracy of the estimates of the POC flux with rivers to the sea is such, that an additional flux of as much as 0.5 G C yr^{-1} could occur.

3. Deposition of organic carbon in inland lakes and reservoirs are estimated to reach 0.25 Gt C yr^{-1} [233].

4. Eutrophic lakes acquire a PCO_2 below that of atmosphere and become small sinks for atmospheric CO_2. In addition to organic deposition, lakes and reservoirs can precipitate carbonates due to increasing pH-values during photosynthesis.

5. Increased input of labile organic matter into rivers causes the riverine PCO_2 to rise. Rivers become sources of CO_2 due to surface degassing. Additional free CO_2 – typically one tenth of the HCO_3^-C – is discharged into the oceans where buffering with solid marine carbonates prevents loss of this CO_2 to the air. The size of this sink, is about 0.04 Gt C yr^{-1} [101].

6. Rivers discharge increasing amounts of phosphate and nitrate into the coastal seas. Rivers from industrialized agricultures already carry nutrient concentrations two orders of magnitudes higher than pristine rivers (Fig. 31). The N/P

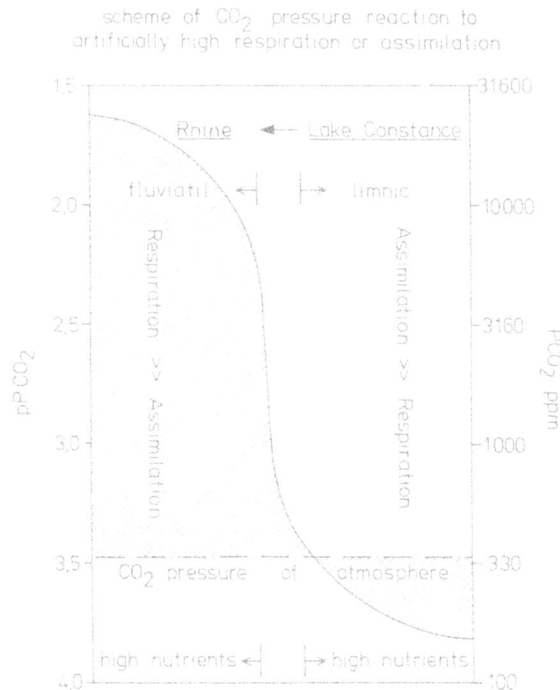

Fig. 29. Scheme of the reaction of PCO_2 in the lacustrine and fluvial environment and the effects of pollution on PCO_2 in these environments [101]

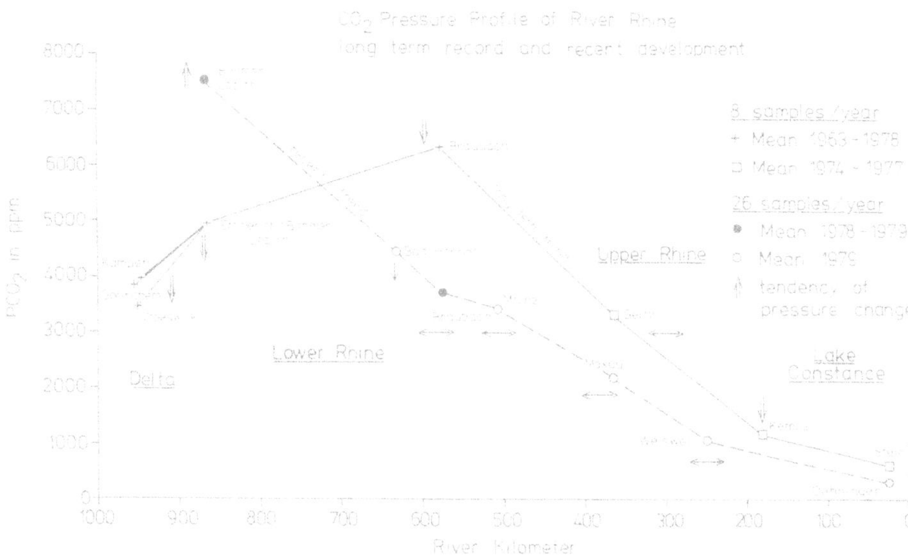

Fig. 30. Longitudinal PCO_2 profile of the River Rhine as evident from long-term mean PCO_2 of 13 hydrochemical monitoring stations. Times for which means apply are indicated. River flow is from right (outflow of Lake Constance) to left (Rhine delta in the Netherlands) [101]

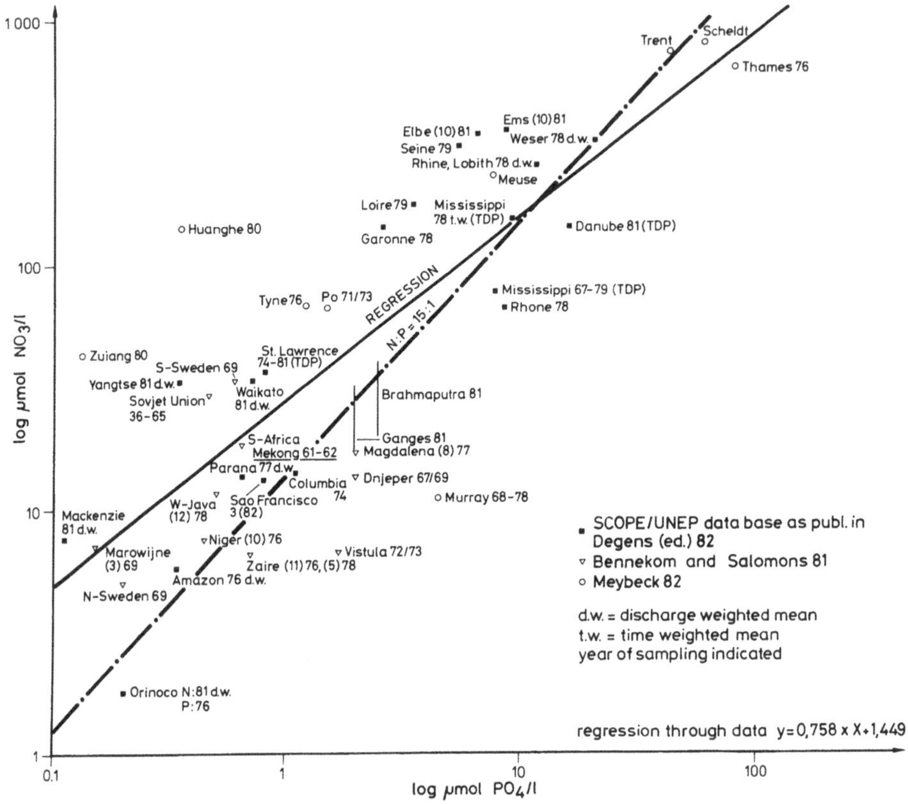

Fig. 31. Scatter plot of mean nitrate versus mean phosphate concentrations for various world rivers. 76 = average of 1976, (10)81 = single sample Oct. 1981, t.w. = time weighted mean, d.w. = discharge weighted man, TDP = Total Dissolved Phosphate, TDN = Total Dissolved Nitrogen (inorganic)

ratio discharged is similar to the requirements of marine plankton and could have triggered intensified photosynthesis in the order of 0.1 Gt C yr^{-1}.

Air-Sea-Exchange

Theoretical Concepts

In order to penetrate a gas-liquid-interface a molecule needs energy to overcome surface tension. It may derive this energy from the kinetic energy of either (turbulent) eddy diffusion or molecular diffusion. The role of capillary waves as a reservoir of kinetic energy in liquid-gas transfer has been discussed [239]. Based on empirical evidence the common assumption is that at the phase boundary eddy kinetic energy contributes – if at all – minimally to gas transfer if compared with molecular diffusion. For convenience one therefore defines a strictly laminar layer in which only molecular diffusion transports a gas.

The air-water interface is then considered a composite of a liquid and a gaseous laminar layer or "film". The air above and the water below the respective films are well mixed by fully developed turbulent eddy motion where the relative contribution of molecular diffusion to transport is negligible. Although in reality the transition from laminar to turbulent is certainly continuous, chemical engineers and oceanographers have successfully applied the simplified concept of a sharp discontinuity between a purely molecular diffusive film and a purely turbulent reservoir of the respective phase [240–246].

Modelling approaches from aerodynamic research on flow along a solid-gas boundary which describe the continuous transition have been adapted to the air-sea-exchange problem [239]. Let us remain with the discontinuous film concept and see what flux is to be expected.

The gas-flux F through a laminar film of thickness dz is, according to Fick's first law of diffusion, proportional to the concentration gradient dC/dz across the film:

$$F = D(dC/dz) \tag{49}$$

where D is the coefficient of diffusion of gas in the film. Since D has dimensions L^2T^{-1}, a "transfer velocity" $k = D/dz$ (dimensions LT^{-1}) can be defined. For steady state transport of gas across the two layers of the interface we may then write [246]:

$$F = k_g(C_g - C_{sg}) = k_l(C_{sl} - C_l) \tag{50}$$

where k_g and k_l are the transfer velocities for the gas and liquid phases, respectively. C_g is the concentration of the gas in the well mixed gas phase above the gas film, C_l is the concentration of the gas in the well mixed liquid phase below the liquid film, C_{sg} and C_{sl} are the concentrations of the gas at the boundary of the two films in the gas and the liquid phases, respectively. If the exchanging gas obeys Henry's law, then:

$$C_{sg} = HC_{sl}^{max} \tag{51}$$

where H = [equilibrium concentration in gas phase (g per cm^3 of air)/equilibrium concentration of unionized dissolved gas in liquid phase (g per cm^3 of water)]. The inverse of k is a measure of the "resistance" to gas transfer, and has dimensions (TL^{-1}).

Eliminating C_{sg} and C_{sl} between Eqs. (50) and (51) and after some rearrangements, one arrives at an expression for the overall resistance R to gas transfer through the air-sea-interface which may be written as the sum of apparent series resistances [246]:

$$R = R_L + R_G \tag{52}$$

where R_L and R_G are the liquid and gas film resistance, respectively. They may be expressed as:

$$R_L = 1/k_l = dz_l/D_l, \tag{53a}$$

$$R_G = 1/k_g = dz_g/HD_g. \tag{53b}$$

with dz_l and dz_g as thicknesses of the liquid and gaseous laminar film and D_l and D_g the molecular diffusivities of the gas in the respective films.

In the case of CO_2, transfer between air and sea is liquid resistance controlled [241], i.e. $R_L \gg R_G$. (Examples for $R_G \ll R_L$ would be H_2O, SO_2, NH_3). For gases, R_L depends on their solubility and chemical reactivity in the aqueous phase.

The liquid resistance control on CO_2 exchange between air and sea is evident from the observation that the atmospheric residence time of CO_2 (several years) is so much longer than that of H_2O (several days), although their molecular diffusivities in the gas phase are similar [241].

As for CO_2 $R_L \gg R_G$, a simplified air-sea-interface model, containing a molecular film only in the liquid phase, is sufficient for practical purposes.

With regard to this liquid boundary layer's behavior on a macroscopic scale two concepts are used [243]: the "stagnant film" model and the "film replacement" or "surface renewal" model.

In the first, a stagnant liquid film boundary layer covers the entire liquid surface and only changes its thickness in response to wind and wave motion. In the latter, the fluid is at rest at the interface, but water is being removed from there and replaced by water from inside the liquid phase at random intervals. It was stated [239] that this model seems applicable to the case of little mechanical mixing but unstable stratification, e.g. by cooling of the sea surface, due to evaporation and radiation under conditions of vanishing wind speed and low insolation. According to [247], the difference in the rate of diffusion calculated with both models is not of significance for oceanographic conditions.

Chemical enhancement of oceanic CO_2 gas exchange, resulting from the reaction:

$$CO_2 + H_2O + CO_3^{2-} = 2HCO_3^- \qquad (17\text{–}19)$$

was included in the stagnant film model [241]. The model was further extended by introducing the constraint of electro-neutrality [244]. Chemical enhancement dominates over pure physical transport at film thicknesses greater 400 microns. For the average oceanic thicknesses below 100 microns it is negligible. The average time of transit for a dissolved CO_2 molecule diffusing through a 40 micron thick surface layer is less than 1 second, but the half-time for the reaction of CO_2 and H_2O at 20 °C is about 50 seconds [244]. Therefore there is no appreciable chance of a CO_2 molecule to react with water during the time of passage through the film.

It has been suggested [248] that the presence of carbonic anhydrase may significantly accelerate the reactions. More recent work does not seem to support this suggestion (Enns, quoted in [249]).

Laboratory Experiments

Because CO_2-transfer is liquid film controlled it depends on the liquid film's thickness – apart from the concentration difference between liquid and gas phase. The film thickness is a function of wind speed. Various laboratory experiments have been performed in search of a relation between film thickness and wind speed and a parameterization of observed gas fluxes in terms of measurable physical quantities [242, 250–256].

A roughly quadratic increase of gas transfer with wind speed was observed [242, 245] as well as a distinctly linear one [253]. At the onset of rough waves there is a sudden increase in gas transfer rates [254, 255]. Dampening of waves by a monolayer of oleyl alcohol led to an 80% reduction in gas transfer [253]. For wind speeds greater 9 ms^{-1} the presence of bubbles created by breaking waves is responsible for a jump of gas transfer velocities up to a factor of 3 [256].

Evidently, the laminar film concept of the gas-liquid interface is an oversimplification of nature, and as was stated by Hasse and Liss [239]: "It is safe to say that the physics of gas exchange processes remain largely unresolved."

Field Determinations

The flux of CO_2 between ocean and atmosphere has been determined by various methods:

1. Natural radiocarbon method: A steady state balance for natural radiocarbon in the oceans is assumed, which implies that the net invasion of naturally produced ^{14}C into the ocean just balances the rate of decay of ^{14}C in the sea. In this way Craig [257] calculated a global CO_2 exchange of 20 moles m^{-2} yr^{-1}. Based on an extended data set Broecker and Peng [247] arrived at a value of 19 moles m^{-2} yr^{-1}.

2. Radon method: the evasion rate of ^{222}Rn out of the ocean is determined by measuring the degree of disequilibrium between ^{226}Ra and it's daughter ^{222}Rn in the surface ocean. Correcting for the different diffusion coefficients of ^{222}Rn and CO_2, a global average for the CO_2 exchange rate of 16 moles m^{-2} yr^{-1} is obtained [258] in good agreement with earlier results [259].

3. Bomb radiocarbon method: based on the inventory of bomb-produced ^{14}C and Atlantic Ocean ^{14}C data Stuiver [260] calculated the oceanic uptake of bomb produced ^{14}C. Relating it to CO_2 fluxes yields, after correcting for a preindustrial atmospheric CO_2 level of 290 ppmv, an exchange flux of 20 moles m^{-2} yr^{-1}. Münnich and Roether [372] calculated film thicknesses ranging from 16 to 72 microns for the North Atlantic.

From the available data it was concluded that the average preindustrial CO_2 exchange between atmosphere and ocean was 19 moles m^{-2} yr^{-1}, corresponding to a mean atmospheric residence time of 8 years [249]. The estimated uncertainty is $\pm 20\%$. The average film thickness of the world ocean has been calculated by the ^{14}C-method as 30 microns [247]. Results obtained by the radon method from a GEOSECS profile from the Atlantic range from 24 to 126 microns [258]. Re-evaluation of these data with regard to wind speed indicates a significant increase of gas exchange with wind speed, but to an uncertain extent [261].

The abovementioned methods are indirect methods. The first direct CO_2 flux measurement over the ocean was reported from off the beach of Sable Island [262] by means of the "eddy correlation technique" which has been widely used for water vapor, heat and momentum flux determinations. (When applied to gas exchange measurements, the gases concentrations at a fixed point are correlated with the vertical wind fluctuations. For a discussion of methodological problems with this technique see [373]). Under the particular set of environmental conditions of the experiment a CO_2 flux of 0.8 micromoles m^{-2} s^{-1} out of the ocean was measured, which is equivalent to about 25 moles m^{-2} yr^{-1}. Measurements by the same technique have also been reported from shallow coastal waters at an outer submerged reef near Miami, Florida [263]. Surprisingly high CO_2 fluxes out of the water were

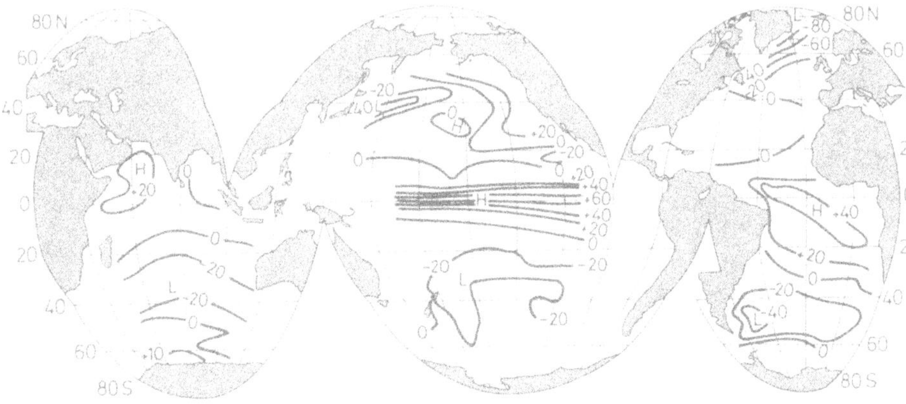

Fig. 32. Global distribution of the difference between the partial pressures of CO_2 in the atmosphere and surface ocean water. Values are given in μatm. H/L indicate regions where the ocean is source/sink for atmospheric CO_2 [264]

found, e.g. 0.3–0.35 mg CO_2 m^{-2} s^{-1} at a wind speed of 10 ms^{-1}, equivalent to 208 moles CO_2 m^{-2} yr^{-1}. This may in part be due to the high partial pressure of CO_2 measured in the water (445 ppm) and the presence of bubbles in breaking waves. The threefold enhancement of gas exchange due to bubbles [256] would account for 60 moles m^{-2} yr^{-1} CO_2 flux if applied to the global average of about 20 moles m^{-2} yr^{-1}. Clearly, more observations are needed before the range of variability in the magnitude of CO_2 fluxes across the air sea interface can be assessed.

At southern latitudes higher than 40 degrees the mean exchange rate is about 50% higher than the global average according to Peng et al. [258]. Low latitude oceans are a source, high latitude oceans are a sink for CO_2. Global distributions of the CO_2 partial pressure differences between atmosphere and ocean are shown in Figure 32.

Assuming an ocean area of $3{,}620 \times 10^{14}$ m^2 and a CO_2 flux of 19 moles m^{-2} yr^{-1}, a global value of $8{,}878 \times 10^{15}$ moles yr^{-1} equivalent to 391×10^{15} g yr^{-1} of CO_2 or 107×10^{15} g yr^{-1} of C would result. It means that the natural exchange of CO_2 across the sea surface is in balance with about 390 gigatons of atmospheric CO_2 entering the oceans and an equal amount of CO_2 returning to the atmospheric. Assuming further that the global emission of anthropogenic CO_2 is 5×10^{15} g yr^{-1} of C, equivalent to about 18×10^{15} g yr^{-1} of CO_2 and assuming finally that about half of it enters the ocean we find that the net flux of anthropogenic CO_2 into the ocean is only in the order of 2% of the natural exchange. It therefore cannot be determined with the aid of the gradient-flux-relationship described above.

Mid-Water Stratification

The ocean water column exhibits a series of mid-water boundaries of physical, chemical or biological nature. They separate water masses and control the flow of matter from surface to sediment. For instance, density boundaries introduced by

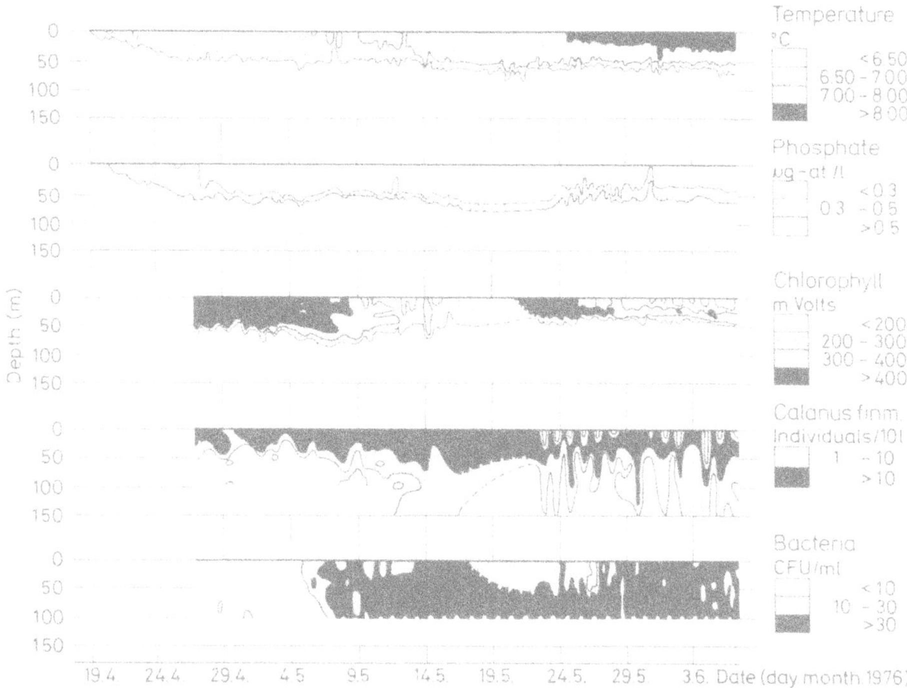

Fig. 33. Depth and time profiles of water temperature, phosphate, chlorophyll, *Calanus finmarchicus*, and colony-forming bacteria (CFU) at a central station of the Fladenground Experiment (Flex 76) positioned in the center of the North Sea (58°55′N; 0°32′E) (after „Meeresforschung in Hamburg". Illustrated documentation of the „Sonderforschungsbereich 94" at the University of Hamburg, FRG, 1977)

temperature and salinity gradients cut the water column into an upper and lower layer and by that set the stage for the onset of a plankton bloom in the euphotic zone, or the development of anoxia in restricted basins such as the Black Sea or the Cariaco Trench. To illustrate the impact of a thermocline on the carbon flux in the ocean, data from an experiment done in the North Sea are presented (Fladenground Experiment FLEX [76, 44]).

In general, phytoplankton growth depends on light intensity, availability of mineral nutrients, turbulent mixing, remineralisation, secondary production and adaptation processes. Optimal conditions for primary production may arise in regions of upwelling or at times a water body stratifies, for example, during the spring season. Figure 33 summarizes the events associated with the formation of a thermocline at a station located in the central region of the North Sea (58°55′N, 0°32′E). Principal factor to start the spring bloom is the development of a thermocline. The interface permits planktonic cells to remain in the euphotic layer for an extended period of time and to generate first a micro- and then a macroenvironment for the orderly growth of a plankton community. In the initial phase massive excretion of DOC by the cells seems to be needed to remove toxic metals from the aquatic environment through metal-organic complexation. Following the first phytoplankton maximum (29 April to May), grazers and bacteria appear in large

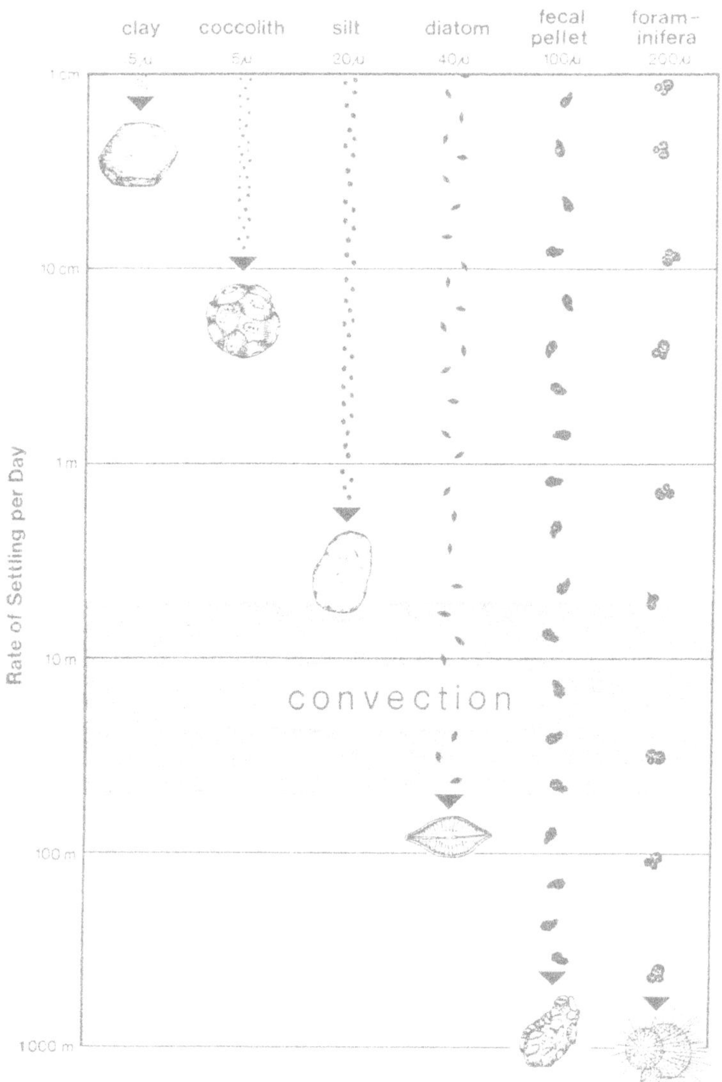

Fig. 34. Settling velocities of different-sized particles in sea water [270]

numbers. Whereas changes in mineral nutrients – here exemplified by phosphate – and variations in phytoplankton density – shown as chlorophyll values – are restricted to the upper layer. Zooplankton and bacteria living on organic matter traverse freely through the thermocline.

The action of grazers (e.g. copepods) will not only involve the uptake of organic matter but the removal of any suspended particle from the euphotic zone be it coccolith, clay mineral or fly ash. They package them in form of fecal pellets which are jetted to the sea floor at velocities of several hundred meters per day (Fig. 34). Another way to transfer organics and minerals to the deep sea is by means of macro

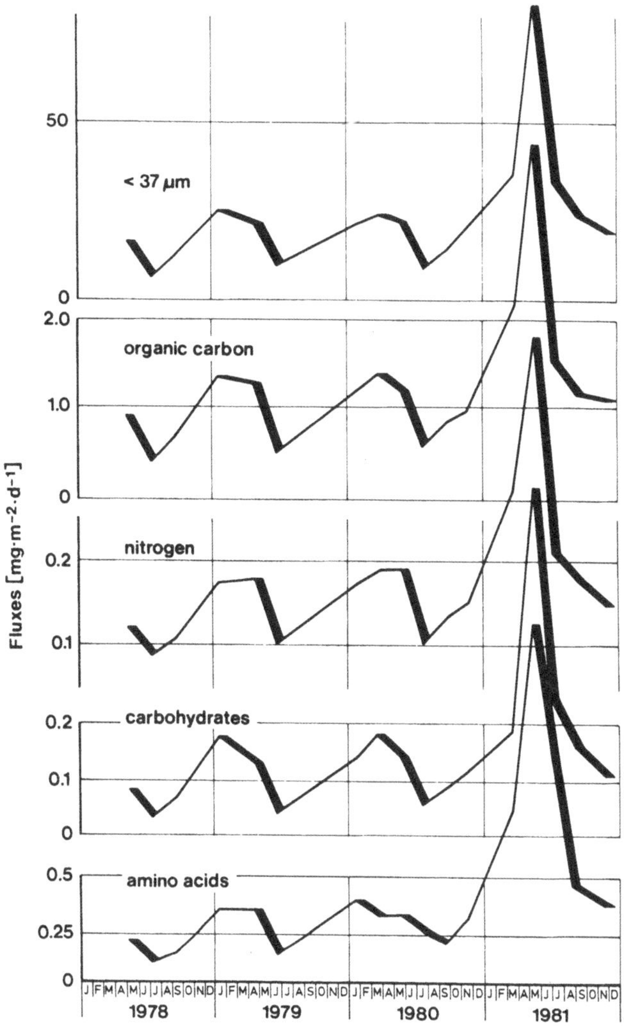

Fig. 35. Fluxes of organic carbon, nitrogen, amino acids and sugars in the $\leqq 37$ μm fraction of materials arriving at a sediment trap set at 3,000 m in the Sargasso Sea [270]

floc formation. In principle, phytoplankton excretes mucus which is instrumental by a kind of "glue-on-effect" in scavenging detritus, whereby macro-aggregates are formed which rapidly sink through density boundaries. Time series sediment traps deployed in various shallow and deep sea environments permit collection of such particles in mid-water [265–270]. Variations in the distribution pattern of sugars, amino sugars, and amino acids indicate seasonality in the fluxes of "fresh" materials. Data from the Sargasso Sea [271, 272] are depicted in Figure 35 and indicate not only seasonal but also interannual variations in the flux of carbon and minerals.

In summary, balanced plankton communities are quite effective in the transfer of suspended particles to the sediment-water interface. For instance, carbonates, clay minerals and organic matter can be rapidly transported from the upper to the lower water layer. It is this transfer mechanism which today regulates the vertical flow of carbon to marine sediments. In case such biological devices do not exist or are retarded, clay-sized particles will stay in the euphotic zone since their settling velocities – according to Stoke's law – amount to only a few cm per day and are thus far below the scale of convection in the upper layer.

Biomineralization

The majority of articles in the field of biomineralization is descriptive in nature. For instance, the morphology of a mollusc shell is depicted in great detail or the intricate relationship between shell organic matrix and mineral phase is studied under the microscope. Other papers deal with the chemical composition of the organic matrix, the type and shape of crystals or the trace metal content of various biominerals. Only a few articles exist on mechanisms that initiate and control mineral deposition and relate crystal growth to distinct physiological and environmental happenings. The subject matter has been thoroughly reviewed at other places [271–273].

In the open sea, carbonate deposition is a biological phenomenon and involves organic membranes as substrate for $CaCO_3$ nucleation and mineral growth. Inorganic precipitation is a rare event and restricted to a few marginal environments. The viewpoint is often expressed that organisms deposit their calcium carbonate along the familiar route:

$$Ca^{2+} + 2HCO_3^- \rightarrow CaCO_3 + CO_2 + H_2O.$$

In this context respiratory CO_2 is frequently mentioned as the sole or principal source for the carbonate ion in skeletal calcite or aragonite [274]. Stable isotope data, however, argue strongly against this supposition [275–277]. Only in a few species can a significant contribution from respiratory CO_2 be demonstrated [278]. Most invertebrates including marine plankton, and the otoliths in fish [279] have carbonates which are isotopically identical to the $^{13}C/^{12}C$ ratios present in the dissolved bicarbonate pool in their environment. In fact, the distribution of oxygen isotopes of carbonates in many aquatic species reflects the water temperature in the habitat where the organisms live. Thus we must assume that bicarbonate enters the inner system of the cell via permeable membranes where it forms a pool from which $CaCO_3$ is rapidly extracted leaving little opportunity for $^{13}C/^{12}C$ exchange with the isotopically light respiratory CO_2. Alternately, carbonate deposition may proceed through direct contact with the outside environment.

The two enzyme systems frequently mentioned in connection with biomineralization are (i) carbonic anhydrase and (ii) alkaline phosphatase. The first one – among other physiological tasks – is regulating the carbonate system whereas the second one is involved in the Ca^{2+} transport process. We will now concentrate on the functional role of carbonic anhydrase in view of its central position in carbonate physiology. For a more general treatment on calcium regulation in calcification see [272].

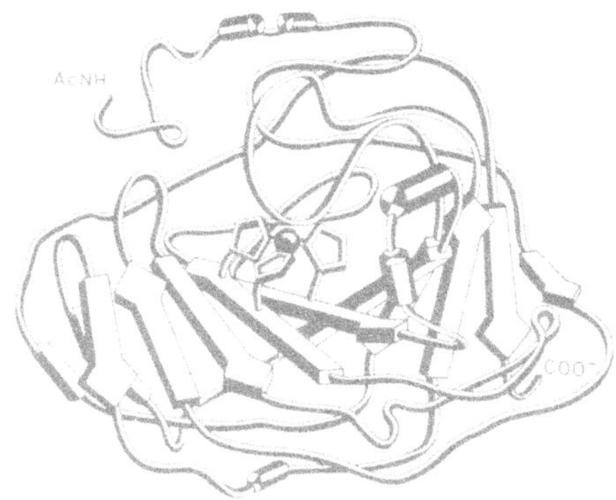

Fig. 36. Idealized drawing of the main chain folding of human erythrocyte carbonic anhydrase C (HCAC). The helices are represented by cylinders and the pleated sheet strands are drawn as arrows in the direction from amino to carboxyl end. The ball supported on three histidyl residues represents the zinc ion [284, 285]

Carbonic anhydrase is a zinc-containing enzyme [280, 281] with a molecular weight of approximately 30,000 daltons [282]. It can occur in the form of multiple isoenzymes which differ in their specific activity [281, 283]. A schematic drawing of the molecule is shown in Figure 36.

The enzyme is quite versatile in that it catalyses the interconversion of CO_2 and HCO_3^- in solution, the hydration of aliphatic aldehydes [286], pyridinecarboxylaldehydes and pyruvic acid [287], and acts as esterases with respect to certain monoesters of carboxylic acids and diesters of carbonic acid [288, 289].

The catalytic mechanism of CO_2 hydration-dehydration by carbonic anhydrase represents the focal issue of the present discussion. We have to consider two aspects: (i) the mode of binding of the CO_2 substrate at the active site, and (ii) the physical-chemical state of ligands on the zinc ion.

A direct coordination of CO_2 to the zinc can be excluded. Based on infrared studies it was suggested that the CO_2 molecule is bound to the active site of carbonic anhydrase in an unstrained fashion and loosely bound to a hydrophobic cavity [290]. Since the physical properties of N_2O and CO_2 are perfectly matched, both molecules should thus compete for binding to the active site. This, however, is not the case which implies that in addition to true substrate binding also nonproductive binding of CO_2 is present. It is noteworthy that no general class of competitive inhibitors of the CO_2 hydration activity of carbonic anhydrase has been found to date. All potent inhibitors of enzymic activity become directly coordinated to the zinc atom. This indicates that CO_2-binding is rather specific, perhaps involving a stereochemical adjustment such as binding of a CO_2 molecule which is known to increase rates of hydration by several orders of magnitude.

There is general agreement that carbonic anhydrase activity is linked to the zinc ion and its ligands. At the active site water is bound to the zinc but the state and involvement of the water in the actual catalysis is a matter of controversy. According to a widely accepted opinion the zinc-bound water is ionized and the activity is a function of pH [291].

The basic form (ZnOH) is active in the hydration reaction, while the acid form ($ZnOH_2^+$) is active in the dehydration reaction. The zinc-bound water participates in both the electron donor and acceptor requirements of the reaction. HCO_3^- is considered to be the substrate in the dehydration reaction and proton transfer must accompany the breaking of the C-O bond.

This currently accepted viewpoint is in conflict with proton relaxation data [292] which indicate that only the high-pH form of the enzyme has a water ligand and that there is no bound water or OH^- at low pH. The idea has been advanced [292] that only the high-pH form catalyzes both the hydration and dehydration reaction. Furthermore on kinetic grounds there is strong evidence against HCO_3^- as a substrate in the dehydration reaction. Instead, it was proposed that dehydration of HCO_3^- catalyzed by carbonic anhydrase proceeds by the initial formation of a complex between the high-pH form of the enzyme and a neutral H_2CO_3 molecule which displaces the H_2O on the metal. Thus, the reaction that is reversibly catalyzed reads:

$$CO_2 + H_2O \leftrightarrows H_2CO_3.$$

The idea of H_2CO_3 as substrate for carbonic anhydrase is strongly supported by inhibitor binding studies [293, 294]. Small anion or sulfonamide inhibitors are linked to the zinc ion at high pH. The complex picks up a proton and the inhibitor is bound in a neutral form.

The strongest support for the involvement of carbonic anhydrase in the regulation of mineral deposition comes from studies on carbonic anhydrase inhibitors [295]. In the presence of inhibitors such as Diamox or sulfanilamide (Fig. 37) enzymic activities are strongly reduced and result in a decrease in shell thickness of eggs [296] or malformation in calcareous tests of the kind shown in Figure 38. In the presence of carbonic anhydrase inhibitors, rates of calcification are strongly affected as shown in studies on corals [298].

These studies indicate that the actual role of carbonic anhydrase in calcification does not rest on its ability to hydrate CO_2 but to remove carbonic acid from the site of calcification:

$$Ca^{2+} + 2HCO_3^- \rightarrow Ca(HCO_3)_2 \rightarrow CaCO_3 + H_2CO_3.$$

The following conclusions can be drawn from these data:

Carbonate deposition involves interaction of Ca^{2+} and HCO_3^- resulting in the formation of an unstable intermediary product $Ca(HCO_3)_2$. This product will break down into $CaCO_3$ and carbonic acid (H_2CO_3). As long as calcium is not the limiting factor, the rate of formation of calcium carbonate will depend on the rate by which carbonic acid is removed from the calcification site. In the presence of carbonic anhydrase, calcification is significantly increased, due to the formation of a complex between the high-pH form and a neutral H_2CO_3 molecule, which is the substrate for carbonic anhydrase [292].

Whereas enzymes regulate the interactions of Ca^{2+} and dissolved carbonates, well-defined proteins and polysaccharides are involved in transportation of Ca^{2+} and nucleation of crystals [272]. Species-specific proteins also serve as templates for the oriented epitaxial growth of biominerals. In a way, the morphology of a calcareous shell is a macroscopic expression of the molecular organisation of the shell

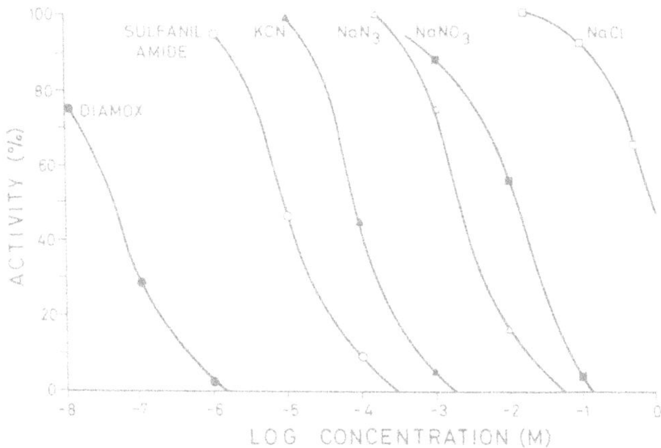

Fig. 37. Inhibition of carbonic anhydrase in crude enzyme preparation from *Serraticardia maxima* by various compounds [295]

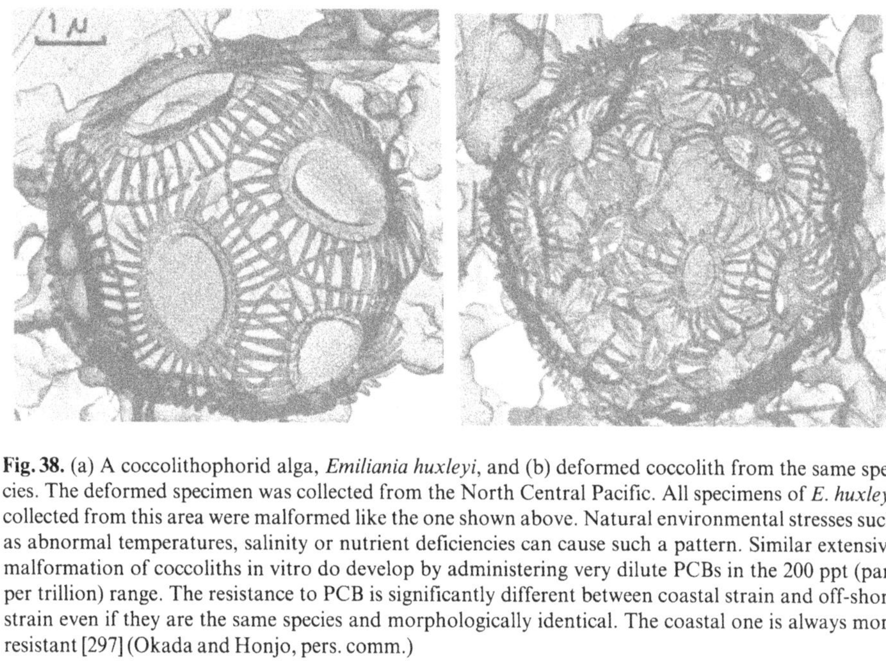

Fig. 38. (a) A coccolithophorid alga, *Emiliania huxleyi*, and (b) deformed coccolith from the same species. The deformed specimen was collected from the North Central Pacific. All specimens of *E. huxleyi* collected from this area were malformed like the one shown above. Natural environmental stresses such as abnormal temperatures, salinity or nutrient deficiencies can cause such a pattern. Similar extensive malformation of coccoliths in vitro do develop by administering very dilute PCBs in the 200 ppt (part per trillion) range. The resistance to PCB is significantly different between coastal strain and off-shore strain even if they are the same species and morphologically identical. The coastal one is always more resistant [297] (Okada and Honjo, pers. comm.)

organic matrix whose conformation is highly dependent on the availability of metal ions and the chemistry and physics of the surrounding environment (e.g. temperature, salinity).

In conclusion, during calcification bicarbonate ions of the sea are transformed to bio-carbonates without carbon isotope fractionation and without the release of CO_2 and water. In contrast, photosynthesis will lead to a ^{12}C-enrichment in the organic matter and a corresponding depletion in the CO_2.

This partitioning of carbon isotopes seems to be the prime reason behind $\delta\,^{13}C$ fluctuations observed in DSDP cores from the late Cretaceous to the Present. First indications for pronounced $\delta\,^{13}C$ variations came from studies on selected foraminifera (reviewed in [299–301]). This work has been extended to include bulk carbonates and the measured $\delta\,^{13}C$ values are depicted in Figure 39. A lightening across the Cretaceous-Tertiary boundary and one near the Paleocene-Eocene boundary is apparent. For most of the Eocene and Oligocene $\delta\,^{13}C$ stays close to +2 per mil. At about 15 million years before present, the $\delta\,^{13}C$ curve falls off by almost 3 per mil to its modern value of close to 0 per mil.

It is tempting to relate these changes to the global carbon budget, but the system is too intricate to be on safe grounds. One thing we know for sure: biological calcification is generally unaccompanied by carbon isotope fractionation between the marine bicarbonate pool and the shell material. Consequently, the carbon isotope pattern of the marine bicarbonate is conservatively transferred to bulk carbonate. Thus, $\delta\,^{13}C$ fluctuations in the about 39,000 Gt dissolved inorganic carbon of world ocean can only be brought about by changes in the rate of organic matter removal, addition both on land and in the sea, or variations in quantity and ^{13}C content of the CO_2 streaming-off from crust and mantle.

We stated previously that juvenile CO_2 has a $\delta\,^{13}C$ close to -5 per mil (Fig. 6). This CO_2 becomes globally partioned into "limestone" and "organic carbon" with $\delta\,^{13}C$ values of about 0 and -25 per mil, respectively. Photosynthesis is the prime reason behind the partitioning. In order to raise $\delta\,^{13}C$ of dissolved carbonates in modern ocean from 0 to +1 per mil, about 1,500 Gt of organic carbon has to be removed from the exogenic cycle. In contrast, a lowering to -1 per mil requires an oxidation of an equal amount of sedimentary organic matter to CO_2. Although the activity of organisms is responsible for carbon isotope fractionation, actually it is the deposition or oxidation rate of organic matter into or from the geological column which counts. Since oceanic anoxic events or peat deposition on land are effective mechanisms for the burial of C_{org}, high $\delta\,^{13}C$ values recorded in marine $CaCO_3$ at certain times during the upper Cretaceous or Tertiary may find their explanation.

Unfortunately we do not know rate and carbon isotope signal of the endogenic CO_2 recharging the dissolved carbonate pool in the Tertiary-Quaternary ocean at any one time. For the moment we can, therefore, only state that even though we do not yet fully understand the meaning of the fluctuating $\delta\,^{13}C$ curve shown in Figure 39 the ups and downs indicate global episodes caused by processes such as peat formation on land, oceanic anoxic events, changes in rates of burial, mass extinctions, tectonic pulses, or climatic impacts.

For most of the Phanerozoic, biomineralization has been the chief mechanism for the deposition of marine carbonates. However, in the Precambian sea, only inorganic carbonate precipitation has occured and carbon in calcite, aragonite or dolomite should have been deposited in isotopic equilibrium with sea water bicarbonate making the carbonate heavier by a few per mil. The $\delta\,^{13}C$ curve prepared by Schidlowski [302] for the Vendian-Cambro-Ordovician time period shows the expected trend. Extending this curve into older rock formations, however, indicates that the isotopic composition of carbonate carbon stays within a narrow range of 5 per mil ($+0.5 \pm 2.6$ per mil) for the entire geologic record (3.5 Ga),

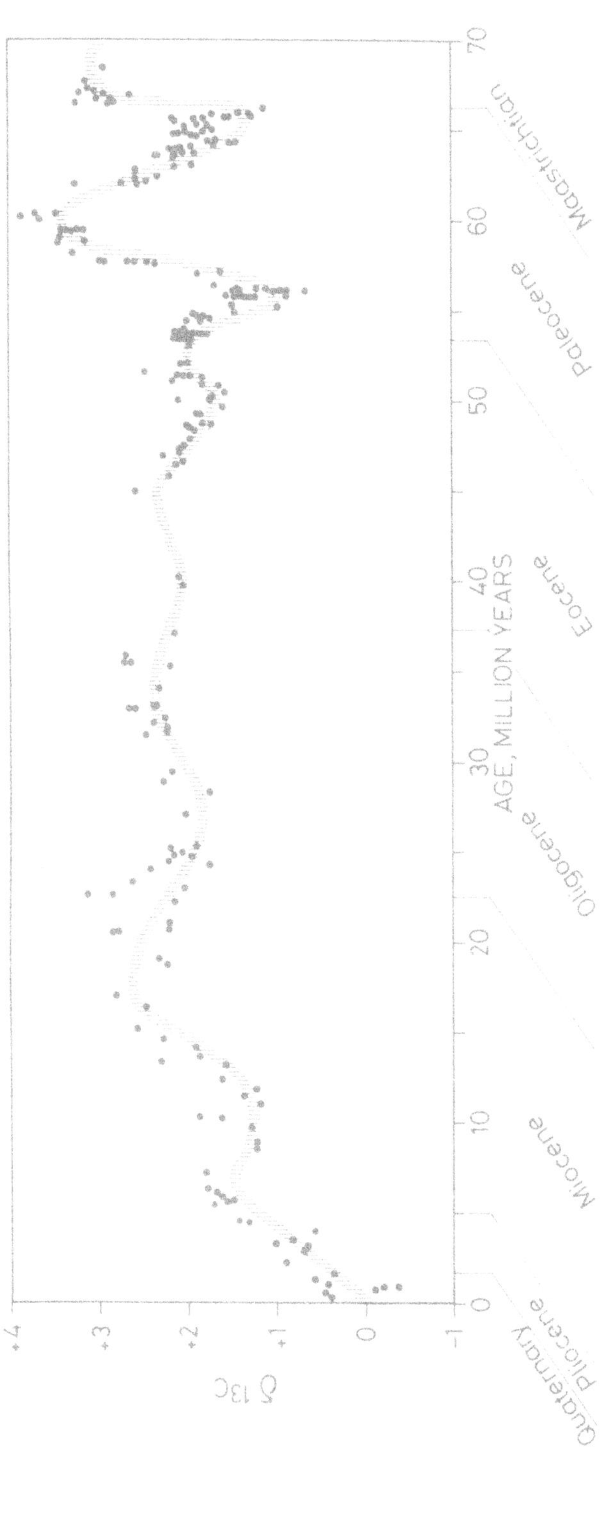

Fig. 39. Carbon isotopic composition of bulk sediment (carbonate fraction) from DSDP Site 525, with data from Site 528 between 25 and 45 million years (Walvis Ridge, Cape Basin) [299]

whereas C_{org} exhibits a large scatter (-26 ± 6 per mil). Some tentative interpretations are offered in [302].

Sediment-Water Interface

The sea floor comprises two thirds of the Earth's surface but it is largely a *terra incognita*. Results of marine geologists on types of sediments and processes occurring at depth are at best a rough approximation of the "real" submarine world. This is so because the sea bottom is sampled at large intervals only [303, 304].

Especially the organic processes presently going on between the water column and the underlying sediments are rather poorly understood. The most prominent factors at the sediment-water boundary are:

(i) the so called benthic layer, a few cm thick where organic matter remineralisation is quite efficient, and

(ii) the upper few centimers of modern sediment in which burrowing organisms and microbial communities play an essential part in the turn-over of organic matter.

Microbial remineralisation of organic matter results in the release of nutrients in high concentrations and characteristic biogeochemical patterns emerge with sediment depth [305].

Interstitial waters in the deep sediments somehow reflect the organic activities displayed at this boundary [306] showing that up to 200 mg C l^{-1} – largely metal complexes – are found in here. Their molecular weight – in contrast to sea water DOC – is greater than 50,000 attesting to the fact, that polymerization of organic matter by biological activitiy and mineral-organic interaction is one of the key elements in fixing organic carbon to the sediment column. This intriguing aspect of a "marine soil" requires more scientific attention, particularly in the light of the circumstance, that this is the check-point where the decision is made as to what and of how much of an organic or inorganic carbon species enters the stratigraphical column.

Chemoclines

Not only physical but also chemical interfaces mark the oceanic water column. Lysocline, carbonate compensation depth (CCD) or oxicline are three prominent boundaries controlling the vertical flow of carbon in the marine habitat [36, 81, 369]. In the ocean, the concentration of the carbonate ion (CO_3^{2-}) decreases with depth which causes an increase in the solubility of calcite (Fig. 40) [81]. The intersect marks the carbonate compensation depth (CCD), below which infalling calcite detritus dissolves. The level at which calcite totally disappeares, is called the lysocline. Above the CCD the seawater is supersaturated with calcite, below the CCD it is undersaturated.

Actually, we are dealing with kinetic boundaries which adjust to the amount of infalling calcareous detritus, degree of upwelling or volcanic emanations. CCD occupies positions several thousand meters below sea level (commonly 3 to 4 km). Since aragonite ($CaCO_3$) is more soluble than calcite ($CaCO_3$) its dissolution starts much earlier. It is of interest to note that CCD can be as much as 5 km below the

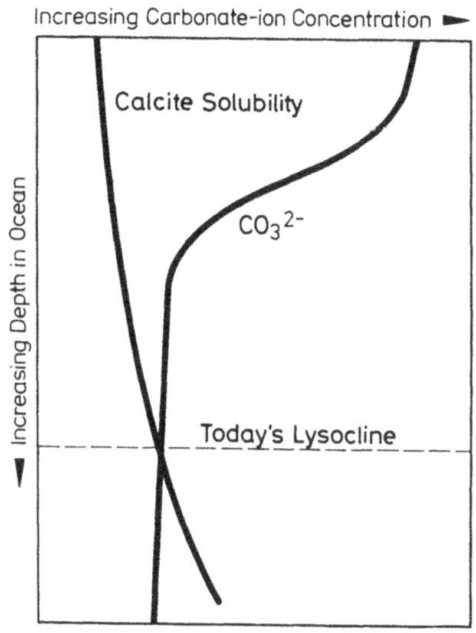

Fig. 40. Calcite solubility in world ocean in relation to CO_3^{-2} content. Today's lysocline is positioned at about 3 km in the Central Atlantic [81]

ocean surface, for instance in the highly productive equatorial Pacific. The consequence to the type of bottom sediments that accumulate are depicted in Figure 41 [370].

The oxicline separating oxic from anoxic waters is often associated with well-developed pycnoclines which physically stratify a water body in mid-water due to temperature or salinity differentials. Such conditions are only rarely encountered in modern ocean (e.g. Black Sea, Cariaco Trench, some Norwegian Fjords) but in meromictic lakes they are quite prominent (e.g. Lake Kivu, Lake Tanganyika). Frequently, CCD is close to thermo-haloclines, as is the case in the Black Sea or some deep East African Rift lakes [369, 371]. Carbonate particles that fall below this phase boundary dissolve, whereas those that settle in shallow areas above the pycnocline remain intact. Scanning electron micrographs of chemically precipitated calcites from the Black Sea sediment above and below the O_2-H_2S interface show this process (Fig. 42) [371].

In the past, the ocean has experienced many rises and falls of pycnoclines, lysoclines, CCD's and oxiclines [81, 369]. These changes have significantly affected the global carbon budgets in many facetious ways. Their presence and level of position appear to be the most determining factors controlling the CO_2 system in the course of geological time.

In the early stages of crustal development, temperatures on Earth, and consequently the PCO_2 in the atmosphere, might have been considerably higher than today [31, 54]. Thus, the question is raised as to why Earth has been spared the "runaway-greenhouse effect" Venus has experienced. The atmosphere of Venus is principally composed of CO_2 (96%) and has a partial pressure of about 99 times that of the Earth which corresponds to a pressure one finds at 1 km in the ocean. The

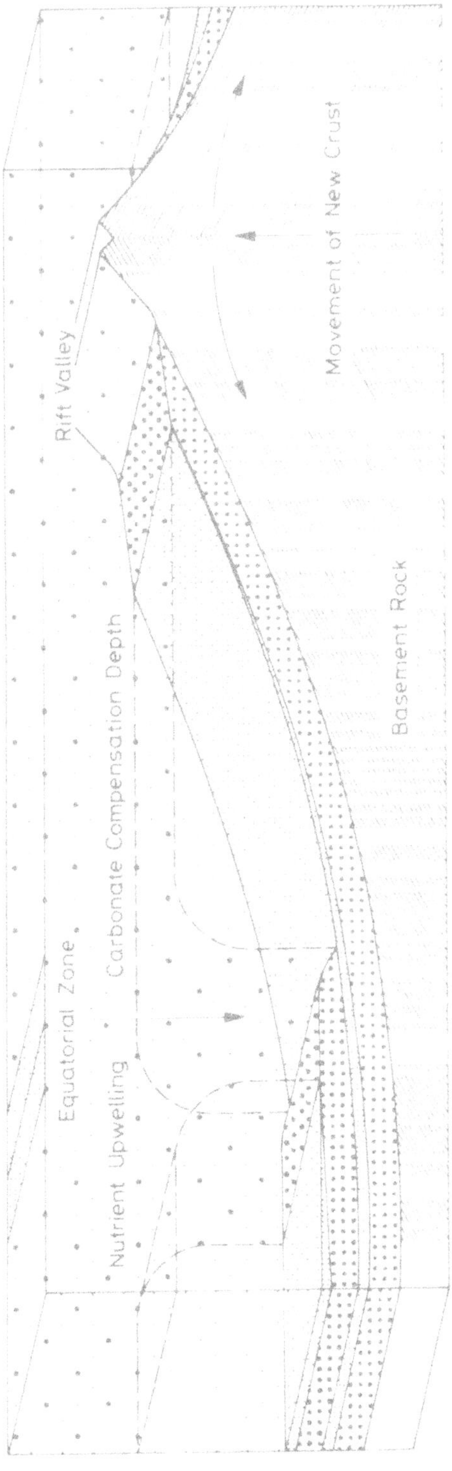

Fig. 41. Dynamic model (schematic) showing the effect of the carbonate compensation depth (CCD) on the kinds of bottom sediments that accumulate. Note the movement of new crust which is indicated by a vertical striation pattern reflecting magnetic reversals. Sea floor spreading will move areas into and out of the influence sphere of the CCD. In case euxinic conditions develop, due to density stratification, the lysocline and CCD will gradually move upward towards the pycnocline and cause dissolution of infalling $CaCO_3$ detritus (see Fig. 42)

Fig. 41. Scanning electron micrographs of chemically precipitated calcites (right) sedimented above Black Sea pynocline and (left) their appearance below pycnocline; note onset of dissolution. Size of individual crystals about 5–10 μm [371].

CO_2 content has a mass of 130×10^{21} g C which would equal all the carbon contained in the Earth's crust and upper mantle down to a depth of 120 km [36]. This suggests that Venus has conservatively collected all emanating carbon in its atmosphere in the form of CO_2 almost in the same way as Earth has accumulated ^{40}Ar over the past 4 billion years. In contrast to Venus, however, carbon on Earth is constantly recycled principally because our planet, having just the right distance to the sun, could retain an ocean, whereas Venus has lost its, and furthermore life is present on Earth. Organisms seem to be chiefly responsible for the return of organic and inorganic carbon to the crustal compartment, whereas volcanism will recharge sea, air and life with carbon and feed the biogeochemical cycle.

We have previously stressed the significance of H_2O and CO_2 in magmatic processes (see p. 133). It thus appears that in modern and ancient oceans a series of "chemical homeostats" exist that are driven by mineral equilibria, physical-chemical feedbacks, and living processes. In conjunction with magmatic processes they were, are, and will be the key elements keeping up the dynamics of our planet Earth.

Environmental Scenarios

Carbon Cycle Modelling

Carbon cycle models have mainly been designed to simulate the movement of anthropogenically released CO_2 through the global environment. The fraction that remains in the air causes the observed increase of CO_2 and is referred to as "air-borne fraction". It's evolution under given rates of man made CO_2-releases to the atmosphere is the principal output required from the models.

Since the beginning of their development [257, 307–309] the box model approach has been applied, whereby the global environment is represented by a few "boxes" (e.g. the ocean, the atmosphere, the biosphere). Fluxes between them vary in proportion to "reservoir size", i.e. compound mass in a box. The proportionality factors are termed "exchange coefficient". Their inverse (reservoir size/flux) is referred to as the compound's residence time in a box. For a comprehensive state of the art report on carbon cycle models see [314].

Model-runs start typically with an assumed pre-industrial steady state carbon cycle (constant atmospheric CO_2-concentration). Then the response to an anthropogenic perturbation is simulated and expressed as the resulting increase of CO_2 in the air. This may then be checked against the observed global increase of atmospheric CO_2, e.g. the Mauna Loa record (see p. 157). The Mauna Loa record itself evolved as the combined effect of:

a) pre-industrial level of CO_2 in the air
b) history of CO_2-release by fossil fuel combustion
c) history of CO_2-release by forest clearing and changing land use
d) uptake of man-made CO_2 by the ocean.

Since only b) is accurately known, models may be tuned to reproduce the Mauna Loa record by independent variation of a), c), and d). This freedom in parameter choice allows, theoretically, the air-borne fractions predicted to vary between 38 and 74% [310].

Lack of observations and the heterogeneity of the continental biosphere make it very difficult to quantify c) for the past and the present (see p. 154). For modelling efforts related to this problem see e.g. [118, 311, 312]. A reliable estimate of d) by realistic ocean carbon cycle models would allow to calculate the biospheric source/sink behaviour as difference between observed CO_2-release from fossil fuels and ocean uptake calculated by the model. For data on a) see chapter "carbon in the air".

Many models have been proposed for numerical simulation of the ocean's uptake capacity of man-made CO_2. In most of them the ocean is one-dimensional and vertically subdivided into the boxes: surface layer, intermediate water, deep water. The exchange coefficients describing mass flux between them correspond to characteristic time scales or "time constants" of oceanic mixing. Exchange with one single atmospheric box is mediated by a constant or temperature dependent buffer factor (see p. 147). A detailed account of this class of models is given in [313].

Their basic disadvantage is that buffer factors and exchange coefficients are not known a priori. They are usually determined by tuning the models themselves to observed distributions of oceanic tracers (like ^{14}C and tritium) and to the input of fossil CO_2 to the atmosphere during the last 100 years. The predictive capability of these models is therefore limited, considering also that the continuous ocean is represented by a few boxes.

A first attempt to reproduce the continuous character of ocean mixing is the "box-diffusion-model" of Oeschger et al. [120]. They introduced vertical eddy diffusion as a mechanism of CO_2 transport into the deep sea. Further sophistications have been proposed [315–319, 122].

Fig. 43. Surface ocean circulation as simulated by the model of Maier-Reimer [320]

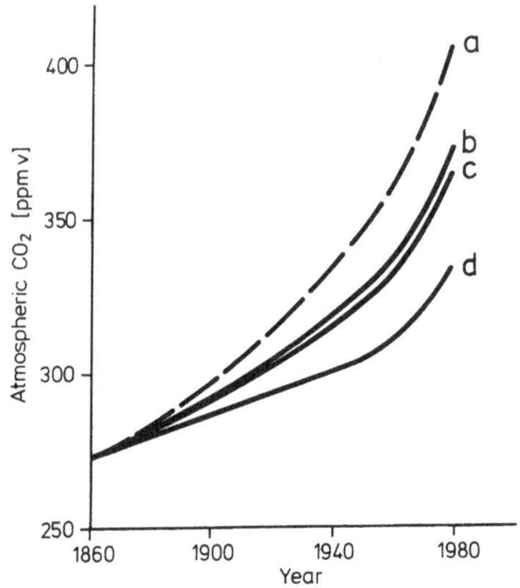

Fig. 44. Simulations of atmospheric CO_2-increase calculated with the model of Maier-Reimer [320]: a) Integrated anthropogenic input (fossil fuel source according to Rotty [321], biospheric source hypothetically assumed as $+1$ Gt C yr^{-1}). b) Atmospheric retention without convective mixing in the ocean circulation model. The resulting air-borne fraction is 73%. c) Atmospheric retention with convective mixing in the model. The resulting air-borne fraction is 65%. d) Same as c) but stabilizing effect of salinity in polar latitudes is neglected. The resulting air-borne fraction is 47%.

Maier-Reimer [320] made a major step recently to overcome the drawbacks of oceanic box models. He employed a general circulation model of the ocean, which is complex enough to reproduce the general features of oceanic circulation (Fig. 43) and simple enough to allow numerical integrations on climatic time scales. The velocity field of the general circulation model advects carbon dioxide as a passive tracer. Thus, a significant improvement is achieved concerning the representation of oceanic water mixing and circulation. With the velocity field provided, it is in principle possible to describe the local rate of change of a carbon species at any (grid-) point of the (model-) ocean according to the tracer distribution equation:

$$\frac{\partial C}{\partial t} = K \Delta C - V \cdot \nabla C + Q - S \qquad (54)$$

C concentration of a carbon species
t time
K eddy diffusivity
V velocity
Q,S local sources and sinks from chemical and biological processes.

The effective time constants – previously prescribed in box models as exchange coefficients – can now be dynamically determined. First experiments [320] indicate that the simulated decay of atmospheric CO_2 following an instantaneous CO_2-input can be described as superposition of a few exponential functions with decay times of roughly 1, 10, and 500 years. They correspond to the observed characteristic time scales of mixing in the ocean: equilibration with atmospheric changes occurs within a few months in the warm surface layer, within about 10 years in intermediate waters above the main thermocline, and within a few hundred years in deep waters.

The decay times also depend on the magnitude of the instantaneous input. For a sudden doubling of atmospheric CO_2 the main time constant is 160 years. In case of an instantaneous quadrupling the invasion of CO_2 into the ocean is more effectively blocked for chemical reasons, and the main time constant becomes then 280 years. These "empirical Greenfunctions" are a major advantage over the linear box models. They could be used to construct a new class of simple, non-linear box models (Maier-Reimer, pers. communication).

In it's present state the model predicts an air-borne fraction of 67% if a hypothetical biospheric source of 1 Gt C yr^{-1} is added to Rotty's fossil fuel input function (Fig. 44).

The incorporation of chemical and biological processes into the transport model will be the next important step towards a realistic ocean carbon cycle model. Such a model will have to be in agreement with the physical oceanographer's view on the dynamics of water transport and what chemists and biologists can say about carbon transfer reactions. From such a model, reliable predictions of the air-borne fraction may be expected in the future.

Radiation Balance

Carbon dioxide contributes to the "greenhouse-effect" of the atmosphere because it strongly absorbs at wavelengths between 14 and 16 μm, thus intercepting terrestrial radiation which would otherwise be lost directly to space (e.g. [322]). The greenhouse effect was first described in 1863 by Tyndall [323], who pointed out, that water vapor transmits a major fraction of the incident sunlight, but strongly absorbs thermal radiation from the Earth. The contribution of CO_2 to maintaining the Earth's heat balance was noted in 1897 by Arrhenius [324]. In 1899 Chamberlain [325] proposed atmospheric CO_2 as a controlling agent in the origin of ice ages. Arrhenius estimated that a doubling of atmospheric CO_2 would produce a global warming of about 6 °C. Concern about adverse climatic effects due to fossil fuel CO_2-emissions has been raised in 1938 by Callendar [326] and in 1941 by Flohn [327].

In 1921 Defant [328], developed probably one of the first mathematical treatments of global climate, followed in 1926 by Angström [329] and in 1930 by Milankovitch [330]. Carbon dioxide was not yet considered in climate models, however, prior to the work of Plass in 1956 [331], followed by Manabe and Möller [332] and Möller [333]. Since then a wealth of climate models including CO_2-response studies have been designed with an increasing degree of complexity and sophistication. For summaries and reviews see [334–337].

The general picture emerging from most models is, that a doubling of CO_2 would lead to a global warming of 1.5–3.5 °C, with almost negligible warming in the tropics and a 7–8 °C warming in polar regions. In the stratosphere the reverse is expected: slight cooling in the polar and more pronounced cooling in the tropical stratosphere.

Two positive feedbacks contribute to these results. First, the ice-albedo-feedback, introduced into modelling by Budyko [338] and Sellers [339] enhances the warming at high latitudes because the diminishing white surfaces of snow and ice during warming decrease re-radiation to space (=albedo) thereby adding to the warming. Second, the water-vapor-feedback, described in detail by Ramanathan [340], enhances warming, because CO_2-warming increases atmospheric water vapor, which then itself exerts an additional greenhouse effect. Regarding the ice-albedo-feedback it should be noted that the weak absorption of CO_2 between 1–3 μm wavelength may lead to a weakening of the radiative energy necessary for the onset of snow melt [341]. In high polar latitudes this may prolong the period of high snow reflectance, equivalent to a cooling, especially in early summer. The quantitative significance of this negative feedback has yet to be established.

The general picture emerging from the models should be regarded as rather preliminary mainly for two reasons. First, most models, like e.g. [342, 344], compute the fast reacting atmosphere's equilibrium response to a fixed CO_2-increase, treating the ocean merely as an evaporating swamp. However, CO_2 is increasing continuously. In addition, the transient and highly nonlinear interactions between atmosphere and the more slowly varying parts of the climate system, namely the oceans and the cryosphere, are the essential characteristics of climate dynamics. Hasselmann [345] pointed out, that the heat capacity of the deeper layers of the ocean could delay the approach to equilibrium of a CO_2-perturbed climate system by many decades. For recent approaches to this problem see [346–352, 374]. Second, the response of clouds to a greenhouse warming is not known. They may enhance or compensate the warming. For example, a decrease of global cloud cover by 2% would lower global mean temperature by 2 °C [353]. Cloud feedback must be regarded as a key unknown in climate modelling.

In spite of their weakness, the models give indications where a CO_2-induced temperature signal might first emerge from the noise of natural variability. A close look at polar and stratospheric records, however, has not revealed any clear CO_2-induced temperature signal so far [354–359]. In the stratosphere, for example, Labitzke and Naujokat [375] found a cooling between 1962 and 1972, but a warming since then (Fig. 45), indicative of the volcanic emission control on stratospheric temperature trends. Recent calculations [376] indicate that, by monitoring the outgoing long-wave flux for small intervals in the 15-micrometer spectral region, changes in stratospheric temperatures due to double atmospheric carbon dioxide are large enough to be detected above the various sources of noise.

Grassl [362], inspired by a theoretical approach proposed by Paltridge [360], designed a thermodynamic model in which entropy is maximized by meridional heat flow in atmosphere and ocean. Variable cloudiness is implicitely accounted for. The extremum condition by which entropy is maximized is, although not deducable from basic physical laws, supported by observation [361]. A CO_2-doubling experiment revealed no particular sensitivity of polar temperatures but slight warming in the tropics. These surprising results appear as the combined effect of less cloudiness and weakened poleward transport of heat.

Assuming that the greenhouse warming of man-made CO_2 is not counterbalanced by negative feedbacks in the climate system, the main effect will be a regional redistribution of an enhanced hydrologic cycle. The regional effects on the water

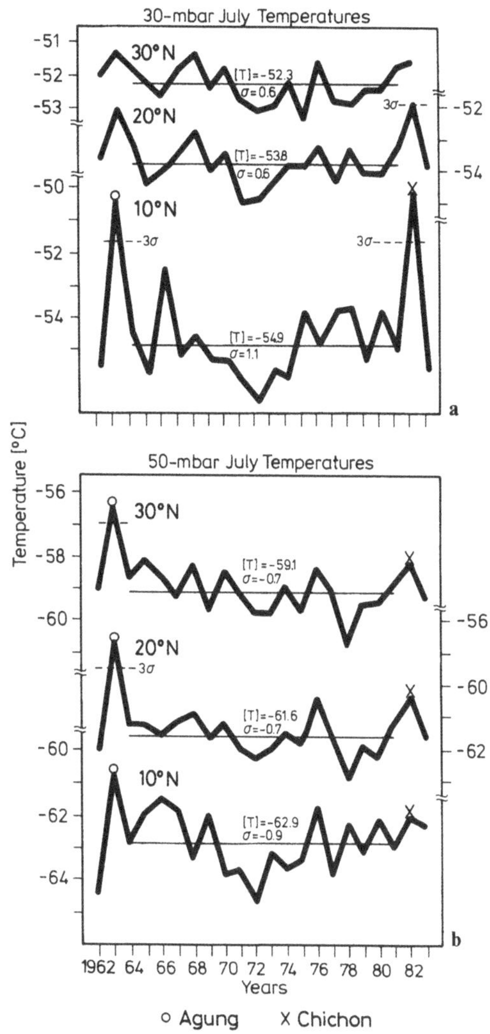

Fig. 45. a) Zonal mean 30-mbar temperatures (°C) during July at 10, 20, and 30°N, for the period 1962 through 1983. The 18 year average (T) is for the period 1964–1981. b) Same as a), but for the 50-mbar level. From Labitzke and Naujokat [375]

cycle appearing in the model of Manabe et al. [363], illustrate that the water resource availability of many regions might be seriously affected. For reasons of continuity, while some regions will suffer from water deficit, others will profit from enhanced water supply by the hydrologic cycle. How, precisely, such a regional pattern might evolve, can not be credibly predicted by the models available at present.

A CO_2-warming of climate might increase methane abundance in the atmosphere [364]. Since both, methane and ozone (which is generated in the troposphere from reactions with methane) exhibit greenhouse effects [365, 366], this may constitute but another positive feedback.

Outlook

The Odyssey of carbon dioxide through "heaven and earth" has finally come to a close. Starting from its probable origin in outer space by disproportionation of carbon monoxide into a carbyne and a CO_2, many other routes can be envisioned along which carbon dioxide is generated in the course of time. Precursors could be crystalline carbon (e.g. graphite) or a series of abiogenic organic molecules. Point of origin for the required oxygen are minerals (e. g. magnetite) or water.

In the solid Earth, carbon dioxide is not just a neutral trace molecule without functions but in combination with H_2O it has a profound impact on mineral equilibria in mantle and crust. In the realm of solidus-liquidus relationships, for instance in a granitic or peridotitic magma, addition of trace quantities of carbon dioxide will lead to a polymerization, i.e. crystallization, whereas admixture of small amounts of water will cause a depolymerization, i.e. rock melting under identical P-T-conditions.

The bulk of carbon dioxide presently discharged by volcanoes or hydrothermal vents comes from recycled marine limestone, organic matter, or sea water bicarbonate to name a few principal sources. Mantle-derived or juvenile CO_2 is only a minor contributor. Rate of CO_2 release from the geosphere is estimated at about 0.2 Gt C yr^{-1}.

At the surface of the Earth, carbon dioxide is the key molecule controlling the carbon cycle in air-water-life systems. The annual turnover amounts to more than 200 Gt C largely a result of biological activities (photosynthesis-respiration) and CO_2 exchange through the air-sea interface. Compared to this figure, the eventual return of carbon to the geosphere is 1,000 times less and about matches the volume of volcanic CO_2 emission. It is concluded that the pristine carbon cycle is a kind of steady state system in which carbon dioxide is the mediating force at all critical points of intersection.

This quasi steady state system attained after billions of years of endogenic and exogenic interactions is temporarily put out of "equilibrium" by the action of man: combustion of fossil fuel and deforestation. At present ca. 6.3–10 Gt C yr^{-1}. are emitted into the air in the form of CO_2. This is close to 150 times the volume of CO_2 discharged by land volcanoes. The excess in CO_2 is in part digested by the marine carbon pool and in part by the soil humus deposit. Yet, one half to one third of human induced CO_2 remains air-borne and causes a rise in the atmospheric CO_2 level at an annual rate of approx. 1 ppm.

The future increase will depend on the amounts of CO_2 released by meeting rising energy demands and by the fraction taken up in the world ocean. How energy needs will develop and to what extent they will be met by burning of fossil fuels is a matter of speculation. Prognostic modelling of the oceanic carbon cycle and it's interaction with climate should become feasible in the years to come. An important step in this direction is the model of Maier-Reimer. For any given scenario of CO_2-increase in the air, only a range of probable climatic responses is what, in principle, may be predicted. Before that range narrows down to a meaningful value, a comprehensive theory of multi-component system with highly non-linear interactions on a wide spectrum of time scales has yet to be developed. Key unknowns in present climate models are cloud-feedback and long-term atmosphere-ocean-cryosphere interactions.

Currently, most modellers tend to expect a global warming trend, with pronounced regional and seasonal rearrangements of the hydrologic cycle, if CO_2 were doubled. Obviously, this would threat some and please others. Important negative feedbacks may have been omitted, but to be on the safe side, worst case scenarios have to be taken serious, until revision renders them obsolete.

We like to emphasize what the modelling community is well aware of: a model is a model is a model! Consequently it is up for revision at any time, because ingrained in a *good* model is – following Karl Popper – it's refutability.

Acknowledgements

The project entitled "The Integrated Carbon Cycle" presently carried out at the SCOPE/UNEP International Carbon Unit, University of Hamburg, jointly with the Max-Planck-Institut für Meteorologie (Prof. K. Hasselmann), Hamburg, and the Faculty of Biological Sciences (Prof. H. Lieth), University of Osnabrück has been supported by the following agencies: Bundesministerium für Forschung und Technologie, Deutsche Forschungsgemeinschaft, Umweltbundesamt, Gesamtverband des Deutschen Steinkohlebergbaues, Beirat für Umweltforschung der Stadt Hamburg, Shell Grants Committee, European Community Commission, Scientific Committee on Problems for the Environment (SCOPE), United Nations Environment Program (UNEP), and the University of Hamburg. The authors are grateful to the agencies for their generous funding of a rather involved issue. We have profited greatly from discussions with numerous visitors at the SCOPE/UNEP International Carbon Unit. In particular we like to thank Ingo Aselmann, Robert Bacastow, Bert Bolin, Anders Björkström, Werner Deuser, Wolfgang Dreybroth, Paul Duvigneaud, John Edmond, Hermann Flohn, Döke Eisma, Jürgen Fischer, Erik Galimov, Gan Wei-Bin, Hartmut Grassl, Klaus Hasselmann, Susumo Honjo, Venu Ittekkot, Michail Ivanov, Gundolf Kohlmaier, Karin Labitzke, Jerry Leenheer, Helmut Lieth, Alexander Lisitzin, Jerry Olson, Ernst Maier-Reimer, Michel Meybeck, Wim Mook, Ralph Rotty, Dominique Raynaud, Jeff Richey, Maurits La Riviere, Wolfgang Roether, Thomas Rosswall, Nicholas Shakleton, Uli Siegenthaler, Gilbert White, and George Woodwell.

References

1. Cameron, A.G.W.: Space Sci. Rev. *15*, 121 (1973)
2. Ahrens, L.H. (ed.): Origin and Distribution of the Elements, p. 909, Pergamon Press, Oxford–New York–Toronto–Sydney–Paris–Frankfurt 1979
3. Blitz L.: Scient. Amer. *246*/4, 72 (1982)
4. Mann, A.P.C., Williams, D.A.: Nature *283*, 721 (1980)
5. Salpeter, E.W.: In: Analyse extraterrestrischen Materials [Kiesl, W., Malissa, H. (eds.)], pp. 203–212, Springer-Verlag, Wien–New York 1974
6. Matheja, J., Degens, E.T.: Structural Molecular Biology of Phosphates, p. 180, Gustav Fischer Verlag, Stuttgart 1971
7. Weiss, A.H., et al.: J. Catal. *16*, 332 (1970)
8. Mizutani, H., et al.: Origins of Life *6*, 513 (1975)

9. Solomon, P. M., Edmunds, M.G. (eds.): Clouds in the Galaxy, p. 300, Pergamon Press, Oxford–New York–Toronto–Sydney–Paris–Frankfurt 1980
10. Suess, H.E.: Origins of Life *6*, 9 (1975)
11. Morris, M., Richard, L.J.: Ann. Rev. Astron. Astrophys. *20*, 517 (1982)
12. Anders, E.: Phil. Trans. R. Soc. London A *285*, 23 (1977)
13. Hartmann, W.K.: Icarus *33*, 50 (1978)
14. Smith, J.V.: Miner. Mag. *43*, 1 (1979)
15. Urey, H.C.: Astrophys. J. Suppl. *1*, 147 (1954)
16. Ringwood, A.E.: Origin of the Earth and Moon, p. 295, Springer-Verlag, New York–Heidelberg–Berlin 1979
17. Smith, J.V.: J. Geol. *90*, 1 (1983)
18. Urey, H.C.: Astrophys. J. *124*, 623 (1956)
19. Brett, R.: Science *153*, 60 (1966)
20. Hayatsu, R., et al.: Science *209*, 1515 (1980)
21. Mason, B.: Geochim. Cosmochim. Acta *30*, 23 (1966)
22. Nagy, B., et al.: Ann. N.Y. Acad. Sci. *93*, 27 (1961)
23. Kaplan, I.R., et al.: Geochim. Cosmochim. Acta *27*, 805 (1963)
24. Clayton, R.N.: Science *140*, 192 (1963)
25. Whittaker, A.G., et al.: Science *209*, 1512 (1980)
26. Lewis, R.S., et al.: J. Geophys. Res. *82*, 779 (1977)
27. Potter, A.E., Del Duca, B.: Icarus *3*, 103 (1964)
28. Whittaker, A.G.: Science *200*. 763 (1978)
29. Brandt, W.: Scient. Amer. *68*/3, 91 (1968)
30. Freund, F., et al.: Geochim. Cosmochim. Acta *44*, 1319 (1980)
31. Walker, J.C.G.: J. Atmos. Sci. *32*, 1248 (1975)
32. Mysen, B.O.: Rev. Geophys. Space Phys. *15*, 351 (1977)
33. Wyllie, P.J.: J. Geol. *85*, 187 (1977)
34. Wong, H.K., Degens. E.T.: Tectonophysics *95*, 191 (1983)
35. Engel, A.E.J., et al.: Bull. Geol. Soc. Am. *85*, 843 (1974)
36. Degens, E.T.: Environ. Intern. *2*, 401 (1979)
37. Taylor, H.P., et al.: Geochim. Cosmochim. Acta *31*, 407 (1967)
38. Craig, H.: Geochim. Cosmochim. Acta *3*, 53 (1953)
39. Degens, E.T.: In: Organic Geochemistry [Eglinton, G., Murphy, M.T.J. (eds.)], pp. 304–329, Springer-Verlag, New York–Heidelberg–Berlin 1969
40. Schwarcz, H.P.: In: Handbook of Geochemistry [Wedepohl, K.H. (ed.)], II-1, 6-B-1-6-B-16, Springer-Verlag, New York–Heidelberg–Berlin 1969
41. Hathaway, J.C., Degens, E.T.: Science *165*, 690 (1969)
42. Ballard, R.D.: Geo *12*, 6 (1979)
43. Lupton, J.E., Craig, H.: Science *214*, 13 (1981)
44. Bolin, B., et al. (eds.): The Global Carbon Cycle, p. 491, SCOPE Rep. *13*, J. Wiley & Sons, Chichester–New York–Brisbane–Toronto 1979
45. Jacoby, G,. (ed.): Proc. Int. Meeting Stable Isot. Tree-Ring Res., p. 150, U.S. Dep. Energy Conf. 790518, Washington, D.C. 1980
46. Freyer, H.D.: Tellus *31*, 124 (1979)
47. Stuiver, M.: Science *199*, 253 (1978)
48. Degens, E.T. (ed.): Transport of Carbon and Minerals in Major World Rivers, Part I, p. 766, Mitt. Geol.-Pal. Inst. Univ. Hamburg *52* (SCOPE/UNEP Sonderbd.) 1982
49. Graßl, H., et al.: Naturwissenschaften, in press (1984)
50. Wasserburg, G.J., et al.: Phil. Trans. R. Soc. London A *285*, 7 (1977)
51. Schubert, G., Covey, C.: Scient. Amer. *245*/1, 44 (1981)
52. Walker, J.C.G., et al.: J. Geophys. Res. *75*, 3558 (1970)
53. Holland, H.D.: The Chemistry of the Atmosphere and Oceans, p. 35, J. Wiley & Sons, New York–Chichester–Brisbane–Toronto 1978
54. Sagan, C., Mullen, G.: Science *177*, 52 (1972)
55. Edmond, J.M., Von Damm, K.: Sci. American *248*/4, 70 (1983)
56. McDuff, R.E., Edmond, J.M.: Earth Plan. Sci. L. *57*, 117 (1982)
57. Henley, R.W., Ellis, A.J.: Earth-Science Rev. *19*, 1 (1983)

58. Einsele, G., et al.: Nature *283*, 441 (1980)
59. Spooner, E.T.C., Bray, C.J.: Nature *266*, 808 (1977)
60. Spiess, F.N., et al.: Science *207*, 1421 (1980)
61. Malahoff, A. (ed.): Polymetallic Sulfides, p. 91, Mar. Techn. Soc. J. *16*, No. 3, 1982
62. Lewis, B.T.R.: Science *220*, 151 (1983)
63. Buch, K.: Meeresforschung *1*, 15 (1951)
64. Hansson, I.: Deep-Sea Res. *20*, 461 (1973)
65. Mehrbach, C., et al.: Limnol. Oceanogr. *18*, 897 (1973)
66. Weiss, R.F.: Mar. Chem. *2*, 203 (1974)
67. Ingle, S.E.: Mar. Chem. *3*, 301 (1975)
68. Edmond, J.M., Gieskes, J.M.: Geochim. Cosmochim. Acta *34*, 1261 (1970)
69. Edmond, J.M. Deep-Sea Res. *17*, 737 (1970)
70. Broecker, W.S., et al.: Science *206*, 409 (1979)
71. Takahashi, T., et al.: In: Carbon Cycle Modelling [Bolin, B. (ed.)], pp. 159–199, SCOPE Rep. *16*, J. Wiley & Sons, Chichester–New York–Brisbane–Toronto 1981
72. Brewer, P.J.: In: Carbon Dioxide, Science and Consensus, II, pp. 93–122, Conf. 820970, U.S. Dept. Energy, Washington, D.C. 1983
73. Bradshaw, A.L., et al.: Earth Planet. Sci. L. *55*, 99 (1981)
74. Stuiver, M., et al.: Science *219*, 849 (1983)
75. Broecker, W.S.: Chemical Oceanography, p. 214, Harcourt Brace & Jovanovich Inc., New York 1974
76. Honjo, S.: In: The Fate of Fossil Fuel Co_2 in the Oceans [Andersen, N.R., Malahoff, A. (eds.)], pp. 269–294, Plenum Press, New York 1977
77. Martinez, L., et al.: Science *221*, 152 (1983)
78. Berner, W., et al.: Nature, *275*, 53 (1979)
79. Neftel, A., et al.: Nature *295*, 220 (1982)
80. Broecker, W.S.: Geochim. Cosmochim. Acta *46*, 1689 (1982)
81. Broecker, W.S.: Scient. Am. *249*/3, 100 (1983)
82. Sundquist, E., et al.: Science *204*, 1203 (1979)
83. Weiss, R.F., et al.: Nature *300*, 511 (1982)
84. Mopper, K., Degens, E.T.: In: The Global Carbon Cycle [Bolin, B., et al. (eds.)], pp. 293–316, SCOPE Rep. *13*, J. Wiley & Sons, Chichester–New York–Brisbane–Toronto 1979
85. Cauwet, G.: Oceanol. Acta *1*, 99 (1978)
86. de Vooys, C.G.N.: In: The Global Carbon Cycle [Bolin B., et al. (eds.)], pp. 259–292, SCOPE Rep. *13*, J. Wiley & Sons, Chichester–New York–Brisbane–Toronto 1979
87. Jenkins, W.J.: Nature *300*, 246 (1982)
88. Riley, G.A.: J. Mar. Res. *6*, 54 (1946)
89. Shulenberger, E., Reid, J.L.: Deep-Sea Res. *28* A, 901 (1981)
90. Li, W.K.W., et al.: Science *219*, 292 (1983)
91. Harvey, G.R.: Mar. Chem. *12*, 333 (1983)
92. Degens, E.T., Ittekkot, V.: In: Transport of Carbon and Minerals in Major World Rivers, Part II [Degens E.T., et al. (eds.)], Mitt. Geol.-Paläont. Inst. Univ. Hamburg *54*, (SCOPE/UNEP Sonderbd.) (in press)
93. Williams, P.M., et al.: Nature *224*, 256 (1969)
94. Devlin, R.M., Barker, A.V.: Photosynthesis, Univ. of Massachusetts, New York 1971
95. Calvin, M., Baasham, J.A.: The Photosynthesis of Carbon Compounds, p. 127, Benjamin, New York 1962
96. Hatch, M.D., Slack, C.R.: Ann. Rev. Plant Physiol. *21*, 141 (1970)
97. Goudriaan, J., Ajtay, G.L.: In: The Global Carbon Cycle [Bolin, B., et al. (eds.)], pp. 237–249, SCOPE Rep. *13*, J. Wiley & Sons, Chichester–New York–Brisbane–Toronto 1979
98. de Wit, C.T.: Photosynthesis of Leaf Canopy, Agricult. Res. Rep. 663, Pudoc, Wageningen 1965
99. Lehninger, A.L.: Biochemie, p. 747, Verlag Chemie, Weinheim 1975
100. Stanier, R.Y., et al.: General Microbiology, p. 873, MacMillan Press, London 1972
101. Kempe, S.: In: Transport of Carbon and Minerals in Major World Rivers, Part I [Degens, E.T. (ed.)], pp. 91–332, Mitt. Geol.-Paläont. Inst. Univ. Hamburg *52*, (SCOPE/UNEP Sonderbd.) 1982
102. Stumm, W., Morgan, J.J.: Aquatic Chemistry, 2nd Ed., p. 780, J. Wiley & Sons, Chichester–New York–Brisbane–Toronto 1981

103. Ajtay, G., et al.: In: The Global Carbon Cycle [Bolin, B., et al. (eds.)], pp. 129–181, SCOPE Rep. *13*, J. Wiley & Sons, Chichester–New York–Brisbane–Toronto 1979
104. Bolin, B., et al.: In: The Global Carbon Cycle [Bolin, B., et al. (eds.)], pp. 1–56, SCOPE Rep. *13*, J. Wiley & Sons, Chichester–New York–Brisbane–Toronto 1979
105. Olson, J.: In: Carbon Dioxide Review: 1982 [Clark, W.C. (ed.)], pp. 388–398, Oxford Univ. Press, New York 1982
106. Matthews, E.: J. Climate Appl. Meteorol. *22*, 474 (1983)
107. Esser, G., et al.: In: Transport of Carbon and Minerals in Major World Rivers, Part I [Degens E.T. (ed.)], pp. 39–58, Mitt. Geol.-Paläont. Inst. Univ. Hamburg *52* (SCOPE/UNEP Sonderbd.) 1982
108. Bohn, H.L.: Soil Sci. Soc. Amer. J. *40*, 468 (1976)
109. Schlesinger, W.H.: Ann. Rev. Ecol. Syst. *8*, 51 (1977)
110. Post, W.M., et al.: Nature *298*, 156 (1982)
111. Whittaker, R.H., et al.: In: Primary Productivity of the Biosphere [Lieth, H., et al. (eds.)], pp. 305–328, Ecol. Stud. *14*, Springer-Verlag, Berlin 1975
112. Aselmann, I.H., Lieth, H.: In: Transport of Carbon and Minerals in Major World Rivers, Part II [Degens, E.T., et al. (eds.)], Mitt. Geol.-Paläont. Inst. Univ. Hamburg *54* (SCOPE/UNEP Sonderbd.) (in press)
113. Bolin, B.: Science *196*, 613 (1977)
114. Woodwell, J.M., et al.: Science *199*, 141 (1978)
115. Hampicke, U.: In: The Global Cycle [Bolin, B., et al. (eds.)], pp. 219–236, SCOPE Rep. *13*, J. Wiley & Sons, Chichester–New York–Brisbane-Toronto 1979
116. Bramryd, T.: In: The Global Carbon Cycle [Bolin, B., et al. (eds.)], pp. 183–218, SCOPE Rep. *13*, J. Wiley & Sons, Chichester–New York–Brisbane–Toronto 1979
117. Seiler, W., Crutzen, P.: Climatic Change *2*, 207 (1980)
118. Moore, B., et al.: In: Carbon Cycle Modelling [Bolin, B. (ed.)], pp. 365–385, SCOPE Rep. *16*, J. Wiley & Sons, Chichester–New York–Brisbane–Toronto 1981
119. Houghton, R.A., et al.: Ecol. Monogr. *53*, 235 (1983)
120. Oeschger, H., et al.: Tellus *27*, 168 (1975)
121. Freyer, H.D., Belacy, N.: J. Geoph. Res. *88* (C11), 6844 (1983)
122. Peng, T.H., et al.: J. Geophys. Res. *88*, C6, 3609 (1983)
123. Cooper, C.F.: In: Carbon Dioxide Review: 1982 [Clark, W.C. (ed.)], pp. 298–320, Oxford Univ. Press, New York 1982
124. Sionit, N., et al.: Crop Sci. *20*, 687 (1980)
125. Degens, E.T., et al.: In: Nitrogen, Phosphorus, and Sulphur: Global Cycles [Svennson, B.H., Söderlund, R. (eds.)], pp. 185–191, SCOPE Rep. 7, Ecological Bulletins, Stockholm 1976
126. Freyer, H.D.: In: The Global Carbon Cycle [Bolin, B., et al. (eds.)], pp. 79–99, J. Wiley & Sons, Chichester–New York–Brisbane–Toronto 1979
127. Ehhalt, D.H.: Tellus *26*, 58 (1974)
128. Seiler, W.: Tellus *26*, 116 (1974)
129. Freyer, H.D.: In: The Global Carbon Cycle [Bolin, B., et al. (eds.)], pp. 101–128, SCOPE *13*, J. Wiley & Sons, Chichester–New York–Brisbane–Toronto 1979
130. Goldberg, E.D. (ed.): Atmospheric Chemistry, Rep. Dahlem Workshop, May 2–7, 1982, p. 384, Springer-Verlag, Berlin–Heidelberg–New York 1982
131. Crutzen, P.J.: In: The Major Biogeochemical Cycles and their Interactions [Bolin, B., Cook, R.B. (eds.)], pp. 67–112, SCOPE Rep. *21*, J. Wiley & Sons, Chichester–New York–Brisbane 1983
132. Galbally, I.E., et al.: In: Cycle of Carbon, Nitrogen, Sulfur and Phosphorus in Terrestrial and Aquatic Ecosystems [Freeney J.R., Galbally, I.E. (eds.)], pp. 1–9, Springer-Verlag, Berlin–Heidelberg–New York 1982
133. Duce, R., et al.: Rev. Geophys. Space Phys. *21*, 921 (1983)
134. Keeling, C.D.: In: Carbon Dioxide, Science and Consensus, pp. II 3–62, CONF 820970, U.S. Dept. Energy, Washington, D.C. 1983
135. Bacastow, R.B.: Nature *261*, 116 (1976)
136. Newell, R.E., Weare, B.C.: Geophys. Res. L. *4*, 1 (1977)
137. Rust, B.W., et al.: J. Geophys. Res. *84*, 3115 (1979)
138. Pearman, G.I., et al.: In: Carbon Dioxide and Climate: Australian Research [Pearman, G.I. (ed.)], Australian Academy of Science, Canberra 1980
139. Bacastow, R.B., et al.: Science *210*, 66 (1980)
140. Pearman, G.I., Hyson, P.: J. Geophys. Res. *86*, 9839 (1981)

141. Gillette, D.A.: Atmos. Envir. *16*, 2537 (1982)
142. Dörr, H., et al.: J. Geophys. Res. *88*, 1309 (1983)
143. Schnell, R.C., et al.: Geophys. Res. *86*, 5364 (1981)
144. Garratt, J.R., Pearman, G.I.: Atmos. Envir. *7*, 1257 (1973)
145. McRae, J.E., Graedel, T.E.: J. Geophys. Res. *84*, 5011 (1979)
146. Bischof, W., Bolin, B.: Tellus *18*, 155 (1966)
147. Georgii, H.W., Jost, D.: Nature *221*, 1040 (1969)
148. Mauersberger, K., Finstad, R.: Geophys. Res. L. *7*, 873 (1980)
149. Ehhalt, D.H., et al.: J. Atmos. Sci. *32*, 163 (1975)
150. Volz, A., et al.: In: Proc. WMO Conf. on Regional and Global Observation of Atmosph. Pollution Relative to Climate, Boulder, Co. 8/1979
151. Bischof, W., et al.: Nature *288*, 347 (1980)
152. Halter, B.C., Peterson, J.T.: Atmosph. Environm. *15*, 1391 (1981)
153. Callendar, G.S.: Quart. J. Roy. Met. Soc. *66*, 395 (1940)
154. Preining, O.: Atmosph. Environm. *17*, 1595 (1983)
155. Stuiver, M.: Science *202*, 1109 (1978)
156. Barnola, J.M., et al.: Nature *303*, 410 (1983)
157. Delmas, J.R., et al.: Nature *284*, 155 (1980)
158. Lorius, C., Raynaud, D.: In: Carbon Dioxide, Text on Current Views and Developm. in Energy/Climate Res. [Bach, W. et al. (eds.)], pp. 145–176, Reidel, Dordrecht (in press)
159. McElroy, M.B.: Nature *302*, 328 (1983)
160. Berger, W.: Naturwiss. *69*, 87 (1982)
161. Aharon, P.: In: Cycling of Carbon, Nitrogen, Sulfur, and Phosphorus in Terrestrial and Aquatic Ecosystems [Freney, J.R., Galbally, I.E. (eds.)], pp. 119–132, Springer-Verlag, Berlin–Heidelberg–New York 1982
162. Pedersen, T.F.: Geology *11*, 16 (1983)
163. Müller, P.J., Suess, E.: Deap-Sea Res. *26*, 1347 (1979)
164. Neftel, A., et al.: Ann. Glaciol. (in press)
165. Siegenthaler, U.: Paper presented at the IAMAP-meeting, Aug. 20–27, Hamburg 1983
166. Junge, C.E., Czeplak, G.: Tellus *20*, 422 (1968)
167. Machta, L.: Bull. Am. Meteorol. Soc. *53*, 402 (1972)
168. Pearman, G.I., Hyson, P.: Geophys. Res. *85*, 4457 (1980)
169. Pearman, G.I., et al.: J. Geophys. Res. *88*, 3581 (1983)
170. Fung, I., et al.: J. Geophys. Res. *88*, 1281 (1983)
171. Suess, H.E.: In: Nuclear Processes in Geological Settings [National Academy of Sciences (ed.)], pp. 52–56, National Research Council Publication 1955
172. Shell Briefing Service: Weltenergie – Daten und Fakten, p. 12, Deutsche Shell AG, Hamburg (1983)
173. Barr, T.N.: Science *214*, 1087 (1981)
174. Brown, L.R.: Science *214*, 995 (1981)
175. Degens, E.T.: In: Transport of Carbon and Minerals in Major World Rivers, Part I [Degens, E.T. (ed.)], pp. 1–12, Mitt. Geol.-Pal. Univ. Hamburg *52* (SCOPE/UNEP Sonderbd.) 1982
176. Landsberg, H.H.: Science *218*, 973 (1982)
177. Abelson, P.H.: Science *213*, 605 (1981)
178. Abelson, P.H.: Science *215*, 1349 (1982)
179. Bostel, J., et al.: KFA Jülich Spec. Publ. *156*, p. 137, 1982
180. Hammond, A.L.: Science *197*, 564 (1977)
181. Maugh II, T.H.: Science *216*, 1213 (1982)
182. Richardson, S.W., White, A.A.L.: Nature *286*, 103 (1980)
183. McDonald, A.: Energy in a Finite World, p. 68, Int. Inst. Appl. Syst. Anal., Laxenburg, Austria 1981
184. Exxon Background Series: World Energy Outlook, Exxon Corp., p. 32, 1980
185. Energy 83: Proc. Conf. Hamburg, p. 778, Hamburg Messe und Congress GmbH 1983
186. Wilson, C.L.: Coal – Bridge to the Future, p. 247, Ballinger Publ. Co., Cambridge, Mass. 1980
187. Stobaugh, R., Yergin, D. (eds.): Energy Future, p. 353, Random House, New York 1979
188. Meyer-Abich, K.M., et al. (eds.): Wirtschaftspolitische Steuerungsmöglichkeiten zur Einsparung von Energie durch alternative Technologien, Part III, p. 1111 (plus Appendix, p. 441), University Essen 1978
189. Isaacs, J.D., Achmitt, W.R.: Science *207*, 265 (1980)

190. Colombo, U.: Science *217*, 705 (1982)
191. Kerr, R.A.: Science *218*, 668 (1982)
192. Sassin, W.: Scient. Americ. *243*/3, 107 (1980)
193. Commoner, B.: A Nearly Perfect Fuel, pp. 66–95, The New Yorker (May 2) 1983
194. Kerr, R.A.: Science *207*, 1455 (1980)
195. Bell, P.R.: In: Carbon Dioxide Review: 1982 [Clark, W.C. (ed.)], pp. 401–406, Oxford Univ. Press, New York 1982
196. von Huene, R., Aubouin, J. (eds.): Joides J. VIII, 2. p. 100, 1982
197. Degens, E.T., et al.: Geol Rdsch. *62*, 245 (1973)
198. Deuser, W.G., et al: Science *181*, 51 (1973)
199. Baross, J.A., Deming, J.W.: Nature *303*, 423 (1983)
200. Stetter, K.O.: Nature *300*, 258 (1982)
201. Abelson, P.H.: Science *221*, 815 (1983)
202. Orr Jr., F.M.: J. Petrol. Techn. (July), 1285 (1983)
203. Kempe, S.: In: The Global Carbon Cycle [Bolin, B., et al. (eds.)], pp. 343–377, J. Wiley & Sons, Chichester–New York–Brisbane–Toronto 1979
204. Garrels, R.M., Mackenzie, F.T.: Evolution of Sedimentary Rocks, p. 397, W.W. Norton & Comp., New York 1971
205. Garrels, R.M., Christ, C.L.: Solutions, Minerals, and Equilibria, p. 450, Freeman, Cooper & Comp., San Francisco 1965
206. Drever, F.I.: The Geochemistry of Natural Waters, p. 388, Prentice-Hall, Englewood Cliffs, N.F. 1982
207. Wigley, T.M.L.: Canad. J. Earth Sci. *8*, 468 (1971)
208. Dester, D.R.: Ion Association of Sodium, Magnesium and Calcium with Sulfate in Aqueous Solution, p. 116, Ph. D. Thesis, Oregon State Univ. 1969
209. Reardon, E.J., Langmuir, D.: Geochim. Cosmochim. Acta *40*, 549 (1976)
210. Wigley, T.M.L.: Canad. J. Earth Sci. *10*, 306 (1973)
211. Wigley, T.M.L.: Geochim. Cosmochim. Acta *37*, 1397 (1973)
212. Jacobson, R.L., Langmuir, D.: Geochim. Cosmochim. Acta *38*, 301 (1974).
213. Lafon, J.M.: Geochim. Cosmochim. Acta *34*, 935 (1970)
214. Kempe, S.: Ann. Speleol. *30*, 699 (1975)
215. Wigley, T.M.L.: Brit. Geomorph. Res. Group Techn. Bull. *20*, 1 (1977)
216. Langmuir, D.: Geochim. Comochim. Acta *35*, 1023 (1971)
217. Fischbeck, R.: N. Jb. Miner. Abh. *126*, 269 (1976)
218. Bögli, A.: Z. Geomorph. Suppl. *2*, 4 (1960)
219. Bögli, A.: Karsthydrographie und Physische Speläologie, p. 292, Springer-Verlag, Berlin 1978
220. Wigley, T.M.L., Plummer, L.N.: Geochim. Cosmochim. Acta *40*, 989 (1976)
221. Plummer, L.N., et al.: Am. J. Sci. *278*, 179 (1978)
222. Plummer, L.N., et al.: In: Chemical Modelling in Aqueous Systems [Jeune, E.A. (ed.)], pp. 538–572, ALS Symp. Ser. *93*, 1979
223. Dreybrodt, W.: Chem. Geol. *31*, 245 (1981)
224. Dreybrodt, W.: Chem. Geol. *32*, 221 (1981)
225. Livingstone, D.A.: U.S. Geol. Surv. Prof. Pap. *440* G, 1 (1963)
226. Kempe, S.: In: The Global Carbon Cycle [Bolin, B., et al. (eds.)], pp. 317–342, J. Wiley & Sons, Chichester–New York–Brisbane–Toronto 1979
227. Meybeck, M.: In: Dissolved Loads of Rivers and Surface Water Quantity Quality Relationships, pp. 173–192, Proc. Hamburg Symp. Aug. 1983, IHHS Publ. *141*, 1983
228. Baumgartner, A., Reichel, E.: The World Water Balance, p. 179, R. Oldenbourg Verlag, München–Wien 1975
229. Balazs, D.: Proc. 7th Intern. Congr. Speleology, Sheffield, 13 (1977)
230. Kempe, S.: J. Geoph. Res. (in press) (1984)
231. Martins, O.: Anorganische und Organische Geochemie des Nigers im Abflußjahr 1980/1981, p. 140, Ph. D. Thesis, Geol.-Paläont. Inst. Univ. Hamburg 1983
232. Nordin, C.F., Meade, R.H.: In: Flux of Organic Carbon by Rivers to the Oceans, pp. 173–219, Conf. 8009140, U.S. Dept. Energy, Washington, D.C. 1981
233. Mulholland, P.J.: In: Flux of Organic Carbon by Rivers to the Oceans, pp. 142–172, Conf. 8009140, U.S. Dept. Energy, Washington, D.C. 1981
234. Meybeck, M.: Amer. J. Sci. *282*, 401 (1982)

235. Schlesinger, W., Melack, F.M.: Tellus, *33*, 172 (1981)
236. Degens, E.T. (ed.): Transport of Carbon and Minerals in Major World Rivers, Part I, p. 766, Mitt. Geol.-Paläont. Inst. Univ. Hamburg *52* (SCOPE/UNEP Sonderbd.) 1982
237. Degens, E.T., et al. (eds.): Transport of Carbon and Minerals in Major World Rivers, Part II, Mitt. Geol.-Paläont. Inst. Univ. Hamburg *54* (SCOPE/UNEP Sonderbd.), (in press)
238. Meade, R.H.: J. Geol. *90*, 235 (1982)
239. Hasse, L., Liss, P.S.: Tellus *32*, 470 (1980)
240. Whitmann, W.S.: Chem. Metall. Engng. *29*, 146 (1923)
241. Bolin, B.: Tellus *12*, 274 (1980)
242. Kanwisher, J.: Deep-Sea Res. *10*, 195 (1963)
243. Danckwerts, P.V.: Gas-liquid Reactions, p. 276, McGraw-Hill, New York–Düsseldorf–London 1970
244. Quinn, J.A., Otto, N.C.: J. Geophys. Res. *76*, 1539 (1971)
245. Liss, P.: Deep-Sea Res. *20*, 221 (1973)
246. Liss, P., Slater, P.G.: Nature *247*, 181 (1974)
247. Broecker, W.S., Peng, T.H.: Tellus *26*, 21 (1974)
248. Berger, R., Libby, W.F.: Science *164*, 1395 (1969)
249. Bolin, B., et al.: Carbon Cycle Modelling [Bolin, B. (ed.)], pp. 1–28, SCOPE Rep. *16*, J. Wiley & Sons, Chichester–New York–Brisbane–Toronto 1981
250. Downing, A.L., Truesdale, G.A.: J. Appl. Chem. *7*, 590 (1957)
251. Sigiura, Y., et al.: J. Mar. Res. *21*, 11 (1963)
252. Hoover, T.E., Berkshire, D.C.: J. Geophys. Res. *74*, 456 (1969)
253. Broecker, H.C., et al.: J. Mar. Res. *36*, 595 (1978)
254. Jähne, B., et al.: Tellus *31*, 321 (1979)
255. Flothmann, D., et al.: Naturwiss. *66*, 49 (1979)
256. Merlivat, L., Memery, L.: J. Geophys. Res. *88*, 707 (1983)
257. Craig, H.: Tellus *9*, 1 (1957)
258. Peng, T.H., et al.: J. Geophys. Res. *84*, 2471 (1979)
259. Roether, W., Dromer, B.: Paleoph. *116*, 476 (1978)
260. Stuiver, M.: J. Geophys. Res. *85*, 2711 (1980)
261. Deacon, E.L.: Bound. Lay. Meteorol. *21*, 31 (1981)
262. Jones, E.P., Smith, S.D.: J. Geophys. Res. *82*, 5990 (1977)
263. Wesely, M.L., et al.: J. Geophys. Res. *87*, 8827 (1982)
264. Takahashi, T., et al.: In: Carbon Dioxide, Science and Consensus, pp. II 123–145, Conf-820970, U.S. Dept. Energy, Washington, D.C. 1983
265. Honjo, S., et al.: Science *216*, 516 (1982)
266. Honjo, S.: Science *218*, 883 (1982)
267. Deuser, W.G., et al.: Deep-Sea Res. *28*, 495 (1981)
268. Deuser, W.G., et al.: Science *219*, 388 (1983)
269. Ittekkot, V., et al.: Deep-Sea Res. (in press)
270. Degens, E.T., Ittekkot, V.: J. Geol. Soc. London (in press)
271. Westbroek, P., De Jong, E.W. (eds.): Biomineralisation and Biological Metal Accumulation, p. 533, D. Reidel Publishing Company, Dordrecht–Boston–London 1983
272. Degens, E.T.: In: Topics in Current Chemistry *64*, pp. 1–112, Springer-Verlag, Berlin–Heidelberg–New York 1976
273. Krampitz, G., Witt, W.: In: Topics in Current Chemistry *78*, pp. 57–144, Springer-Verlag, Berlin–Heidelberg–New York 1979
274. Steemann-Nielsen, E.: Physiol. Plant. *19*, 236 (1966)
275. Deuser, W.G., Degens, E.T.: Nature *215*, 1033 (1967)
276. Emrich, K., et al.: Earth Plan. Sci. L. *8*, 363 (1970)
277. Mook, W.G., et al.: Earth Plan. Sci. L. *22*, 169 (1974)
278. Keith, M.L., Weber, J.N.: Science *150*, 498 (1965)
279. Degens, E.T., et al.: Mar. Biol. *2*, 105 (1969)
280. Maren, T.H.: Physiolog. Rev. *47*, 595 (1967)
281. Carter, M.J.: Biol. Rev. *47*, 405 (1972)
282. Coleman, J.E.: J. Biol. Chem. *242*, 5212 (1967)
283. Edsall, J.: Ann. N.Y. Acad. Sci. *68*, 41 (1968)
284. Liljas, A., et al.: Nature New Biol. *235*, 131 (1972)

285. Kannan, K.K., et al.: Cold Spring Harbor Symp. Quant. Biol. *36*, 221 (1972)
286. Pocker, Y., Meany, J.E.: Biochem. *4*, 2535 (1965)
287. Pocker, Y., Meany, J.E., Biochem. *6*, 239 (1967)
288. Verpoorte, J.A., et al.: J. Biol. Chem. *242*, 4221 (1967)
289. Pocker, Y., Guilbert, L.J.: Biochem. *13*, 70 (1974)
290. Wang, J.H.: Proc. Nat. Acad. Sci. USA *66*, 874 (1970)
291. Khalifah, R.G.: J. Biol. Chem. *246* (1971)
292. Koenig, S.G., Brown, R.D.: Proc. Nat. Acad. Sci. USA *69*, 2422 (1972)
293. Taylor, P.W., Burgen, A.S.V.: Biochem. *10*, 3859 (1971)
294. Taylor, P.W., et al.: Biochem. *10*, 3866 (1971)
295. Okazaki, M.: Botanica Marina *15*, 133 (1972)
296. Newton, I., Bogan, J.: Nature *249*, 582 (1974)
297. Honjo, S.: In: Research in the Sea [Fey, P.M. (ed.)], p. 2, Woods Hole Oceanographic Inst., Woods Hole Mass 1974
298. Goreau, T.E.: Biol. Bull. *116*, 59 (1959)
299. Shackleton, N.J.: J. Geol. Soc. London (in press)
300. Shackleton, N.J., Hall, M.A.: DSDP 74 (in press)
301. Shackleton, N.J.: Progr. Oceanogr. *11*, 199 (1982)
302. Schidlowski, M.: In: Mineral Deposits and the Evolution of the Biosphere [Holland, H.D., Schidlowski, M. (eds.)], pp. 103–122, Springer-Verlag, Berlin–Heidelberg–New York 1982
303. Seibold, E., Berger, W.H.: The Sea Floor, p. 288, Springer-Verlag, Berlin–Heidelberg–New York 1982
304. Heezen, B.C., Hollister, C.D.: The Face of the Deep, p. 659, Oxford Univ. Press, New York 1971
305. Jörgensen, B.B.: In: The Major Biogeochemical Cycle and their Interactions [Bolin, B., Cook, R.B. (eds.)], pp. 477–509, SCOPE Rep. *21*, J. Wiley & Sons, Chichester–New York–Brisbane–Toronto–Singapore 1983
306. Degens, E.T., Ittekkot, V.: In: Transport of Carbon and Minerals in Major World Rivers, Part II [Degens, E.T., et al. (eds.)], Mitt. Geol.-Paläont. Inst. Univ. Hamburg *54* (SCOPE/UNEP Sonderbd.), (in press)
307. Eriksson, E., Welander, P.: Tellus *8*, 155 (1955)
308. Revelle, R., Suess, H.E.: Tellus *1*, 18 (1957)
309. Bolin, B., Eriksson, E.: Rossby Memorial Vol., p. 130, Rockfeller Inst. Press, New York (1959)
310. Oeschger, H., Heimann, M.: J. Geophys. Res. *88*, 1258 (1983)
311. Emanuel, W.R., et al.: In: Carbon Cycle Modelling [Bolin, B. (ed.)], pp. 335–353, SCOPE Rep. *16*, J. Wiley & Sons, Chichester–New York–Brisbane–Toronto 1981
312. Kohlmaier, G.H.: Radiat. Environm. Biophys. *19*, 67 (1981)
313. Bacastow, R., Björkström, A.: In: Carbon Cycle Modelling [Bolin, B. (ed.)], pp. 29–79, SCOPE Rep. *16*, J. Wiley & Sons, Chichester–New York–Brisbane–Toronto 1981
314. Bolin, B. (ed.): Carbon Cycle Modelling, p. 390, SCOPE Rep. *16*, J. Wiley & Sons, Chichester–New York–Brisbane–Toronto 1981
315. Björkström, A.: In: The Global Carbon Cycle [Bolin, B. (ed.)], pp. 403–457, SCOPE Rep. *13*, J. Wiley & Sons, Chichester–New York–Brisbane–Toronto 1979
316. Viecelli, et al.: Climatic Change *3*, 281 (1981)
317. Bolin, B., et al.: Tellus *35* B, 206 (1983)
318. Siegenthaler, U.: J. Geophys. Res. *88*, 3599 (1983)
319. Kratz, G., et al.: Radiocarbon (in press)
320. Maier-Reimer, E.: J. Geophys. Res. (submitted)
321. Rotty, R.M.: J. Geophys. Res. *88*, 1301 (1983)
322. Bolle, H.J.: In: The Handbook of Environmental Chemistry [Hutzinger, O. (ed.)], Vol. 1, Part B, pp. 131–305, Springer-Verlag, Berlin 1982
323. Tyndall, J.: Philosoph. Mag. *4*, 200 (1863)
324. Arrhenius, S.: Philosoph. Mag. *41*, 237 (1897)
325. Chamberlain, T.C.: J. Geol. *7*, 545 (1899)
326. Callendar, G.S.: Quart. J. Roy. Met. Soc. *64*, 223 (1938)
327. Flohn, H.: Zeitschr. f. Erdkunde *9*, 13 (1941)
328. Defant, A.: Geogr. Ann. *3*, 209 (1921)
329. Angström, A.: Gerlands Beitr. Geophys. *15*, 1 (1926)

330. Milankovitch, M.: In: Handbuch der Klimatologie [Köppen, W., Geiger, R. (eds.)], Vol. 1, Part A, pp. 176. Borntraeger Verlag, Berlin 1980
331. Plass, G.N.: Tellus *8*, 140 (1956)
332. Manabe, S., Möller, F.: Monthly Weather Rev. *89*, 503 (1961)
333. Möller, F.: J. Geophys. Res. *68*, 3877 (1963)
334. Saltzman, B.: Advances in Geophysics *20*, 183 (1978)
335. Clark, W.C. (ed.): Carbon Dioxide Review: 1982, p. 469, Oxford Univ. Press, New York 1982
336. National Research Council: Carbon Dioxide and Climate: A Second Assessment, p. 72, National Academy Press, Washington, D.C. 1982
337. Schlesinger, M.E.: In: Carbon Dioxide, Science and Consensus, pp. III 3–139, Conf. 820970, U.S. Dept. Energy, Washington, D.C. 1983
338. Budyko, M.I.: Tellus *21*, 611 (1969)
339. Sellers, W.D.: J. Appl. Meteorol. *8*, 392 (1969)
340. Ramanathan, V.: J. Atmosph. Sci. *38*, 918 (1981)
341. Choudhury, B., Kukla, G.: Nature *280*, 668 (1979)
342. Manabe, S., Wetherald, R.T.: J. Atmos. Sci. *37*, 99 (1980)
343. Manabe, S., Stouffer, R.J.: J. Geophys. Res. *85*, 5529 (1980)
344. Washington, W.M., Meehl, G.A.: J. Geophys. Res. *88*, 6600 (1983)
345. Hasselmann, K.: In: Man's Impact on Climate [Bach, W., et al. (eds.)], pp. 43–55, Elsevier, New York 1979
346. Hunt, B.G., Wells, N.C.: J. Geophys. Res. *84*, 787 (1979)
347. Hoffert, M.I., et al.: J. Geophys. Res. *84*, 6667 (1980)
348. Thompson, S.L., Schneider, S.H.: Nature *290*, 9 (1980)
349. Schneider, S.H., Thompson, S.L.: J. Geophys. Res. *86*, 3135 (1981)
350. Offenborn, W., Grassl, H.: Contr. Atmosph. Phys. *54*, 107 (1981)
351. Bryan, K., et al.: Science *215*, 56 (1982)
352. Hasselmann, K.: Progr. Oceanogr. *11*, 69 (1982)
353. Schneider, S.: Atmos. Sci. *29*, 1413 (1972)
354. Naujokat, B.: J. Geophys. Res. *86*, 9811 (1981)
355. Kukla, G., Gavin, J.: Science *214*, 497 (1981)
356. Labitzke, K.: J. Met. Soc. Japan *60*, 124 (1982)
357. Thompson, S.L. Schneider, S.H.: Nature *295*, 645 (1982)
358. McCracken, M.C.: In: Carbon Dioxide, Science and Consensus, pp. V 3–47, Conf.-820970, U.S. Dept. Energy, Washington, D.C. 1983
359. Zwally, H.J., et al.: Science *220*, 1005 (1983)
360. Paldridge, G.: Quart. J. Roy. Met. Soc. *101*, 475 (1975)
361. Mobbs, S.D.: Quart. J. Roy. Met. Soc. *108*, 535 (1982)
363. Manabe, S., et al.: Climatic Change *3*, 347 (1981)
364. Hameed, S., Cess, R.D.: Tellus *35*B, 1 (1983)
365. Fishman, J. et al.: Nature *282*, 818 (1979)
366. WMO (World Meteorological Organisation): Ozone Research and Monitoring Project, Rep. *14*, Geneva 1982
367. Keeling, C.D., et al.: Tellus *82*, 538 (1976)
368. Suess, H.E.: Science *122*, 415 (1955)
369. Degens, E.T., et al.: In: Energy – Present and Future Options, Vol. 1 (Merrick, D., Marshall, R. (eds.)], pp. 93–186, J. Wiley & Sons, Chichester–New York–Brisbane–Toronto 1981
370. Heezen, B.C., McGregor, D.: Scient. Amer. *229*/5, 102 (1973)
371. Degens, E.T., Stoffers, P.: In: Time- and Strata-Bound Ore Deposits [Klemm, D.D., Schneider, H.F. (eds.)], pp. 25–45, Springer-Verlag, Berlin–Heidelberg 1977
372. Münnich, K.O., Roether, W.: In: Radioactive Dating and Methods of Low Level Counting, pp. 93–104, I.A.E.A. (International Atomic Energy Agency), Vienna 1967
373. Webb, E.K., et al.: Quart. J. Roy. Met. Soc. *106*, 85 (1980)
374. Adem, J., Garduno, R.: Geofisica International *21*, 309 (1982)
375. Labitzke, K., Naujokat, B.: Beitr. Phys. Atmosph. *11*/83, 1 (1983)
376. Kiehl, J.T.: Science *222*, 504 (1983)
377. Shackleton, N.J., et al.: Nature *306*, 319 (1983)
378. Kempe, S., Degens, E.T.: [subm. to Chem. Geol. (1984)]

Subject Index

www.ingramcontent.com/pod-product-compliance
Lightning Source LLC
Chambersburg PA
CBHW081952130526
45046CB00024B/2508
* 9 7 8 3 6 6 2 1 5 2 3 5 5 *